AF439071

# Invertebrate Animals (Metazoa: Invertebrata) from the Bulgarian Black Sea Coast

*Zdravko Hubenov*

Copyright © 2025 Zdravko Hubenov
Copyright © 2025 Generis Publishing

All rights reserved. This book or any portion thereof may not be reproduced or used in any manner whatsoever without the written permission of the publisher except for the use of brief quotations in a book review.

Title: **Invertebrate Animals (Metazoa: Invertebrata) from the Bulgarian Black Sea Coast**

ISBN: 979-8-89248-898-3

Author: Zdravko Hubenov

Cover image: www.pixabay.com

Publisher: Generis Publishing
Online orders: www.generis-publishing.com
Contact email: info@generis-publishing.com

# Invertebrate Animals (Metazoa: Invertebrata) from the Bulgarian Black Sea Coast

**Zdravko Hubenov**

National Museum of Natural History, Bulgarian Academy of Sciences, 1 Tsar Osvoboditel Blvd., 1000 Sofia; e-mail: zkhubenov@abv.bg; zhubenov@nmnhs.com

# Contents

# Abstract

The investigations of the Black Sea territory for the last 2 centuries are generalized in this work. A total of 19 types, 39 classes, 123 orders, 477 families and 1590 species are known from the Bulgarian Black Sea. They include 1059 species (66.6%) marine and marine-brackish forms and 531 species (33.4%) freshwater-brackish, freshwater and terrestrial forms, connected with water. Five types (Nematoda, Rotifera, Annelida, Arthropoda and Mollusca) have a high species richness (over 100 species). Of these, the richest in species are Arthropoda (813 species – 51.1%), Annelida (175 species – 11.0%) and Mollusca (155 species – 9.7%). The rest 14 types include from 1 to 38 species. There are some well-studied regions (over 200 species recorded). Firstly is the vicinity of Varna (601 species) where the investigations continue more than 100 years. The aquatory of the towns Nesebar, Pomorie, Burgas and Sozopol (from 220 to 274 species) and the region of the Cape Kaliakra (230 species) are well-studied. Of the coastal basins most studied are the lakes Durankulak, Ezerets-Shabla, Beloslav, Varna, Pomorie, Atanasovsko, Burgas, Mandra and the firth of Ropotamo River (known to 100 species). The vertical distribution has been analyzed for 805 species (76.0%) – marine and marine-brackish forms. The great number of species is found from 0 to 25 m on sand (396 species) and rocky (257 species) bottom. The groups of stenohypo- (52 species – 6.5%), stenoepi- (465 species – 58.1%), meso- (115 species – 14.4%) and eurybathic forms (168 species – 21.0%) are represented. The marine and marine-brackish species are divided into 162 zoogeographical categories, combined into 4 main groups and 16 subgroups. The main portion of the Black Sea fauna has an Atlantic-Mediterranean origin and represents the impoverished Atlantic-Mediterranean fauna (740 species – 69.9%). Cosmopolitan, Atlantic-Indian, Atlantic-Pacific, endemic and Caspian relict forms are represented. The benthic (115 species – 97.5%) and marine (114 species – 96.6%) forms of the Black Sea endemics (118 species – 11.2%) predominate. The brackish endemics (11 species – 9.3%) most often are Caspian relicts. The main portions of the Caspian relicts (41 species – 3.9%) are benthic brackish forms (38 species – 92.7%). The freshwater-brackish, freshwater and terrestrial forms, connected with water, are divided into 80 zoogeographical categories, combined into 2 groups and 5 subgroups. Typical for the coast is the prevalence of the species, distributed in Palearctic and beyond it (296 species – 58.3%). Species, distributed only in Palearctic but in more than one subregion (79 species

– 15.5%) and species, distributed within one Palearctic subregion (126 species – 24.8%) are represented – Eurosiberian (55 species – 10.8%) and Mediterranean (71 species – 13.9%). A short characteristic of the planktonic and benthic cenoses is done and some coastal basins are scrutinized. An attention is paid to the invasive immigrants that changed the Black Sea communities during the last 60 years. The species of economic and conservation importance are discussed.

**Key words**: Black Sea coast, faunistic diversity, invertebrates, zoogeography, alien species

# Introduction

The Bulgarian Black Sea fauna had been studied for more than 118 years (Chichkoff 1907, 1908, 1912, 1924). A vast material of faunistic data, concerning Bulgarian Black Sea, was accumulated. During the last 70 years, the coast is under a drastic anthropogenic impact and large landscape changes. Considerable changes in the Black Sea cenoses are caused by some invasive species, introduced in the last 100 years (Cvetkov & Marinov 1986; Konsulov 1998; Gomoiu et al. 2002; Todorova 2005, 2011, 2015, 2017; Todorova et al. 2008a, 2008b, 2009, 2022, 2023; Konsulova et al. 2010; Todorova & Panayotova 2011, 2015; Todorova & Moncheva 2013; Karachle et al. 2017; Todorova & Milkova 2017; Băncilă et al. 2022). The dynamic natures of the fauna, economic and social importance of wildlife and biodiversity conservation require periodic updating of the faunistic diversity of Bulgaria.

The published catalogues of the Bulgarian Black Sea fauna (Valkanov, 1957a; Valkanov & Marinov, 1964; Marinov & Golemansky, 1989; Marinov, 1990; Konsulov & Konsulova, 1993) have not a systematic character and taxa of the genus and species group (the families are not presented) are in alphabetical order. These works are published in Bulgarian and the faunistic analyses there are too short. The published generalized studies in English by Konsulov (1998) and Konsulov & Konsulova (1998) are similar to the works of Marinov (1990) and Konsulov & Konsulova (1993). A part of the names used is out-of-date and needs to be updated. There is a contemporary systematic view for some taxonomic groups, included in the monograph series Fauna of Bulgaria (Polychaeta – Marinov, 1977; Harpacticoida – Apostolov & Marinov, 1988), in survey papers (Mollusca – Wilke, 1996; Hubenov, 2005b, 2007a, 2007b, 2015, 2024; Uzunova 2016) or in dissertations (Nematoda – Stoykov, 1980; Crustacea: Malacostraca – Uzunova, 2006). There is a lack of zoogeographical characteristic of the fauna except Polychaeta, Harpacticoida, Malacostraca and Mollusca. The submitted zoogeographical analyses of the groups mentioned above are done according to their origin or areography (different principle) and are difficult to compare. Commonly the benthos hydrobionts are scrutinized separately from the plankton forms. A generalized zoogeographical work on the Bulgarian Black Sea fauna lacks.

The aim of this work is to present the Bulgarian marine invertebrate fauna as well as to analyze the taxonomic diversity, the level of study and some zoogeographical and ecological features of the Black Sea invertebrates.

# Approach, material and methods

The investigations of the Black Sea territory for the last 2 centuries are generalized in this work. The paper generalizes the works of Caspers (1951), Valkanov (1957a), Valkanov & Marinov (1964), Marinov & Golemansky (1989), Marinov (1990), Konsulov & Konsulova (1993, 1998), Konsulov (1998) and Golemansky (2007). Data from 839 publications and the dissertations of Konsulov (1991), Kamburska (2004), Todorova (2005), Uzunova (2006), Trayanova (2008) and Todorova (2014) are included. Currently some coastal wetlands have been investigated in connection with their management plans (Durankulak Lake, Shabla Lake, Pomorie Lake, Atanasovsko Lake and protected area Poda). These investigations are also included in the work.

The categories of type class and order are used (an exception is made for supertype Arthropoda because of the structure of the superior taxonomic categories). All water (marine, brackish and freshwater) and many terrestrial invertebrate animals, connected with the coast and coastal basins are included.

In the numbering of the localities (Table 1), for convenience, the old numbering used in catalogues of Valkanov (1957a), Valkanov & Marinov (1964) and Marinov & Golemansky (1989) is presented. In many cases, the information on the coastal basins, given in these publications is outdated (before 1957). Today, a part of the brackish basins along the coast do not exist in its original form. They are converted into bays, harbors, dams, have no connection with the sea or are drained. For most of the smaller basins there is a lack of present-day faunistic investigations. Twenty new localities, which have no equivalent in the previous catalogues, are included. The number of known species in the separate localities is presented as well. It shows mainly their level of study and to a less extent, the actual species diversity. The localities of the species are presented in Abbreviations used. The coastal brackish basins are included.

For each species are given: 1) recent scientific name; 2) synonyms and names under which it is reported for Bulgaria; 3) distribution in the Bulgarian Black Sea aquatory; 4) depth at which it was established; 5) areographical characteristic; 6) brief ecological data; 7) conservation status; 8) literature sources. For introduced taxa, the natural range is indicated first, followed by the secondary (anthropogenic range). The invasive species are marked.

For the marine species, the depth to which they are established in the Bulgarian Black Sea is given. Species, for which there are no data from the Bulgarian coast, data from other regions of the sea are presented. When information in the Bulgarian literature differs significantly from the one, reported for other parts of the sea, the respective foreign data for the Black Sea are presented, after the Bulgarian data. In freshwater Mollusca, the presented depth refers to the whole country.

For species that inhabit both fresh and salt waters, an areographical categorization for seas and freshwater basins is presented. The categorization for the freshwaters is given in brackets. Some taxa, distributed both in the sea and freshwaters (Supercosmopolitan) are analyzed both to the marine and freshwater forms. The brackish species are included to marine or freshwater forms according to the fact whether they are marine-brackish or freshwater-brackish. When the reports of species distribution are discrepant, a second categorization is presented. An attention is paid to the immigrants and invasive forms that had changed considerably the Black Sea communities in XX century.

There is no unanimity among the experts about the zoogeographical status of the Black Sea, which is either considered as independent subregion or is unified with the Mediterranean Sea (and Lusitanean Atlantic subregion). The zoogeographical scheme used here (Table 4) is based on the works of Guryanova (1964), De Lattin (1967), Golikov & Starobogatov (1968, 1972), Starobogatov (1970), Mordukhay-Boltovskogo (1972), Golikov (1982), Nesis (1982), Riedl (1983), Bănărescu (1990), Abbott & Dance (1991), Elder & Pernetta (1991), Bruyne (2003), Hook (2008), Earle & Glover (2009). The zoogeographical categorization of species is done on the basis of data of their distribution, taken from the literature and the newest electronic issues (Tables 4 and 5).

The presented ecological data are taken from the Bulgarian literature. Only if there are no data from Bulgaria, foreign data are included for the corresponding species. The conservation value of taxa is determined regarding to their populations inhabiting Bulgaria. For local endemics, 100 % of their populations are localized in Bulgaria, therefore they are given the highest conservation category (world importance). This category also includes regional endemics because of their restricted distribution and species from the IUCN Red List. Taxa of European importance include Black Sea endemics as well as the species from Bern Convention and Habitats Directive. Relicts and rare taxa (if not listed under other category) form the group of national importance. The species, included in

Black Sea Red Data Book (Dumont et al., 1999), Red Data Book of Bulgaria (Biserkov & Golemanski, 2011, 2015), European and IUCN Red List are marked.

The literature references do not include all publications addressed to the corresponding species from the Bulgarian coast (to 9 references quoted). Most often the first record of taxa is given, its inclusion in catalogues and some new or important literature data. Under updating of the names and  specifying of the species distribution, some electronic issues are used: Antarctic Invertebrates, CLEMAM (Check List of European Marine Mollusca), DAISIE (Delivering Alien Invasive Species Inventories for Europe), EOL (Encyclopedia of Life), ERMS (European Register of Marine Species), EUNIS biodiversity database, Fauna Europaea, Global Invasive Species Database, Global Names Index, ITIS (Integrated Taxonomic Information System), Marine Planktonic Copepods, Marine Species Identification Portal, MarLIN (The Marine Life Information Network), NARMS (North Atlantic Register for Marine Species), NeMys, NEOBANIS (European Network on Invasive Alien Species), PESI (A Pan-European Species directories Infrastructure), PlanktonNet Image, OBIS (Ocean Biogeographic Information System), The World of Copepods, World Polychaeta Database, WoRMS (World Register of Marine Species).

# Unexplored territories and literature data

Despite the prolonged hydrobiological investigations and good knowledge of the Bulgarian  Black Sea fauna as a whole, unexplored areas still remain. The possible reasons for this fact are as follows: lack of specialists on many taxonomic groups; great loading of the specialists with environmental or conservation projects, therefore the time for faunistic research is insufficient; periodic standard surveys of the fixed number of monitoring stations, a relative remoteness of natural science centers; a poor attendance by many zoologists in comparison with other regions or change in coastal communities as a result of anthropogenic impact. Most of the literature data related to these regions are fragmentary, outdated, concern separated systematic groups or are scattered in different works which are not specially referred to them.

Today, the most poorly investigated territories in regard to many groups is the southern coast (south of Cape Maslen Nos) and the coastal zone with a depth less than 10 m, where the oceanographic ships rarely enter. Some of the coastal basins were explored long ago so the investigations do not reflect the recent condition of their fauna.

Weaknesses in the literature data which limit the obtaining of equivalent information for the comparison of the territories include: different levels of study of individual taxa; insufficient research of many groups in the corresponding areas; a lack of exact localities for the part of the recorded species; existence of rich synonymy; outdated data; a lack of generalized investigations for most of the groups; significant differences in the number of taxa in the separate areas; unexplored territories; prolonged periods of data accumulation for most regions; predominance of ecological studies versus those of fauna; independent review of benthos and plankton forms. These weaknesses lead to the following 5 problems: 1) Continuous supplementation of an existing historical list of fauna. As a result, species diversity in a given area is higher than in reality. 2) Incomparability of data in terms of time periods. Data comparisons between two areas very often cover different periods as it is not possible to study all taxonomic groups and territories simultaneously. 3) Incomparability of benthos – plankton data. Many studies are look at either benthos only or plankton only, despite the fact that most taxa have both a benthic and planktonic stages. 4) Incomplete reporting of anthropogenic influences, successional and landscape changes on the composition of the communities along the coast. A number of well-studied brackish basins in the past no longer exist or have changed. 5) Prioritization of research in areas under monitoring or environmental protection legislation.

# Abbreviations used

**Taxa**: [ ] – names and synonyms under which the species are recorded for Bulgaria; **D** – distribution, **E** – ekological data, **R** – References, **V** – depth, **Z** – zoogeographical categories.

**Distribution**: 1-29 – Localities in Black Sea (**1-10** – North Black Sea, **11-22** – South Black Sea), **32-54** – Localities along the coast, **58-96** – Coastal basins, **figures** – numbers of the localities in Table 1, **Depth** – dash before the figure (-7, -12) indicates the latter as maximum depth, dash after the figure (7-, 12-) indicates the latter as minimum depth, **?** – uncertain data or lack of data, * – outdated information and significantly altered habitat versus time of collection of the material.

**Localities of the species**: * – highly altered habitat to the time of collection of the material; [ ] – old geographical names and old data, before 1957.

**Sea localities: 1** – Durankulak [Blatnitsa] (Durankulak north – Durankulak Lake – Krapets); **2** – Shabla (Shabla Lake – Shabla Tuzla – Cape Shabla); **3** – Cape Kaliakra (Rusalka – Bolata – Cape Kaliakra); **4** – Kavarna (Cape Chairburun – Cape Chirakman – Cape Kalkanburun); **5** – Balchik (Balchik Tuzla – Balchik); **6** – Batova (Albena – Kranevo); **7** – Varna (Golden Sands – Evksinograd – Varna – Cape Galata – Pasha Dere River); **8** – Kamchiya (Cape Ilandzhik – Camping Ray – Kamchiya River – Shkorpilovtsi); **9** – Byala (Cape Cherni – Byala – Dvoynitsa River – Obzor); **10** – Cape Emine (Irakli – Cape Emine – Cocketrice sandy bank); **11** – Nesebar (Elenite – Sunny Beach – Nesebar – Ravda); **12** – Pomorie (Aheloy – Pomorie – Camping Evropa – Cape Lahna); **13** – Burgas (Saraphovo – Burgas – Kraymorie – Park Rosenets – Cape Chukalya); **14** – Sveta Anastasiya Island [Bolshevik Island]; **15** – Chernomorets (Cape Atiya – Chernomorets – Cape Chervenka [Cape Hrisotira]); **16** – Sozopol (Camping Gradina – Sozopol – Kavatsite – Dyuni); **17** – Cape Maslen Nos (Alepu Marsh – Ropotamo River – C. Maslen Nos – Stomoplo Marsh); **18** – Primorsko (Stomoplo Marsh – Primorsko – Dyavolska Reka River); **19** – Kiten [Urdoviza] (International Youth Centre – Kiten – Karaagach River – Lozenets); **20** – Tsarevo [Michurin, Vasiliko] (Cape Arapya – Tsarevo – Varvara); **21** – Ahtopol (Varvara – Achtopol – Veleka River); **22** – Sinemorets (Sinemorets – Silistar River – Rezovo); **23** – *Zostera* overgrowths (0-6 m); **24** – Rocky sublittoral, *Cystoseira* and other algae, *Mytilus* (from 0.5-1 m to 15-25 m); **25** – Sandy sublittoral (1-25 m); clean sand – to 17 m, with *Branchiostoma* – to 20 m; **26** – Coastal silt (from 15-20 m to 30-40 m); dominated by *Melinna* – to 25-30 m; **27** – *Mytilus* silt (from 15-20 m to 60-80 m); **28** – *Phaseolina* silt (from 65 m to 140-180 m); **29** – Black Sea, pelagic in front of the Bulgarian coast.

**Localities along the sea coast**: **32** – Lithotelms, Shabla – Cape Kaliakra, Varna; **33** – Lithotelms, Ravda, Sozopol – Cape Maslen Nos; **34** – Basins of Varna Aquarium; **35** – Subterranean (ground) waters of sandy beach, interstitial, mesopsamal; **36** – Sunny Beach, coastal zone, sand bottom and floating algae; **37** – Arkutino, coastal zone; **38** – Camping (Residence) Perla, coastal zone, sand bottom and floating algae; **39** – Kiten, coastal zone, sand bottom and floating algae; **40** – Lozenets, coastal zone; **41** – Small saltwater marshes along the coast; **42** – Small freshwater marshes along the coast; **43** – Mouths of small streams; **44** – Temporary salty puddles and floods around the coastal basins; **45** – Rocks along the entire coast, rocky supralittoral; **46** – Algae washed ashore along the coast, supralittoral; **47** – Salty soils around coastal basins; **48** – Terrestrial coastal zone with halophilic plants (to 50-100 m from the sea); **49** – Sea coastal zone; littoral (medio- or pseudolittoral); **50** – Rocky littoral (medio- or pseudolittoral, enteromorpha zone); **51** – Sandy littoral (medio- or pseudolittoral); **52** – Sandy supralittoral; **53** – Springs and wells with brackish water along the coast, Durakulak – Cape Kaliakra; **54** – Springs along the coast, Sozopol – Cape Maslen Nos.

**Coastal basins (lakes, swamps, firths and river floods)**: **58** – Durankulak [Blatnitsa] Lake: [0-5‰, 3.4 km², depth 4 m], 1-4‰, average salinity 2‰; **59** – Ezerets Lake: [1-2‰, average salinity 1.6‰], 0.58-0.79‰, 0.72 km², depth 9.0 m; **60** – Shabla Lake: [0.1-2‰, 0.6-1.6‰], 0.52-0.60‰, 0.79 km², depth 9.5 m; **61** – *Schabla Tuzla: [10-30‰], 22-200‰, 0.19 km², depth 0.6 m; **62** – Nanevska Tuzla [Tauk Liman]: 1-90‰ (often about 20‰) 0.10 km², depth 0.3 m; **63** – *Bolata River Mouth: 0.1‰; **64** – *Balchik Tuzla: [80-150‰], 35-160‰, 0.14 km², depth 0.5-0.8 m; **65** – Batova River Mouth and Swamp: 0.03-6‰, depth 0.5-1 m; **66** – *Golden Sands Marshes: [0-60‰]; **67** – *Sindel [Sultanlar] Swamp: [0‰]; **68** – *Beloslav [Devnya, Gebedzhe] Lake: 0.1-15.6‰, 3.90 km², depth 3.5 m, sea canal 1923; **69** – *Varna Lake: [5-14‰], 6.5-8-16.8‰, 17.40 km², depth 19 m, sea canals 1909, 1976; **70** – Pasha Dere [Chatal Dere, Novata Voda] River Mouth: [0-7‰]; **71** – Kamchiya River Mouth Swamps: [0.4-0.7‰], average 0.1‰; **72** – Fandakliyska Reka [Shkorpilova] River Mouth: [0.1‰]; **73** – *Dvoynitsa [Cherta, Suha Kamchiya] River Mouth; **74** – *Hadzhiyska River

Mouth [Nesebar Marsh]: [1-10‰]; **75** – Aheloy River Mouth; **76** – Pomorie Lake: 30-70 to 140‰, 8.50 km², depth 1.4 m; **77** – Atanasovsko Lake: 1-250‰, average 50-60‰, 16.90 km², depth 0.3-0.8 m; **78** – *Burgas [Vaya] Lake: [9-20‰], 1.8-45‰, average 10.6‰, 27.60 km², depth 1.3 m; **79** – *Mandra Dam [Mandra Lake to 1963]: [0.1-12‰, max. 30‰, 14.00 km², depth 1.1-5 m]; **80** – Uzungeren-Poda Complex: 0.1-32‰, 3.12 km²; **81** – Tsiganski Skelet Marsh [Chengene Skele Marsh]: [7-20‰]; **82** – Alepu Marsh: 4-11-27‰ (usually 3.5-7.2‰), 0.14 km², depth 0.6-1 m; **83** – Arkutino Marsh: 0.1-1‰, 0.03 km², depth 0.5 m; **84** – Ropotamo River Mouth: 5-15‰; **85** – Stomoplo Marsh: [2-25‰], 1.5-4‰, 6-14‰, 0.06 km², depth 0.5 m; **86** – *Dyavolsko Blato Swamp: [1-20‰], 6-14‰, 0.80 km², depth 1 m; **87** – *Dyavolska Reka River Mouth: depth 4 m; **88** – Karaagachka Reka [Kitenska, Oryashka] River Mouth and Swamp: [5-15‰]; **89** – Tsarevska Reka [Michurinska (small)] River: [3‰]; **90** – Izgrevsko Dere [Michurinska Reka (great)] River: [5-10‰]; **91** – Puddles and mouths of streams between Tsarevo and Ahtopol; **92** – Veleka River Mouth: [0-0.5‰]; **93** – Butamyata [Potamyata] River Mouth: [12‰]; **94** – Silistar River Mouth: [5-15‰]; **95** – Rezovska Reka [Rezvaya] River Mouth: [0-1.45‰]; **96** – Black Sea coastal lakes and swamps.

**Zoogeographical categories** (the abbreviations in brackets refer to the freshwater and terrestrial species): **aam** – Arctic-Atlantic-Mediterranean, **aami** – Arctic-Atlantic-Mediterranean-Indian, **aamip** – Arctic-Atlantic-Mediterranean-Indo-Pacific, **aaminp** – Arctic-Atlantic-Mediterranean-Indo-North Pacific, **aamni** – Arctic-Atlantic-Mediterranean-North Indian, **aamswp** – Arctic-Atlantic-Mediterranean-Southwest Pacific, **aanambp** – Arctic-Antarctic-Atlantic-Mediterranean-Boreal Pacific, **aanamip** – Arctic-Antarctic-Atlantic-Mediterranean-Indo-Pacific, **aannam** – Arctic-Antarctic-North Atlantic-Mediterranean, **ab** – Amphiboreal, **abam** – Arctic-Boreal Atlantic-Mediterranean, **abambp** – Arctic-Boreal Atlantic-Mediterranean-Boreal Pacific, **abap** – Arctic-Boreal Atlantic-Pontian, **abapbp** – Arctic-Boreal Atlantic-Pontian-Boreal Pacific, **abapnep** – Arctic-Boreal Atlantic-Pontian-Northeast Pacific, **ace** – Arctic-Circumeuropean, **acem** – Arctic-Circumeuropean-Mauritanian, **acmnz** – Arctic-Celtic-Mediterranean-New Zealand, **acp** – Arctic-Celtic-Pontian, **adep** – Adriatic-Aegean-Pontian, **adp** – Adriatic-Pontian, **adpc** – Adriatic-Pontian-Caspian, **am (am)** – Atlantic-Mediterranean, **ami** – Atlantic-Mediterranean-Indian, **aminp** – Atlantic-Mediterranean-Indo-North Pacific, **aminwp** – Atlantic-Mediterranean-Indo-Northwest Pacific, **aminz** – Atlantic-Mediterranean-Indo-New Zealand, **amip** – Atlantic-Mediterranean-Indo-Pacific, **amiswp** – Atlantic-Mediterranean-Indo-Southwest Pacific, **amiwp** – Atlantic-

Mediterranean-Indo-West Pacific, **amj** – Atlantic-Mediterranean-Japonic, **amnei** – Atlantic-Mediterranean-Northeast Indian, **amnep** – Atlantic-Mediterranean-Northeast Pacific, **amni** – Atlantic-Mediterranean-North Indian, **amnp** – Atlantic-Mediterranean-North Pacific, **amnz** – Atlantic-Mediterranean-New Zealand, **amp** – Atlantic-Mediterranean-Pacific, **amrs** – Atlantic-Mediterranean-Red Sea, **amrsp** – Atlantic-Mediterranean-Red Sea-Pacific, **amswp** – Atlantic-Mediterranean-Southwest Pacific, **amwi** – Atlantic-Mediterranean-West Indian, **amwp** – Atlantic-Mediterranean-West Pacific, **anam** – Arctic-North Atlantic-Mediterranean, **anaminp** – Arctic-North Atlantic-Mediterranean-Indo-North Pacific, **anamip** – Arctic-North Atlantic-Mediterranean-Indo-Pacific, **anamnep** – Arctic-North Atlantic-Mediterranean-Northeast Pacific, **anamnp** – Arctic-North Atlantic-Mediterranean-North Pacific, **anamp** – Arctic-North Atlantic-Mediterranean-Pacific, **anamrs** – Arctic-North Atlantic-Mediterranean-Red Sea, **anap** – Arctic-North Atlantic-Pontian, **anapnep** – Arctic-North Atlantic-Pontian-Northeast Pacific, **anclm** – Antarctic-Celtic-Lusitanian-Mediterranean, **anpip** – Antarctic-Pontian-Indo-Pacific, **antami** – Antarctic-Atlantic-Mediterranean-Indian, **antamip** – Antarctic-Atlantic-Mediterranean-Indo-Pacific, **antamp** – Antarctic-Atlantic-Mediterranean-Pacific, **ap** – Atlantic-Pontian, **api** – Atlantic-Pontian-Indian, **apswp** – Atlantic-Pontian-Southwest Pacific, **(ase)** – Atlantic-South European, **(atm)** – Afrotropical-Mediterranean, **baap** – Boreal-Antiboreal Atlantic-Pontian, **bam** – Boreal Atlantic-Mediterranean, **bambp** – Boreal Atlantic-Mediterranean-Boreal Pacific, **bami** – Boreal Atlantic-Mediterranean-Indian, **bamnep** – Boreal Atlantic-Mediterranean-Northeast Pacific, **bamswp** – Boreal Atlantic-Mediterranean-Southwest Pacific, **bap** – Boreal Atlantic-Pontian, **bapbp** – Boreal Atlantic-Pontian-Boreal Pacific, **bapp** – Boreal Atlantic-Pontian-Pacific, **cacpnz** – Carolinian-Celtic-Pontian-New Zealand, **calm** – Carolinian-Lusitanian-Mediterranean, **calp** – Carolinian-Lusitanian-Pontian, **cb** – Circumboreal, **cbm** – Circumboreal-Mediterranean, **cbma** – Circumboreal-Mediterranean-Australian, **cclm** – Carolinian-Celtic-Lusitanian-Mediterranean, **ccp** – Carolinian-Celtic-Pontian, **ce** – Circumeuropean, **cem** – Circumeuropean-Mauritanian, **cg** – Circumglobal, **clm** – Celtic-Lusitanian-Mediterranean, **clmi** – Celtic-Lusitanian-Mediterranean-Indian, **clmm** – Celtic-Lusitanian-Mediterranean-Mauritanian, **clmnei** – Celtic-Lusitanian-Mediterranean-Northeast Indian, **clmnwi** – Celtic-Lusitanian-Mediterranean-Northwest Indian, **clmnz** – Celtic-Lusitanian-Mediterranean-New Zealand, **clmrs** – Celtic-Lusitanian-Mediterranean-Red Sea, **clmwi** – Celtic-Lusitanian-Mediterranean-West Indian, **clp** – Celtic-Lusitanian-Pontian, **clpnz** – Celtic-Lusitanian-Pontian-New Zealand, **cm** – Celtic-Mediterranean, **cp** – Celtic-Pontian, **cpc** – Celtic-

Pontian-Caspian, **cpj** – Celtic-Pontian-Japonic, **cpnei** – Celtic-Pontian-Northeast Indian, **cpnz** – Celtic-Pontian-New Zealand, **cpwp** – Celtic-Pontian-West Pacific, **(cse)** – Central and South European, **(csee)** – Central and Southeast European, **(cseea)** – Central and Southeast European-Anatolian, **(cseeit)** – Central and Southeast European-Iran-Turanian, **(csena)** – Central and South European-North African, **cst** – Circumsubtropical, **(dp)** – Disjunct Palaearctic, **(e)** – European, **(ea)** – European-Australian, **eam** – East Atlantic-Mediterranean, **eami** – East Atlantic-Mediterranean-Indian, **eamip** – East Atlantic-Mediterranean-Indo-Pacific, **eamiswp** – East Atlantic-Mediterranean-Indo-Southwest Pacific, **eamp** – East Atlantic-Mediterranean-Pacific, **eamrs** – East Atlantic-Mediterranean-Red Sea, **eamswi** – East Atlantic-Mediterranean-Southwest Indian, **eamwi** – East Atlantic-Mediterranean-West Indian, **(ean)** – European-Anatolian, **(Eb)** – Balkan endemic, **(Ebg)** – Bulgarian endemic, **(eca)** – European-Central Asian, **(eit)** – European-Iran-Turanian, **(El)** – Local Bulgarian endemic, **em (em)** – East Mediterranean, **(emca)** – East Mediterranean-Central Asian, **(ena)** – European-North African, **(Ep)** – Pontian endemic, **ep** – Aegean-Pontian, **(Er)** – Regional Bulgarian endemic, **(esca)** – Eurosiberian-Central Asian, **(et)** – European-Turanian, **(ewca)** – European-West Central Asian, **(h)** – Holarctic, **(ha)** – Holarctic-Australian, **ham** – Holatlantic-Mediterranean, **(hat)** – Holarctic-Afrotropical, **(hata)** – Holarctic-Afrotropical-Australian, **(hn)** – Holarctic-Neotropical, **(hna)** – Holarctic-Neotropical-Australian, **(hnat)** – Holarctic-Neotropical-Afrotropical, **(hnata)** – Holarctic-Neotropical-Afrotropical-Australian, **(hno)** – Holarctic-Neotropical-Oriental, **(hnoa)** – Holarctic-Neotropical-Oriental-Australian, **(ho)** – Holarctic-Oriental, **(hoa)** – Holarctic-Oriental-Australian, **(hoes)** – Holoeurosiberian, **hom (hom)** – Holomediterranean, **(hop)** – Holopalaearctic, **(hpt)** – Holarctic-Paleotropical, **(hpta)** – Holarctic-Paleotropical-Australian, **(hptn)** – Holarctic-Paleotropical-Neotropical, **i** – introduced species (immigrants), **j** – Japanese, **K (k)** – Cosmopolitan, **kclm** – Caribbean-Celtic-Lusitanian-Mediterranean, **klm** – Caribbean-Lusitanian-Mediterranean, **kmm** – Caribbean-Mediterranean-Mauritanian, **lm** – Lusitanian-Mediterranean, **lmi** – Lusitanian-Mediterranean-Indian, **lmm** – Lusitanian-Mediterranean-Mauritanian, **lmmg** – Lusitanian-Mediterranean-Mauritanian-Guinean, **lmmwi** – Lusitanian-Mediterranean-Mauritanian-West Indian, **lmnei** – Lusitanian-Mediterranean-Northeast Indian, **lmnz** – Lusitanian-Mediterranean-New Zealand, **lmsa** – Lusitanian-Mediterranean-South African, **lmwi** – Lusitanian-Mediterranean-West Indian, **lmwiwp** – Lusitanian-Mediterranean-West Indo-West Pacific, **lmwp** – Lusitanian-Mediterranean-West Pacific, **lp** – Lusitanian-Pontian, **m** –

Mediterranean, (**mca**) – Mediterranean-Central Asian, **miwp** – Mediterranean-Indo-West Pacific, **mj** – Mediterranean-Japonic, **mmgt** – Mediterranean-Mauritanian-Guinean-Tasmanian, **mni** – Mediterranean-North Indian, **mnz** – Mediterranean-New Zealand, **mrs** – Mediterranean-Red Sea, (**mwca**) – Mediterranean-West Central Asian, (**na**) – North American, **nam** – North Atlantic-Mediterranean, **namep** – North Atlantic-Mediterranean-East Pacific, **nami** – North Atlantic-Mediterranean-Indian, **namim** – North Atlantic-Mediterranean-Indo-Malayan, **naminz** – North Atlantic-Mediterranean-Indo-New Zealand, **namip** – North Atlantic-Mediterranean-Indo-Pacific, **namiwp** – North Atlantic-Mediterranean-Indo-West Pacific, **namj** – North Atlantic-Mediterranean-Japonic, **namnei** – North Atlantic-Mediterranean-Northeast Indian, **namnep** – North Atlantic-Mediterranean-Northeast Pacific, **namni** – North Atlantic-Mediterranean-North Indian, **namnp** – North Atlantic-Mediterranean-North Pacific, **namnz** – North Atlantic-Mediterranean-New Zealand, **namp** – North Atlantic-Mediterranean-Pacific, **namrs** – North Atlantic-Mediterranean-Red Sea, **namrsnep** – North Atlantic-Mediterranean-Red Sea-Northeast Pacific, **namsp** – North Atlantic-Mediterranean-South Pacific, **namsep** – North Atlantic-Mediterranean-Southeast Pacific, **namswp** – North Atlantic-Mediterranean-Southwest Pacific, **namwi** – North Atlantic-Mediterranean-West Indian, **namwp** – North Atlantic-Mediterranean-West Pacific, **nap** – North Atlantic-Pontian, **napnei** – North Atlantic-Pontian-Northeast Indian, **neamal** – Northeast Atlantic-Mediterranean-Aleutian, **neamep** – Northeast Atlantic-Mediterranean-East Pacific, **neaminz** – Northeast Atlantic-Mediterranean-Indo-New Zealand, **neamj** – Northeast Atlantic-Mediterranean-Japonic, **neamnp** – Northeast Atlantic-Mediterranean-North Pacific, **neamnz** – Northeast Atlantic-Mediterranean-New Zealand, **neamswp** – Northeast Atlantic-Mediterranean-Southwest Pacific, **neamwp** – Northeast Atlantic-Mediterranean-West Pacific, (**nem**) – Northeast Mediterranean, (**nemit**) – Northeast Mediterranean-Iran-Turanian, **nm** (**nm**) – North Mediterranean, (**nmwca**) – North Mediterranean-West Central Asian, **nz** – New Zealand, (**o**) – Oriental, (**om**) – Oriental-Mediterranean, (**omca**) – Oriental-Mediterranean-Central Asian, (**omcaa**) – Oriental-Mediterranean-Central Asian-Australian, **p** – Pontian, (**pat**) – Palearctic-Afrotropical, (**pata**) – Palearctic-Afrotropical-Australian, **pc** – Pontian-Caspian, **pca** – Pontian-Caspian-Aral, **pinz** – Pontian-Indo-New Zealand, (**pm**) – Pontomediterranean, **pnep** – Pontian-Northeast Pacific, (**pno**) – Palearctic-Neotropical-Oriental, (**po**) – Palearctic-Oriental, (**poa**) – Palearctic-Oriental-Australian, (**ppt**) – Palearctic-Paleotropical, (**ppta**) – Palearctic-Paleotropical-Australian, (**ptm**) – Paleotropical-Mediterranean, (**ptmca**) – Paleotropical-

Mediterranean-Central Asian, (**ptsp**) – Paleotropical-South Palearctic, **R** – relict, **Rc** – Caspian relict, (**se**) – South European, (**see**) – Southeast European, (**seea**) – Southeast European-Anatolian, (**seep**) – Southeast European-Pontian, (**seepc**) – Southeast European-Pontian-Caspian, **Sf** – subfossil, **SK (sk)** – Subcosmopolitan, **tam** – Tropical Atlantic-Mediterranean, (**tp**) – Transpalaearctic, (**tpo**) – Transpalaearctic-Oriental, **vck** – Virginian-Carolinian-Caribbean, **vclm** – Virginian-Celtic-Lusitanian-Mediterranean, (**wces**) – West and Central Eurosiberian, (**wcp**) – West and Central Palaearctic, (**wcpo**) – West and Central Palaearctic-Oriental, (**wes**) – West Eurosiberian, (**wesa**) – West Eurosiberian-Anatolian, (**wp**) – West Palearctic, (**wpat**) – West Palearctic-Afrotropical, (**wppt**) – West Palearctic-Paleotropical, + – species known only from shells, ● – occurrence of endemic taxa, **?** – probable category.

**Ecological data**: **ar** – argillophilous, $\alpha$ – $\alpha$-mesosaprobic, $\alpha$-$\beta$ – $\alpha$-$\beta$-mesosaprobic, $\beta$ – $\beta$-mesosaprobic, **B** – brackish, **bt** – benthos, **co** – commensal, **CR** – critically endangered, **cr** – crenobiont, **cs** – coastal silt, **eb** – eurybathic, **ec** – ectoparasite, **eh** – euryhaline, **ep** – epibathic, **epi** – epibiont, **epp** – epipelagic, **et** – eurythermal, **eu** – eurybiont, **gw** – ground-water, **ha** – halophilous or halobiont, **hb** – hypobathic, **if** – interstitial fauna, **is** – invasive species, **L** – freshwater, **l** – littoral zone (medio-, pseudolittoral, intertidal), **lr** – rocky littoral, **ls** – sandy littoral, **lt** – rocks or lithophilous, **M** – marine, **mb** – mesobathic, **mc** – *Mytilus* cenosis, **ms** – *Mytilus* silt, **o** – oligosaprobic, **p** – plankton, **pa** – parasite, **pe** – pelophilous, **ph** – algae overgrowth or phytophilous, **phc** – *Phyllophora* coenosis, **phs** – *Phaseolina* silt, **po** – potamophilous, **pp** – pelagic, **ps** – sand or psammophilous, **rh** – rhithrophilous, **ro** – rocky, **s** – silt, **sb** – stenobathic, **sep** – stenoepibathic, **sg** – shells and sand with shells, **shb** – stenohypobathic, **sl** – sublittoral zone (infra- and circalittoral, subtidal), **slc** – *Cystoseira* sublittoral, **slr** – rocky sublittoral, **sls** – sandy sublittoral, **sp** – supralittoral zone (supratidal), **spr** – rocky supralittoral, **sps** – sandy supralittoral, **sw** – stagnant water, **T** – terrestrial, **th** – thermophile, **TL** – terrestrial forms connected with water, **tx** - trogloxene, **x** – xenosaprobic, **zc** – *Zostera* cenosis, **‰** – limiting freshwater level for marine and salinity level for the freshwater forms, ( ) – rarely exception.

**Conservation status**: ■ – Black Sea Red Data Book, ▲ – Red Data Book of Bulgaria, ♦ – European and IUCN Red List. **BA** – Barcelona Convention, SPA/BD Protocol (Annex II – 1999, 2017), Protocol concerning Specially Protected Areas and Biological Diversity in the Mediterranean, **BC** – Bern Convention (II, III); **BSZL** – Black Sea Zoobenthost List IUCN Criteria, **BU** – Bucharest Convention, **DD** – data deficient, **Ei** – European importance, **EN** –

endangered, **EX** – Extinct, **HD** – Habitats Directive, **LC** – least concern, **LR** – lower risk, **N** – national importance, **NE** – not evaluated, **NT** – near threatened, **r** – rare, **VU** – vulnerable, **W** – world importance.

# Results and Discussion

A total of 19 types, 39 classes, 123 orders, 477 families and 1590 species have been known from the Bulgarian Black Sea (Table 3). These taxa include 1059 species (66.6%) marine and marine-brackish forms and 531 species (33.4%) freshwater-brackish, freshwater and terrestrial forms, connected with water. A small number of supercosmopolitan forms (25 species), inhabitants of the marine, freshwater and terrestrial cenoses are scrutinized to both two groups. Five types (Nematoda, Rotifera, Annelida, Arthropoda and Mollusca) have a high species composition (over 100 species). Of these, the richest in species are Arthropoda (813 species – 51.1%), Annelida (175 species – 11.0%) and Mollusca (155 species – 9.79%). The rest 14 types include from 1 to 38 species. The Bulgarian fauna comprises about 70% of the known 2000-2200 species from the Black Sea and Azov Sea (Tables 2 and 3). For individual taxa this percentage varies considerably and depends on the level of study. The species composition varies depending on whether the authors considered only marine and marine-brackish forms or include freshwater-brackish, freshwater and terrestrial forms, related to water. The rich in brackish basins Ukrainian and Russian Black Sea coast is considerably superior to the Bulgarian coast in brackish taxa.

# Register of the marine, brackish, freshwater and terrestrial Invertebrate free-living fauna from the Bulgarian Black Sea coast

## PORIFERA

### DEMOSPONGIAE

#### HADROMERIDA

#### Clionaidae

*Pione vastifica* (Hancock, 1849) [*Cliona pontica, C. stationis, Vioa grantii*]; **D** – 7, 8, 9, 10, 11, 12, 13, 14, 15, 16, 27; **V** – -42; **Z** – namwp; **E** – M, bt, mb, mc; **R** – 32, 249, 389.

#### Suberitidae

*Suberites carnosus* (Johnston, 1842) [*Halichondria, S. domuncula*]; **D** – 3, 7, 16, 34; **V** – 6-180; **Z** – am; **E** – M, bt, eb, ms, phs; **R** – 32, 249, 389.

*Suberites prototypus* Czerniavsky, 1880 [*Prosuberites brevispinus, P. epiphytum*]; **D** – 7, 27, 28; **V** – 3-90; **Z** – ● p; **E** – M, bt, eb, ms, phs; **R** – 32, 84, 249, 389.

#### HALICHONDRIDA

#### Halichondriidae

*Halichondria panicea* (Pallas, 1766) [*H. grossa, Spongia*]; **D** – 7; **V** – 2-65; **Z** – K; **E** – M, bt, eb, lt, ■; **R** – 32, 114, 249, 389.

#### POECILOSCLERIDA

#### Microcionidae

*Clathria cleistochela* (Topsent, 1925) [*Microciona*]; **D** – 7; **V** – 0-5; **Z** – lmm, ? i; **E** – M, bt, ? is; **R** – 84, 249, 389.

## Tedaniidae

*Tedania anhelans* (Lieberkühn, 1859) [*T. nigrescens, Haliclona, Reniera digitata*]; **D** – 5, 6, 23, 24; **V** – -22; **Z** – ? amp; **E** – M, bt, ep, ph; **R** – 32, 67, 249, 389.

## Coelosphaeridae

*Lissodendoryx variisclera* (Swartschewsky, 1905) [*L. dictyonoides*]; **D** – 7; **V** – 8-26; **Z** – ● p; **E** – M, bt, ep, lt; **R** – 32, 249, 389.

## Crellidae

*Crella gracilis* (Alander, 1942) [*Kowalewskyella, Yvesia*]; **D** – 3; **V** – 14-60; **Z** – cp; **E** – M, bt, eb, lt; **R** – 32, 249, 389.

## Mycalidae

*Mycale syrinx* (Schmidt, 1862) [*Esperia lorenzii, E. muscoides*]; **D** – 7, 34; **V** – 0-87; **Z** – lmm; **E** – M, bt, eb, ro, s; **R** – 32, 249, 389.

HAPLOSCLERIDA

## Petrosiidae

*Petrosia ficiformis* (Poiret, 1789) [*P. clavata, P. dura, Reniera boutschinskii*]; **D** – 3, 5, 24; **V** – -40; **Z** – kmm; **E** – M, bt, mb, slc, slr; **R** – 32, 249, 389.

## Chalinidae

*Chalinula limbata* (Montagu, 1818) [*Halichona, Haliclonissa, Pachychalina*]; **D** – 7; **Z** – clmm; **E** – M, bt; **R** – 32, 249, 389.

*Haliclona aquaeductus* (Schmidt, 1862) [*H. alba, Adocia, Reniera*]; **D** – 5, 11, 12, 13, 34; **V** – -100; **Z** – lm; **E** – M, bt, eb, ro; **R** – 32, 249, 389.

*Haliclona cinerea* (Grant, 1826) [*H. palmata, Adocia, Reniera, Spongia*]; **D** – 5; **V** – -75; **Z** – aamip; **E** – M, bt, eb, phc; **R** – 32, 249, 389.

*Haliclona flavescens* (Topsent, 1893) [*Reniera*]; **D** – 7; **V** – -35; **Z** – hom; **E** – M, bt, ro; **R** – 84, 249, 389.

*Haliclona grossa* (Schmidt, 1864) [*Adocia, Reniera*]; **D** – 12, 13; **Z** – em; **E** – M, bt; **R** – 32, 249, 389.

*Haliclona implexa* (Schmidt, 1868) [*H. informis, Adocia, Reniera curiosa*]; **D** – 1, 2, 3, 4, 5; **V** – -85; **Z** – mmgt; **E** – M, bt, eb, ro; **R** – 32, 249, 389.

*Haliclona inflata* (Schmidt, 1868) [*Adocia, Reniera*]; **D** – 5, 7, 12, 13; **Z** – adp; **E** – M, bt; **R** – 32, 389.

*Haliclona irregularis* (Czerniavsky, 1880) [? *Ulosa stuposa*, ? *Haliclonissa digitata*]; **D** – 24, 49; **V** – -22; **Z** – ? ● p, ? clm; **E** – M, bt, zc; **R** – 32, 249, 389.

*Haliclona simulans* (Johnston, 1842) [*Adocia densa, Isodictya pallida, Reniera*]; **D** – 5; **Z** – clmm; **E** – M, bt; **R** – 32, 249, 389.

*Haliclona tubulifera* (Swartschewsky, 1905) [*Adocia, Reniera*]; **D** – 3; **Z** – ● p; **E** – M, bt; **R** – 32, 249, 389.

## Spongillidae

*Ephydatia fluviatilis* (Linnaeus, 1759); **D** – 34, *68; **Z** – (hpta); **E** – L, 2.5‰, bt; **R** – 32, 374, 389.

DENDROCERATIDA

## Dysideidae

*Dysidea fragilis* (Montagu, 1818) [*Spongia, Spongelia*]; **D** – 1, 2, 3, 4, 5, 7; **V** – -40; **Z** – clmm; **E** – M, bt, mb, ps, ro; **R** – 32, 84, 389.

CALCAREA

LEUCOSOLENIDA

## Sycettidae

*Sycon ciliatum* (Fabricius, 1780) [*S. coronata*]; **D** – 23, 24, 27, 28, 49, 50; **V** – 0-180; **Z** – aamswp; **E** – M, bt, eb, ph, ro; **R** – 198, 249.

# CNIDARIA

## HYDROZOA

### CAPITATA

#### Hydridae

*Hydra viridissima* Pallas, 1766 [*Chlorohydara*]; **D** – 71, 88, 93; **Z** – (k); **E** – L, 5‰, bt; **R** – 374, 389.

#### Cladonematidae

*Cladonema radiatum* Dujardin, 1843; **D** – 12; **Z** – amip; **E** – M, bt-p; **R** – 91, 249, 389.

#### Corymorphidae

*Corymorpha nutans* Sars, 1835 [*C. sarsi*]; **D** – 7; **V** – 10-100; **Z** – anam; **E** – M, bt-p, eb, r; **R** – 160, 249, 333, 389.

#### Corynidae

*Coryne pusilla* (Gaertner, 1774); **D** – 11, 16; **Z** – namwp; **E** – M, bt-p; **R** – 91, 389.

*Sarsia tubulosa* (Sars, 1835) [*S. mirabilis, Coryne, Syncoryne*]; **D** – 7, 16, 49; **Z** – aanambp; **E** – M, bt-p; **R** – 249, 319, 352, 389.

#### Moerisiidae

*Odessia maeotica* (Ostroumoff, 1896) [*Ostroumovia, Pontia, Moerisia*]; **D** – 7, 71, 78; **Z** – lm, ? Rc; **E** – M-B, 25‰, bt-p, ■; **R** – 297, 299, 300, 319, 373, 374, 376, 383.

*Moerisia inkermanica* Paltschikowa-Ostroumowa, 1925 [*Ostroumovia*]; **D** – *69, 71, *79, 84, 88; **Z** – amip; **E** – M-B-L, 40‰, bt-p; **R** – 249, 317, 374, 376, 386, 389.

# Protohydridae

*Protohydra leuckarti* Greeff, 1870 [*P. squamata*]; **D** – *79, 96; **V** – 0-1; **Z** – cbm; **E** – M, 7‰, bt; **R** – 249, 378, 389.

LEPTOTHECATA (CONICA)

## Aglaopheniidae

*Aglaophenia pluma* (Linnaeus, 1767); **D** – 7, 16; **V** – 0-100; **Z** – amwp; **E** – M, bt, eb; **R** – 91, 249, 295, 389.

## Kirchenpaueriidae

*Kirchenpaueria halecioides* (Alder, 1859) [*Plumularia*]; **D** – 7, 24; **Z** – SK; **E** – M, bt, eb; **R** – 249, 389.

## Blackfordiidae

*Blackfordia virginica* Mayer, 1910 [*Campanulina pontica*]; **D** – 71, *79, 84, *86, 88; **Z** – nami, i; **E** – M, 7‰, bt-p, sl, is; **R** – 160, 249, 374, 389.

## Campanulinidae

*Opercularella lacerata* (Johnston, 1847) [*Campanulina tenuis*]; **D** – 7; **V** – 0-20; **Z** – bam; **E** – M, bt, ep; **R** – 249, 295, 389.

*Phialella quadrata* (Forbes, 1848) [*Campanulina repens*]; **D** – 29, 49; **V** – -10; **Z** – amip, ? SK; **E** – M, bt-p, ep; **R** – 249.

## Sertulariidae

*Sertularella polyzonias* (Linnaeus, 1758); **D** – 3, 7, 12, 16; **V** – 0-100; **Z** – SK, ? K; **E** – M, bt, eb; **R** – 67, 91, 249, 295, 389.

FILIFERA

## Bougainvilliidae

*Calyptospadix cerulea* Clarke, 1882 [*Bougainvillia megas, Garveia franciscana, Perigonimus*]; **D** – *69, 84; **Z** – cst, amip, i; **E** – M-B, bt, eh, l, is; **R** – 102, 249, 252, 295.

33

*Bougainvillia muscus* (Van Beneden, 1844) [*B. ramosa, Perigonimus*]; **D** – 7; **Z** – amip, ? SK; **E** – M, bt-p; **R** – 249, 295, 389.

## Oceaniidae (Clavidae)

*Clava multicornis* (Forskål, 1775) [*C. squamata*]; **D** – 7; **Z** – bam; **E** – M-B, bt, mc, ro; **R** – 295, 389.

*Cordylophora caspia* (Pallas, 1771) [*C. lacustris*]; **D** – 58, *68, *69, *78, *79, 84, 88, 93, 94, 96; **Z** – namnz, i, Rc, ? SK; **E** – B, eh, 0.1-20‰, bt; **R** – 84, 249, 374, 389.

## Eudendriidae

*Eudendrium ramosum* (Linnaeus, 1758) [*Tubularia*]; **D** – 7; **V** – -10; **Z** – SK; **E** – M, bt, eb; **R** – 249, 298, 389.

*Eudendrium* cf. *merulum* Watson, 1985; **D** – 7; **V** – 0-1, 15; **Z** – amiswp; **E** – M, bt, ro, r; **R** – 321.

## Hydractiniidae

*Hydractinia carnea* Sars, 1846 [*Podocoryna*]; **D** – 5, 7, *69; **V** – 0-70; **Z** – amwp; **E** – M, 12‰, bt-p, eb; **R** – 67, 84, 249, 317, 319, 331, 374, 389.

## Rathkeidae

*Rathkea octopunctata* (Sars, 1835); **D** – 7, 29; **Z** – cbma, (? i); **E** – M-B, p (bt), c; **R** – 160, 249, 319, 389.

LIMNOMEDUSAE (LIMNOPOLYPAE)

## Microhydrulidae

*Microhydrula pontica* Valkanov, 1965; **D** – ?; **Z** – clm; **E** – M, 10-20‰, bt; **R** – 391, 392.

## Olindiasidae (Olindiidae)

*Craspedacusta sowerbyi* Lankester, 1880 [*C. ryderi*]; **D** – 34; **Z** – (k), i; **E** – L, is, bt-p; **R** – 386, 389.

? *Olindias phosphorica* (Delle Chiaje, 1841); **D** – 96; **Z** – lm; **E** – M, p, r; **R** – 281.

? *Maeotias marginata* (Modeer, 1791) [*M. inexpectata*]; **D** – ?; **Z** – clm; **E** – M-B, p ; **R** – 281.

PROBOSCOIDA

## Campanulariidae

*Orthopyxis integra* (MacGillivray, 1842) [*Campanularia integriformis*]; **D** – 7, *69; **V** – 1-60; **Z** – amip, ? K; **E** – M, bt, eb, r; **R** – 249, 295, 374, 389.

*Campanularia volubilis* (Linnaeus, 1758) [*C. volubilis* var. *urceolata*, *C. urceolata*]; **D** – 11, 12, 16; **V** – 5-100; **Z** – amp; **E** – M, bt, eb; **R** – 91, 249, 389.

*Clytia hemisphaerica* (Linnaeus, 1758) [*C. johnstoni, Campanularia*]; **D** – 3, 11, 12, 16; **V** – 0-15; **Z** – amp, ? SK; **E** – M, bt-p, ph, ro; **R** – 67, 91, 249, 389.

*Gonothyraea loveni* (Allman, 1859) [*Laomedea, Obelia*]; **D** – 3, 7; **V** – 0-30; **Z** – bamswp, ? K; **E** – M, bt, ep-mb; **R** – 68, 249, 295, 389.

*Hartlaubella gelatinosa* (Pallas, 1766) [*Laomedea bicuspidata, Obelia*]; **D** – 7; **Z** – amp; **E** – M, bt, r; **R** – 84, 249, 389.

*Laomedea angulata* Hincks, 1861 [*Obelia*]; **D** – 7, *69; **Z** – am; **E** – M, 12‰, bt, r; **R** – 84, 249, 389.

*Obelia longissima* (Pallas, 1766) [*Laomedea*]; **D** – 7; **V** – 0-100; **Z** – SK; **E** – M, bt-p, eb; **R** – 84, 160, 249, 389.

SCYPHOZOA

SEMAEOSTOMEAE

## Ulmaridae

*Aurelia aurita* (Linnaeus, 1758); **D** – 1, 2, 3, 4, 5, 6, 7, 8, 9, 10, 11, 12, 13, 14, 15, 16, 17, 18, 19, 20, 21, 22, 29, 69, 84, 88; **V** – 0-100; **Z** – K; **E** – M, p; **R** – 91, 160, 287, 319, 374, 389.

## Rhyzostomatidae

*Rhizostoma pulmo* (Macri, 1778); **D** – 1, 2, 3, 4, 5, 6, 7, 8, 9, 10, 11, 12, 13, 14, 15, 16, 17, 18, 19, 20, 21, 22, 29, *69; **Z** – namrs; **E** – M, p; **R** – 91, 160, 287, 319, 374, 389.



## Kishinouyeidae

*Lucernariopsis campanulata* Lamouroux, 1815 [*Lucernaria*]; **D** – 12, 21, 24; **V** – 0-20; **Z** – clm; **E** – M, bt, ep, slc, ph; **R** – 91, 249, 389.

## ANTHOZOA: HEXACORALIA

### ACTINIARIA

## Actiniidae

*Actinia equina* (Linnaeus, 1758) [*A. equina* var. *zonata*]; **D** – 1, 2, 3, 4, 5, 6, 7, 8, 9, 10, 11, 12, 13, 14, 15, 16, 17, 18, 19, 20, 21, 22, 24, 69; **V** – 0-38; **Z** – eamiwp; **E** – M, bt, ep, lr, slr, sg; **R** – 31, 91, 249, 295, 389.

## Diadumenidae

? *Diadumene lineata* (Verrill, 1869) [? *Actinothoe clavata*]; **D** – 24, 49; **V** – 0-15; **Z** – SK, i; **E** – M-B, bt, 0.5-35‰, is; **R** – 355.

## Sagartiidae

*Actinothoe clavata* (Ilmoni, 1830) [*A. angulicoma, Cyliste viduata*]; **D** – 7, 12, 13, 16, 25, 26, 69, 77; **V** – 1-20; **Z** – clm; **E** – M, bt, ep, mc, sg, s; **R** – 84, 91, 155a, 249, 295, 389.

### CERIANTHARIA

## Cerianthidae

*Pachycerianthus solitarius* (Rapp, 1829) [*Cerianthus vestitus*]; **D** – 2; **V** – 60-175; **Z** – lm; **E** – M, bt, shb, phs; **R** – 249.

# CTENOPHORA

## ATENTACULATA

### BEROIDA

#### Beroidae

*Beroe ovata* Bruguière, 1789; **D** – 1, 2, 3, 4, 5, 6, 7, 8, 9, 10, 11, 12, 13, 14, 15, 16, 17, 18, 19, 20, 21, 22, 29; **Z** – amiwp, i; **E** – M, p, is; **R** – 160, 162, 200.

## TENTACULATA

### CYDIPPIDA

#### Pleurobrachiidae

*Pleurobrachia rhodopis* Chun, 1879 [*P. pileus*, *P. rhdodactyla*]; **D** – 7, 10, 11, 12, 13, 14, 15, 16, 17, 29, *69; **V** – 0-100; **Z** – ● p; **E** – M, p, eb; **R** – 91, 317, 319, 374.

### LOBATA

#### Bolinopsidae

*Bolinopsis vitrea* (L. Agassiz, 1860); **D** – 3, 7, 29; **Z** – cst, amip, ? i; **E** – M, p; **R** – 292, 355.

*Mnemiopsis leidyi* A. Agassiz, 1865 [*Leucothea multicornis*, *Mnemia maccradyi*]; **D** – 1, 2, 3, 4, 5, 6, 7, 8, 9, 10, 11, 12, 13, 14, 15, 16, 17, 18, 19, 20, 21, 22, 29, 69; **V** – 0-25; **Z** – vck, ? K, i; **E** – M, p, is, sb; **R** – 65, 160, 194, 195, 196, 197, 199, 201.

# PLATHELMINTHES

## TURBELLARIA

### ACOELA

#### Convolutidae

*Convoluta hypparchia* Pereyaslawzewa, 1893 [*C. festiva*]; **D** – 7, *69, *79; **Z** – m; **E** – M, bt; **R** – 84, 374.

### Otocelididae

*Otocelis rubropunctata* (Schmidt, 1852) Diesing, 1862; **D** – 7, 25, 26; **V** – -18; **Z** – neamj; **E** – M, bt, pe-ps, ep; **R** – 386.

### Taurididae

*Taurida fulvomaculata* (Ax, 1959) [*Convoluta*]; **D** – 7; **Z** – m; **E** – M, bt, ep, ps; **R** – 392.

MACROSTOMIDA

### Macrostomidae

*Macrostomum appendiculatum* (O. Fabricius, 1826); **D** – 77; **Z** – clm; **E** – ? B-L, 50‰, bt, eh; **R** – 374.

POLYCLADIDA

### Leptoplanidae

*Leptoplana tremellaris* (Müller, 1774) Örsted, 1843; **D** – 5, *69, 84, *86; **Z** – ce; **E** – M, bt; **R** – 67, 374.

### Prosthiostomidae

*Prosthiostomum siphunculus* (Delle Chiaje, 1818) Lang, 1884; **D** – 3; **Z** – clm; **E** – M, bt; **R** – 67, 389.

### Stylochidae

*Stylochus tauricus* Jakubova, 1909 [*Stycoplana*]; **D** – 5; **Z** – m; **E** – M, bt; **R** – 67, 389.

RHABDOCOELA

### Dalyelliidae

*Gieysztoria expedita* (von Hofsten, 1907) [*Dalyellia*]; **D** – *68, 71, 76, *86, 88; **Z** – (pat); **E** – L, 12‰, bt; **R** – 374, 389.

## Karkinorhynchidae

*Baltoplana valkanovi* Ax, 1959; **D** – 7; **Z** – m; **E** – M, 12‰, bt; **R** – 392.

## Polycystididae

*Acrorhynchides reprobatus* (Pereyaslawzewa, 1892) Strand, 1928 [Graff, 1905]; **D** – *69; **Z** – m; **E** – M, bt; **R** – 374, 389.

*Gyratrix hermaphroditus* Ehrenberg, 1831 [*Gyrator*]; **D** – *68, *69, 71; **Z** – SK; **E** – M-L, 5‰, bt, eh; **R** – 374, 389.

*Polycystis naegellii* Kölliker, 1845; **D** – 7, 24; **Z** – clm; **E** – M, bt; **R** – 203, 252.

*Rogneda minuta* Uljanin, 1870 [*Polycystis*]; **D** – *69; **Z** – clm; **E** – M, 12‰, bt, ph-ps; **R** – 374, 389.

*Rogneda polyrhabdota* Ax, 1959; **D** – 7; **Z** – m; **E** – M, bt; **R** – 392.

## Promesostomidae

*Promesostoma marmoratum* (Schultze, 1851) v. Graff, 1882; **D** – *69, *79; **Z** – bam; **E** – M, 8‰, bt; **R** – 374, 389.

Seriata

## Coelogynoporidae

*Coelogynopora biarmata* Steinböck, 1924; **D** – 7, 25, 51; **V** – 1-2; **Z** – clm, ? bam; **E** – M, bt, sep, sls, ps, eh; **R** – 203, 249, 388, 389.

*Coelogynopora tenuiformis* Karling, 1966; **D** – 7, 25, 51; **Z** – pnep; **E** – M, bt, sls, ps, r; **R** – 203, 252.

## Monocelididae

*Archilina endostyla* Ax, 1959; **D** – 7, 25, 51; **V** – 0.5-2; **Z** – m; **E** – M, bt, sep, sls, ps; **R** – 203, 252.

*Monocelis longiceps* (Dugès, 1830); **D** – *69; **Z** – clm; **E** – M, 2‰, bt; **R** – 374, 389.

*Pseudomonocelis ophiocephala* (Schmidt, 1861), Meixner, 1943; **D** – 7, 25, 51; **V** – 1-2; **Z** – m; **E** – M, bt, sep, sls, ps; **R** – 203, 252.

## Otoplanidae

*Archotoplana holotricha* Ax, 1956; **D** – 6, 7, 51; **Z** – m, ? cm; **E** – M, bt, ep, sls, ps; **R** – 203, 252.

*Otoplana bosporana* Ax, 1959; **D** – 7; **Z** – m; **E** – M, bt, ep, sls, ps; **R** – 392.

*Parotoplanella progermaria* Ax, 1956; **D** – 6, 7, 51; **V** – 0-1; **Z** – m, ? lm; **E** – M, bt, sep, ls, sls, ps; **R** – 203, 252.

*Postbursoplana fibulata* Ax, 1955; **D** – 6, 7, 25; **V** – 2; **Z** – m, ? lm; **E** – M, bt, sep, sls, ps; **R** – 203, 252.

*Postbursoplana pontica* Ax, 1959; **D** – 7; **Z** – ● p; **E** – M, bt; **R** – 392.

*Pseudosyrtis subterranea* (Ax, 1951) Ax, 1956 [*Otoplana*]; **D** – 7, 35; **Z** – clm; **E** – M-B, bt, sep, ls, ps; **R** – 389.

*Triporoplana synsiphonioides* Ax, 1956; **D** – 7, 25, 51; **V** – 0.5; **Z** – lm; **E** – M, bt, sep, sls, ps; **R** – 203, 252.

## Dendrocoelidae

*Dendrocoelum lacteum* (Müller, 1774); **D** – 58, *68; **Z** – (h); **E** – L, 2‰, bt, ph; **R** – 206, 374, 389.

## Dugesiidae

*Dugesia polychroa* (Schmidt, 1861) [*Planaria*]; **D** – *68; **Z** – (h); **E** – L, 0.5‰, bt; **R** – 374, 389.

# NEMERTINI

## ANOPLA

### PALAEONEMERTINI

### Cephalothricidae

*Cephalothrix arenaria* Hylbon, 1957; **D** – ?; **Z** – clm; **E** – M, bt, ps; **R** – 249.

*Cephalothrix linearis* (Rathke, 1799) [*C. bioculata*]; **D** – 27, 28; **V** – 40-120; **Z** – bam; **E** – M, bt, shb, s; **R** – 249.

*Cephalothrix rufifrons* (Johnston, 1837); **D** – 27, 28; **V** – 40-120; **Z** – clm; **E** – M, bt, shb, s; **R** – 249.

### Tubulanidae

*Carinina heterosoma* G. I. Müller, 1965; **D** – 26, 27; **V** – 5-70; **Z** – ● p; **E** – M, bt, s, ep-mb; **R** – 249.

### HETERONEMERTINI

### Cerebratulidae

*Cerebratulus marginatus* Renier, 1804; **D** – 27, 28; **V** – 40-120; **Z** – abambp; **E** – M, bt, phs, shb, s; **R** – 249.

*Cerebratulus ventrosulcatus* Bürger, 1892; **D** – 26, 27, 28; **V** – 6-120; **Z** – m; **E** – M, bt, s, eb; **R** – 249.

### Lineidae

*Lineus bilineatus* (Renier, 1804); **D** – ?; **Z** – bambp; **E** – M, bt, sl; **R** – 249.

*Lineus ruber* (O. F. Müller, 1774); **D** – 49; **Z** – nam; **E** – M, bt, sl, mc, ro; **R** – 249.

*Ramphogordius lacteus* Ratke, 1843 [*Lineus*]; **D** – *69; **V** – 2-65; **Z** – clp, ? clm; **E** – M, bt, mb-eb; **R** – 249, 374, 389.

*Notospermus geniculatus* (Delle Chiaje, 1828) [*Lineus*]; **D** – ?; **Z** – lmnz; **E** – M, bt; **R** – 249.

*Micrura fasciolata* Ehrenberg, 1828; **D** – 1, 2, 3, 4, 5, 6, 7, 8, 9, 10, 11, 12, 13, 14, 15, 16, 17, 18, 19, 20, 21, 22, 27, 28; **V** – 18-125; **Z** – clm; **E** – M, bt, mb-eb, s; **R** – 249.

*Pontolineus arenarius* Müller et Scripcariu, 1964; **D** – 25; **V** – 8-16; **Z** – ● p; **E** – M, bt, sep, sls, ps; **R** – 249.

*Pussylineus gabriellae* Corrêa, 1956; **D** – ?; **Z** – m; **E** – M, bt; **R** – 249.

ENOPLA

HOPLONEMERTINI

## Amphiporidae

*Amphiporus bioculatus* McIntosh, 1874; **D** – ?; **Z** – bap, ? bam; **E** – M, bt; **R** – 249.

*Amphiporus lactifloreus* (Johnston, 1828); **D** – ?; **Z** – bam; **E** – M, bt, sl; **R** – 249.

*Zygonemertes maslovskyi* (Czerniavsky, 1880); **D** – 50; **Z** – ● p; **E** – M, bt, lr; **R** – 249.

## Cratenemertidae

*Nipponnemertes pulchra* (Johnston, 1837) [*Amphiporus*]; **D** – ?; **Z** – baap; **E** – M, bt; **R** – 249.

## Emplectonematidae

*Emplectonema gracile* (Johnston, 1837) [*Eunemertes*]; **D** – 3, 24; **V** – 0-10; **Z** – bambp; **E** – M, bt, ep, ro; **R** – 67, 249, 389.

## Ototyphlonemertidae

*Ototyphlonemertes antipai* G. I. Müller, 1968; **D** – 51; **V** – 0.2-1; **Z** – ● p; **E** – M, bt, ls, sep; **R** – 249.

42

## Tetrastemmatidae

*Prostoma graecense* (Böhmig, 1892) [*Tetrastemma*]; **D** – *68, *69; **Z** – (sk); **E** – L-B, 1‰, bt; **R** – 374, 389.

*Tetrastemma bacescui* G. I. Müller, 1962; **D** – 5, 24; **V** – 0.5-8; **Z** – • p; **E** – M, bt, lr, slr, ep; **R** – 249.

*Tetrastemma candidum* (O. F. Müller, 1774) [*Prostoma*]; **D** – 3, 5; **V** – 0-55; **Z** – anamp; **E** – M, bt, mb; **R** – 67, 249, 389.

*Tetrastemma coronatum* (Quantrefages, 1846); **D** – 26, 27; **V** – 4-40-120; **Z** – clm; **E** – M, bt, s, hb-eb; **R** – 249.

*Tetrastemma longissimum* Bürger, 1895 [? nom. dubium]; **D** – ?; **Z** – clm; **E** – M, bt; **R** – 249.

*Tetrastemma melanocephalum* (Johnston, 1837); **D** – 3, 5, 27; **V** – 40-70; **Z** – neaminz; **E** – M, bt, s, mb-eb; **R** – 67, 249.

*Tetrastemma vermiculus* (Quantrefages, 1846) [*Prostomatella*]; **D** – 24; **V** – 0-40; **Z** – nam; **E** – M, bt, slr, mb; **R** – 249.

# GASTROTRICHA

CHAETONOTOIDEA

## Chaetonotidae

*Aspidiophorus mediterraneus* Remane, 1927; **D** – 7, 51; **V** – 0.5-2; **Z** – namsep; **E** – M, bt, sep, ls, eh; **R** – 315, 389, 390.

*Chaetonotus maximus* Ehrenberg, 1831; **D** – 24, 35, 49; **Z** – K; **E** – M (B-L), bt, eh, l, ls; **R** – 249.

*Chaetonotus similis* Zelinka, 1889; **D** – 98; **Z** – (k); **E** – L-B, 8-18‰, bt; **R** – 249.

*Halichaetonotus decipiens* Remane, 1829 [*Chaetonotus*]; **D** – 7, 35; **Z** – nam; **E** – M, gw, ls; **R** – 315, 389.

*Halichaetonotus marinus* (Giard, 1904) [*Chaetonotus pleuracanthus*]; **D** – 16, 35; **Z** – clm; **E** – M, gw, ls, r; **R** – 249, 389.

*Heterolepidoderma marinum* Remane, 1926; **D** – 16, 51; **Z** – clm; **E** – M, bt, l, sl; **R** – 249, 389.

## Xenotrichulidae

*Heteroxenotrichula pygmaea* Remane, 1934 [*Xenotrichula*]; **D** – 7, 35; **Z** – nap, ? SK; **E** – M, gw, ls, eh; **R** – 249.

*Xenotrichula intermedia* Remane, 1934 [*X. beauchampi*]; **D** – 12, 16, 18, 35; **Z** – nam, ? SK; **E** – M, gw, if, ls, eh; **R** – 249, 389, 390.

MACRODASYOIDEA

## Dactylopodolidae

*Dendrodasys ponticus* Valkanov, 1957; **D** – 7, 35; **Z** – pm; **E** – M, gw, if, ls; **R** – 249, 389, 390.

## Macrodasyidae

*Macrodasys africanus* Remane, 1950 [*M. africanus* var. *ponticus* Valkanov, 1957]; **D** – 7, 25, 35, 51; **V** – 0-2; **Z** – api; **E** – M, ep, gw, ls; **R** – 249, 315.

## Thaumastodermatidae

*Acanthodasys aculeatus* Remane, 1927; **D** – 16, 25, 51; **V** – 0-6; **Z** – namni; **E** – M, bt, ep, ls, sls; **R** – 315, 389, 390.

## Turbanellidae

*Turbanella cornuta* Remane, 1925; **D** – 7, 25, 35, 51; **V** – 0-6; **Z** – clm; **E** – M, bt, eh, ep, ls, sls; **R** – 249, 389, 390.

*Turbanella pontica* Valkanov, 1957; **D** – 7, 35, 51; **Z** – ● p; **E** – M (B), gw, ps, ls; **R** – 249, 315, 390.

# NEMATODA (NEMATA)



ENOPLIDA

## Anoplostomatidae

*Anoplostoma viviparum* (Bastian, 1865); **D** – 7, 25, 27, 28; **V** – 10, 120; **Z** – kclm; **E** – M, bt, sl, ps-s, ms-phs; **R** – 121, 344.

## Anticomidae

*Anticoma acuminata* (Eberth, 1863) Stekhoven, 1950; **D** – 7; **Z** – aminz; **E** – M, bt; **R** – 121.

## Enchelidiidae

*Symplocostoma ponticum* Filipjev, 1918; **D** – 7, 25; **V** – 12-40-; **Z** – m; **E** – M, bt, mb, sl, ps-s; **R** – 252, 347.

*Symplocostoma tenuicolle* (Eberth, 1863) [*S. longocolle*]; **D** – 7, 23, 24, 25, 76; **Z** – amswp; **E** – M, bt, eb, slc-zc, ps-s; **R** – 121, 341, 344.

*Polygastrophora hexabulba* (Filipjev, 1918); **D** – 16, 25; **V** – 8-9; **Z** – mnz, ? amnz; **E** – M, bt, ep, sl, ps; **R** – 252, 341, 344.

*Eurystomina ornata* (Eberth, 1863); **D** – 7, 16, 24; **Z** – bam; **E** – M, bt, ph; **R** – 341, 344.

*Eurystomina assimilis* (de Man, 1876); **D** – 7, 23, 24, 25, 27; **Z** – namrs; **E** – M, bt, ph; **R** – 121, 341, 344.

## Enoplidae

*Enoplus littoralis* Filipjev, 1918; **D** – 1, 2, 3, 4, 5, 6, 7, 8, 9, 10, 11, 12, 13, 14, 15, 16, 17, 18, 19, 20, 21, 22, 35, 46, 49, 51; **V** – 0.2-0.4 ; **Z** – clm ; **E** – M, bt, l, ps, gw; **R** – 341, 344.

*Enoplus maeoticus* Filipjev, 1916; **D** – 1, 2, 3, 4, 5, 6, 7, 8, 9, 10, 11, 12, 13, 14, 15, 16, 17, 18, 19, 20, 21, 22, 35, 51; **Z** – em, ? m; **E** – M, bt, l, ps, gw; **R** – 344.

*Enoplus quadridentatus* Berlin, 1853 [*E. euxinus, E. hirtus*]; **D** – 7, 8, 9, 10, 11, 12, 13, 14, 15, 16, 24, 25; **V** – 4, 80; **Z** – clm; **E** – M, bt, ep, slc, ph, ps-s; **R** – 121, 344, 389.

*Enoplus schulzi* Gerlach, 1952; **D** – 7, 13, 35; **Z** – cp; **E** – M, bt, l, gw, if; **R** – 363.

### Ironidae

*Trissonchulus benepapillosus* (Schulz, 1935) [*Dolicholaimus, Syringolaimus*]; **D** – 7, 35; **Z** – namni; **E** – M, bt, l, gw, if; **R** – 362, 386, 389.

*Trissonchulus oceanus* Cobb, 1920 [*Dolicholaimus nudus*]; **D** – 7, 35; **Z** – cm; **E** – M, bt, l, gw, if; **R** – 362.

*Dolicholaimus platonovae* Stoikov, 1979; **D** – 7, 35, 51; **Z** – • p; **E** – M, bt, l, ls, ps, gw; **R** – 343, 344.

*Syringolaimus caspersi* Gerlach, 1951; **D** – 7, 76; **Z** – cp; **E** – M, bt, eh-70‰, s, sls; **R** – 121, 344, 389.

### Leptosomatidae

*Leptosomatides euxinus* Filipjev, 1918; **D** – 3, 28; **V** – 25-, 200; **Z** – cp; **E** – M, bt, shb, s-phs; **R** – 252, 347.

*Leptosomatum sabangense* Steiner, 1915 [*L. bacillatum, Phanoglene*]; **D** – 10, 28; **V** – 0-10-90; **Z** – aanamip; **E** – M, bt, eb, phs, s; **R** – 252, 338, 344.

*Leptosomatum punctatum* (Eberth, 1863); **D** – 3, 28; **V** – 60-100; **Z** – m; **E** – M, bt, hb, s-phs; **R** – 252, 347.

*Pseudocella saveljevi* (Filipjev, 1927) [*Enoploides*]; **D** – 7, 25, 51; **Z** – cp; **E** – M, bt, l, ls, sls; **R** – 344.

### Oncholaimidae

*Viscosia cobbi* Filipjev, 1918; **D** – 16, 20, 25; **V** – 8-10, 150; **Z** – clm; **E** – M, bt, ep-eb, sls, s; **R** – 252, 338, 344.

*Viscosia glabra* (Bastian, 1865); **D** – 7, 25; **V** – 5-10, 120; **Z** – cclm; **E** – M, bt, ep-eb, sls, ps; **R** – 121, 344.

*Viscosia minor* Filipjev, 1918; **D** – 7, 25; **Z** – • p; **E** – M, bt, sls, ps, r; **R** – 344.

*Pontonema zernovi* (Filipjev, 1916) [*Oncholaimus vulgaris, Parancholaimus*]; **D** – 25, 27; **V** – -60, 150; **Z** – • p; **E** – M, bt, eb-mb, ps, s; **R** – 341, 344.

*Oncholaimellus mediterraneus* Stekhoven, 1942; **D** – 13, 35; **Z** – cm, ? clm; **E** – M, bt, l, ps, gw, if; **R** – 362.

*Oncholaimus brevicaudatus* Filipjev, 1918; **D** – 7, 35, 51; **V** – 230; **Z** – calm; **E** – M, bt, l, sl, ps, gw; **R** – 340, 344.

*Oncholaimus campylocercoides* De Koninck & Schuurmans Stekhoven, 1933; **D** – 2, 35, 51; **V** – 80-150; **Z** – cclm; **E** – M, bt, l, ls, ps, gw; **R** – 121, 340, 344.

*Oncholaimus conicauda* Filipjev, 1929; **D** – 7, 35, 51; **V** – 0.3-1; **Z** – clmnwi; **E** – M, bt, 12‰, l, ps, gw; **R** – 252, 341, 344.

*Oncholaimus dujardinii* de Man, 1878; **D** – 3, 7, 25, 35, 51; **V** – 0-8, -230; **Z** – namnz; **E** – M, bt, mb, ls, sls, gw; **R** – 121, 344.

## Oxystominidae

*Oxystomina clavicauda* (Filipjev, 1918); **D** – 16, 25; **V** – 20; **Z** – • p; **E** – M, bt, ep, sls, ps; **R** – 341, 344.

*Oxystomina elongata* (Bütschli, 1874) [*Oxystomatina*]; **D** – 7, 25; **V** – 10-25; **Z** – cclm; **E** – M, bt, ep, sls, ps, s, ar; **R** – 341, 344.

*Halalaimus ponticus* Filipjev, 1922; **D** – 3, 28; **V** – 68, -100; **Z** – • p; **E** – M, bt, hb, s-phs; **R** – 252, 347.

*Nemanema filiforme* (Filipjev, 1918) [*Oxystoma*]; **D** – 10, 27; **V** – 40; **Z** – em; **E** – M, bt, mb, sl, pe, s, ps; **R** – 252, 347.

## Rhabdodemaniidae

*Rhabdodemania pontica* Platonova, 1965; **D** – 27, 28; **V** – 30-150; **Z** – • p; **E** – M, bt, shb, ms, phs; **R** – 344.

## Thoracostomopsidae

*Enoploides amphioxi* Filipjev, 1918; **D** – 25; **V** – 8-9; **Z** – em; **E** – M, bt, eb, eu, ps-pe; **R** – 341, 344.

*Enoploides alexandrae* Uzunov, 1974; **D** – 6, 7, 35; **V** – 0.8; **Z** – • p; **E** – M, bt, l, ps, gw, r; **R** – 344, 361.

*Enoploides brevis* Filipjev, 1918; **D** – 28; **V** – 75-120; **Z** – • p; **E** – M, bt, shb, phs, ms; **R** – 344.

*Enoploides cirrhatus* Filipjev, 1918 [*Brachionus*]; **D** – 7, 12, 16, 25, 27, 28; **V** – 0-60; **Z** – cp; **E** – M, bt, eb, l, sl, eu; **R** – 341, 344.

*Enoploides fluviatilis* Mikoletzky, 1923 [*Brachionus*]; **D** – 58, 59, 60; **Z** – ? p; **E** – M-B, bt, 0.5-2‰; **R** – 338.

*Enoploides hirsutus* Filipjev, 1918; **D** – 10, 27, 28; **V** – 4-100; **Z** – • p; **E** – M, bt, sl, eb, eu, s; **R** – 252, 344.

*Mesacanthion conicum* (Filipjev, 1918) [*Enoplolaimus*]; **D** – 7, 25, 27, 28; **V** – 80; **Z** – • p; **E** – M, bt, sl, eu, ps-pe; **R** – 252, 347.

*Oxyonchus dubius* (Filipjev, 1918) [*Enoplolaimus*]; **D** – 17, 28; **Z** – em; **E** – M, bt, sl, phs; **R** – 252, 347.

## Tobrilidae

*Tobrilus gracilis* (Bastian, 1865) [*Trilobus*]; **D** – *68, *79, 88; **Z** – (e); **E** – L, bt, 3‰; **R** – 374, 389.

## Tripyloididae

*Tripyloides marinus* (Bütschli, 1874); **D** – 7, 23, 28, 33; **V** – 0-100; **Z** – bam; **E** – M, bt, eb, eu, l, sl, s; **R** – 121, 389.

*Bathylaimus australis* Cobb, 1894 [*B. assimilis, B. ponticus*]; **D** – 7, 25; **Z** – am, ? ham; **E** – M, bt, ps; **R** – 121, 344.

*Bathylaimus cobbi* Filipjev, 1922; **D** – 10, 25, 27, 28; **V** – -100; **Z** – • p; **E** – M, bt, eb, ms, phs, ps; **R** – 341, 344.

*Bathylaimus filipjevi* Stoikov, 1976 ?; **D** – 7, 35; **V** – 0.4; **Z** – ● p; **E** – M, bt, ps, ls, gw; **R** – 340, 344.

DORYLAIMIDA

## Dorylaimidae (Qudsianematidae)

*Dorylaimus otmanliensis* Uzunov, 1974; **D** – 13; **Z** – ● p; **E** – M, bt, l, ps, gw, if; **R** – 361.

*Dorylaimus stagnalis* Dujardin, 1845; **D** – 58, 59, 60; **Z** – (e); **E** – L, bt, 0.5-2‰; **R** – 337.

*Eudorylaimus carteri* (Bastian, 1865); **D** – 58, 59, 60; **Z** – (e); **E** – L-T, bt, 0.5-2‰; **R** – 337.

*Eudorylaimus filipjevi* (Gerlach, 1951) [*Dorylaimus*]; **D** – 7, 24; **Z** – ● Er, p; **E** – M, bt, sl, ph; **R** – 121.

*Crocodorylaimus flavomaculatus* (von Linstow, 1876) [*Laimidorus*]; **D** – 58, 59, 60; **Z** – (e); **E** – L, bt, 0.5-2‰; **R** – 337.

CHROMADORIDA

## Ceramonematidae

*Pselionema annulatum* (Filipjev, 1922) [*Ceratonema, Steineria*]; **D** – 3, 28; **V** – 80-100; **Z** – namsp; **E** – M, bt, shb, phs; **R** – 252, 347.

## Chromadoridae

*Chromadorina bioculata* (Schultze, 1857) [*Chromadora*]; **D** – *68; **Z** – cp, (csee); **E** – M-L, bt, 1‰; **R** – 374, 389.

*Chromadora nudicapitata* (Bastian, 1865) [*Ch. quadrilineata*]; **D** – 7, 24; **Z** – amnz; **E** – M, bt, slc, ps; **R** – 121.

*Chromadorita demaniana* Filipjev, 1922; **D** – 25; **V** – 80; **Z** – cp, ? clp; **E** – M, bt, sl, ps; **R** – 121, 389.

*Chromadorita leuckarti* (de Man, 1876); **D** – 7; **V** – -15, -140; **Z** – cp; **E** – M, bt, ep, sl, ps; **R** – 121, 344, 389.

*Dichromadora cephalata* (Steiner, 1916) [*Chromadora cricophana*]; **D** – 7, 12, 25, 35, 51; **V** – 0-2; **Z** – clpnz; **E** – M, bt, eh, l, sl, ps, gw; **R** – 252, 338, 344.

*Hypodontolaimus balticus* (Schneider, 1906); **D** – 11, 35; **Z** – bap; **E** – M, bt, l, sl, ps, gw; **R** – 252, 362.

*Neochromadora poecilosomoides* (Filipjev, 1918) [*Chromadora, Dichromadora*]; **D** – 7, 51; **V** – -140; **Z** – clm, ? vclm; **E** – M, bt, l, ps, slc, zc; **R** – 252, 347.

*Neochromadora sabulicola* (Filipjev, 1918) [*Chromadora*]; **D** – 3, 7, 10, 20, 24, 25; **V** – 120; **Z** – p, ? cp; **E** – M, bt, eb, pe, ph, ps; **R** – 252, 342, 344.

*Ptycholaimellus ponticus* (Filipjev, 1922) [*Hypodontolaimus*]; **D** – 7, 25, 35, 51; **V** – 0.6-1, 50; **Z** – clm, ? cclm; **E** – M, bt, l, sl, ps, gw, r; **R** – 344.

*Spilophorella euxina* Filipjev, 1918; **D** – 7, 25; **V** – 150; **Z** – em; **E** – M, bt, sl, ps, eb; **R** – 121, 389.

## Cyatholaimidae

*Cyatholaimus gracilis* (Ebert, 1863) [*C. demani*]; **D** – 7, 8, 9, 10, 11, 12, 13, 14, 15, 16, 24, 25, 26, 27, 28; **V** – -100, 120; **Z** – clm; **E** – M, bt, eb, ph, ps, s, r; **R** – 121, 344.

*Paracanthonchus caecus* (Bastian, 1865) [*P. abnormis, Cyatholaimus*]; **D** – 3, 7, 10, 13, 20, 27, 35; **V** – 0-73, 150; **Z** – namnz; **E** – M, bt, eb, eu, ph, ps, s; **R** – 342, 344, 362.

## Ethmolaimidae

*Ethmolaimus multipapillatus* Paramonov, 1926; **D** – 76; **Z** – ● p; **E** – M, bt, eh-70‰; **R** – 121, 389.

## Desmodoridae

*Acanthopharynx similis* (Algén, 1932) [*Desmodora, ? Sabatieria ornata*]; **D** – 11, 35; **Z** – pinz; **E** – M, bt, l, ps, gw, if; **R** – 362.

*Chromaspirina pontica* Filipjev, 1918; **D** – 16, 25; **Z** – cm; **E** – M, bt, sls, slc, zc, ps; **R** – 252, 341, 344.

*Desmodora pontica* (Filipjev, 1922); **D** – 3, 10, 25, 27, 28; **Z** – clm; **E** – M, bt, eb, eu, ps, pe, r; **R** – 252, 344.

*Prodesmodora circulata* (Micoletzky, 1913); **D** – 58, 59, 60; **Z** – lm, (e); **E** – M-B, bt, 0.5-2‰; **R** – 337.

*Metachromadora arenaria* Stoikov, 1979; **D** – 25; **V** – 8-9; **Z** – ● p; **E** – M, bt, ep, ps, sls; **R** – 343.

*Metachromadora cystoseirae* Filipjev, 1918 [? Gerlach et Riemann, 1973]; **D** – 24; **Z** – ● p; **E** – bt, sl, slc; **R** – 252, 347.

*Metachromadora macroutera* Filipjev, 1918; **D** – 16, 25; **V** – 8-9, 35; **Z** – ● p; **E** – M, bt, ep-mb, sl, sls; **R** – 252, 338, 344.

*Onyx perfectus* Cobb, 1891; **D** – 7, 25; **V** – 15-17; **Z** – clm; **E** – M, bt, ep, sl, sls; **R** – 252, 347.

*Spirinia parasitifera* (Bastian, 1865) [? *S. zosterae* Filipjev, 1918]; **D** – 7, 24; **V** – 3.5; **Z** – namim; **E** – M, bt, ep, sls, slc, zc; **R** – 252, 344.

*Spirinia sabulicola* (Filipjev, 1918); **D** – 3, 10, 17, 27, 28; **Z** – ● p; **E** – M, bt, eu, ms, phs, r; **R** – 252, 344.

## Desmoscolecidae

*Desmoscolex laevis* Kreis, 1938 [*D. minutus*]; **D** – 7, 78; **V** – 1-100; **Z** – namim; **E** – M, bt, eb, eh, s, phs; **R** – 252, 344, 347.

*Tricoma bacescui* (Paladian & Andriescu, 1963) [*Desmoscolex, Qadricoma*]; **D** – 28; **V** – 75-120; **Z** – ● p; **E** – M, bt, shb, phs, r; **R** – 252, 344.

## Epsilonematidae

*Bathyepsilonema pustulatum* Gerlach, 1952 [*Epsilonema pustulatum ponticum*]; **D** – 7, 35, 51; **V** – 0.2-0.6; **Z** – cm, ? clm; **E** – M, bt, l, ls, gw, r; **R** – 252, 344.

## Leptolaimidae

*Camacolaimus pontolittoralis* Uzunov, 1977; **D** – 11, 35; **Z** – ● p; **E** – M, bt, ls, gw; **R** – 362.

## Meyliidae

*Quadricoma eurycricus* (Filipjev, 1922) [*Desmolorenzenia, Desmoscolex*]; **D** – 10, 11, 25, 28; **V** – 5-90; **Z** – ● p; **E** – M, bt, eb, sls, phs; **R** – 252, 344, 347.

*Quadricoma loricata* Filipjev, 1922 [*Tricoma*]; **D** – 10, 28; **V** – 50-100; **Z** – ● p; **E** – M, bt, hb, phs, s; **R** – 252, 344.

*Quadricoma steineri* Filipjev, 1922 [*Tricoma euxinica*]; **D** – 7, 25; **V** – -50; **Z** – ● p; **E** – M, bt, mb, ps-pe; **R** – 252, 347.

## Monoposthiidae

*Monoposthia costata* (Bastian, 1865); **D** – 23, 24, 25; **Z** – ami; **E** – M, bt, ep, l, sl, ph; **R** – 252, 344, 347.

*Nudora steineri* (Steiner, 1921); **D** – 13, 35; **Z** – cp; **E** – M, bt, l, gw, if, ps; **R** – 252, 362.

## Selachinematidae

*Choanolaimus psammophilus* de Man, 1880; **D** – 11, 35; **Z** – clm; **E** – M, bt, l, gw, if, ps; **R** – 252, 362.

*Halichoanolaimus dolichurus* Ssaweljev, 1912 [*Hypodentolaimus filicauda*]; **D** – 7, 24; **V** – 1-4, 140, **Z** – clm; **E** – M, bt, slc, ph, phs, r; **R** – 252, 344.

*Halichoanolaimus robustus* (Bastian, 1865) [*H. clavicauda*]; **D** – 7, 25; **V** – 6-10, 150; **Z** – neaminz; **E** – M, bt, sep, sls, ps; **R** – 252, 338, 344.

MONHYSTERIDA

### Axonolaimidae

*Axonolaimus ponticus* Filipjev, 1918; **D** – 1, 2, 3, 4, 5, 6, 7, 8, 9, 10, 11, 12, 13, 14, 15, 16, 17, 18, 19, 20, 21, 22, *68, 71, *78, *79, 84, 88, 93, 94; **V** – 0-100; **Z** – clm; **E** – M, bt, eh-1-20‰, sl, eb, eu, ps, pe, ph; **R** – 121, 338, 344, 374, 389.

*Axonolaimus setosus* Filipjev, 1918; **D** – 7, 24, 25; **V** – -40; **Z** – clm; **E** – M, bt, mb, ps, pe, slc; **R** – 252, 344.

## Comesomatidae

*Comesoma stenocephalum* Filipjev, 1918; **D** – 3, 7, 10, 20, 27, 28; **Z** – • p; **E** – M, bt, ms, phs, sls, eu; **R** – 252, 344.

*Sabatieria abyssalis* (Filipjev, 1918); **D** – 10, 27, 28; **V** – 50-100; **Z** – mni; **E** – M, bt, hb, ms, phs, pe; **R** – 252, 338, 344.

*Sabatieria celtica* Southern, 1914 [*S. cupida*]; **D** – 7, 25, 26, 76; **V** – 8; **Z** – amnep, ? ce; **E** – M, bt, sls, s, ps, pe; **R** – 252, 338, 344.

*Sabatieria pulchra* (Schneider, 1906) [*S. vulgaris*]; **D** – 24, 25; **V** – 6-10; **Z** – aam; **E** – M, bt, ep, sl, ps, ph; **R** – 252, 338, 344.

## Linhomoeidae

*Paralinhomoeus filiformis* (Filipjev, 1918) [*Linchomoeus*]; **D** – 25, 28; **V** – 20-100; **Z** – cp; **E** – M, bt, hb, sls, phs; **R** – 252, 347.

*Paralinhomoeus tenuicaudatus* (Bütschli, 1874) [*P. ostraearum*]; **D** – 7, 25, 76; **Z** – clm; **E** – M, bt, sls, ps, phs, r; **R** – 252, 344.

*Terschellingia longicaudata* de Man, 1907 [*T. antonovi*]; **D** – 7, 10, 25, 26; **Z** – aamni; **E** – M, bt, hb, pe, ph, ps; **R** – 252, 338, 344.

*Terschellingia pontica* Filipjev, 1918; **D** – 7, 26, 27, 28; **V** – 15-100; **Z** – • p; **E** – M, bt, eb, sl, s, pe; **R** – 252, 341, 344.

## Monhysteridae

*Monhystera ampullocauda* Paramonov, 1926; **D** – 51; **Z** – • p; **E** – M, bt, ps, ls; **R** – 344.

*Monhystera collaris* Filipjev, 1922; **D** – 7, 26; **Z** – p; **E** – M, bt, sl, pe, s, (sls), r; **R** – 252, 344.

*Monhystera filiformis* Bastian, 1865; **D** – 7, 76; **Z** – clm, ? K; **E** – M, bt, eh-0-70‰, eu; **R** – 3, 121, 344, 389.

*Monhystera parva* (Bastian, 1865) [*M. antarctica, M. heteroparva, M. kessinensis*]; **D** – 7; **Z** – amni; **E** – M, bt; **R** – 121, 389.

*Monhystera rotundicapitata* Filipjev, 1922; **D** – 7, 24; **V** – -10; **Z** – ● p; **E** – M, bt, ep, sls, slc; **R** – 252, 342, 344.

## Sphaerolaimidae

*Sphaerolaimus macrocirculus* Filipjev, 1918; **D** – 10, 25, 27, 28; **V** – 59; **Z** – clm; **E** – M, bt, hb, s, pe, (ps); **R** – 252, 342, 344.

*Sphaerolaimus ostreae* Filipjev, 1918 [*S. maeoticus*]; **D** – 7, 25, 27; **V** – 59; **Z** – em, ? cm; **E** – M, bt, mb, ps, pe, r; **R** – 252, 341, 344.

## Xyalidae

*Daptonema oxycerca* (de Man, 1888) [*Monhystera, Theristus*]; **D** – 7, 25; **Z** – bam; **E** – M, bt, sep, sls; **R** – 121, 389.

*Paramonhystera elliptica* Filipjev, 1918 [*P. setosa, Leptogastrella*]; **D** – 7, 25; **Z** – lm, ? clm; **E** – M, bt, sl, ps, pe, ph; **R** – 121, 389.

*Theristus littoralis* Filipjev, 1922; **D** – 35, 46; **Z** – ● p; **E** – M, bt, l, sp, gw, if; **R** – 252, 363.

*Theristus longicaudatus* Filipjev, 1922; **D** – 17, 20, 25; **Z** – ccp; **E** – M, bt, sls, ps; **R** – 252, 344.

MONONCHIDA

## Mononchidae

*Mononchus truncatus* Bastian, 1865; **D** – 58, 59, 60; **Z** – (e, ? k); **E** – L, 0.5-2‰, bt; **R** – 337.

SECERNENTEA

DIPLOGASTERIDA

## Diplogastridae

*Diplogaster rivalis* (Leyding, 1854); **D** – *68, *69, 93; **Z** – (e); **E** – L, 3‰; **R** – 374, 375, 389.

RHABDITIDA

## Rhabditidae

*Rhabditis marina* Bastian, 1865; **D** – 11, 35; **Z** – namnz, (e); **E** – M (L), bt, l, gw, if; **R** – 252, 362.

TYLENCHIDA

## Criconematidae

*Mesocriconema xenoplax* (Raski, 1952); **D** – 48, 76; **Z** – (ea, ? ha); **E** – T; **R** – 119.

# CEPHALORHYNCHA

KINORHYNCHA

CYCLORHAGIDA

## Centroderidae

*Centroderes spinosus* (Reinhard, 1881) [*Echinoderes*]; **D** – 1, 2, 3, 4, 5, 6, 7, 8, 9, 10, 11, 12, 13, 14, 15, 16, 17, 18, 19, 20, 21, 22; **V** – -128; **Z** – lm; **E** – M, bt, shb, phs, pe; **R** – 229, 249.

## Echinoderidae

*Echinoderes agigens* Bacescu, 1968 [*E. dujardinii*]; **D** – 1, 2, 3, 4, 5, 6, 7, 8, 9, 10, 11, 12, 13, 14, 15, 16, 17, 18, 19, 20, 21, 22, 24, 25, 35; **V** – 1-8, 80; **Z** – m; **E** – M, bt, ep, gw, if, s, slr; **R** – 229, 249.

HOMALORHAGIDA

## Pycnophyidae

*Pycnophyes kielensis* Zelinka, 1928; **D** – 25, 26; **V** – 15-; **Z** – clm; **E** – M, bt, sl, ps-pe, s; **R** – 229, 249.

*Pycnophyes ponticus* (Reinhard, 1881) [*Echinoderes*]; **D** – 1, 2, 3, 4, 5, 6, 7, 8, 9, 10, 11, 12, 13, 14, 15, 16, 17, 18, 19, 20, 21, 22, 26, 27; **V** – 10-40; **Z** – m; **E** – M, bt, mb, s, pe; **R** – 229, 249.

# ROTIFERA (ROTATORIA)



## EUROTATORIA

### ADINETIDA

## Adinetidae

*Adineta barbata* Janson, 1893; **D** – 96; **Z** – (sk); **E** – L, l; **R** – 209, 374.

*Adineta vaga* (Davis, 1873); **D** – 96; **Z** – (k); **E** – L, 10‰; **R** – 106, 209, 389.

### PHILODINIDA

## Habrotrochidae

*Habrotrocha angusticollis* (Murray, 1905); **D** – 96; **Z** – (k); **E** – L; **R** – 209, 373.

## Philodinidae

*Philodina citrina* Ehrenberg, 1832; **D** – 59, 60, *68; **Z** – (k); **E** – L, 2.3‰, l, ph; **R** – 209, 280, 374, 389.

*Philodina roseola* Ehrenberg, 1832 [*Ph. cinnabarina*]; **D** – 33, 60, 96; **Z** – lmnz, (k); **E** – M-L, l, p, ph; **R** – 106, 209, 389.

*Rotaria citrina* (Ehrenberg, 1838); **D** – 96; **Z** – (sk); **E** – L-B, l; **R** – 209.

*Rotaria rotatoria* (Pallas, 1766) [*Rotifer vulgaris*]; **D** – *68, *79, *86, 96; **Z** – (k), ? lmnz; **E** – L-B, 10‰, l, ph; **R** – 209, 374, 389.

*Rotaria tardigrada* (Ehrenberg, 1832); **D** – 49; **Z** – ? lm, (k); **E** – L-B, 18‰, p, l, ph; **R** – 146, 209.

### PLOIMA

## Asplanchnidae

*Asplanchna girodi* de Guerne, 1888; **D** – 78; **Z** – (k); **E** – L-B, p; **R** – 209, 293.

*Asplanchna priodonta* Gosse, 1850; **D** – 77, 78, 79; **Z** – ? ace, (k); **E** – L, 17.3‰, p, eu, sw; **R** – 192, 193, 209, 293.

*Asplanchna sieboldii* (Leydig, 1854); **D** – 59, 60, 78; **Z** – (k); **E** – L, 5‰, p; **R** – 280, 353a.

*Asplanchnopus hyalinus* Harring, 1913; **D** – 77; **Z** – (ho); **E** – L, p; **R** – 155a

### Brachionidae

*Anuraeopsis fissa* Gosse, 1851 [*A. hypelasma*]; **D** – 60, *69, 71, 78; **Z** – ? lmnz, (k); **E** – L-B, 5‰, p; **R** – 209, 280, 353a, 374, 389.

*Brachionus angularis* Gosse, 1851; **D** – 58, 59, 60, *68, *69, 77, 78, 79, 80, 96; **Z** – ham, (k); **E** – L-B, 5‰, p, sw; **R** – 193, 209, 280, 293, 353a, 374, 389.

*Brachionus bennini* Leissling, 1924; **D** – 78; **Z** – (sk); **E** – L, p, sw; **R** – 209, 293.

*Brachionus budapestinensis* Daday, 1885 [*B. similis*]; **D** – 96; **Z** – (ppta, ? k); **E** – L, sw; **R** – 209.

*Brachionus calyciflorus* Pallas, 1776 [*B. amphiceros*, *B. dorcas*, *B. pala*]; **D** – 58, 59, 60, *68, *69, 71, 77, 78, 79, 80, 96; **Z** – ? ham, (k); **E** – L, 5‰, p, sw; **R** – 160, 190, 193, 209, 278, 293, 353a, 374, 389.

*Brachionus diversicornis* (Daday, 1883); **D** – 58, 78; **Z** – clm, (sk); **E** – L, p, sw; **R** – 209, 278, 293, 353a.

*Brachionus falcatus* Zacharias, 1898; **D** – 59; **Z** – (? sk); **E** – L, p, sw; **R** – 280.

*Brachionus forficula* Wierzeyski, 1891; **D** – 60; Z – (ppta); **E** – L, p, sw; **R** – 280.

*Brachionus leydigii* Cohn, 1862; **D** – 59, 60, 96; **Z** – (ppta); **E** – L, p, sw; **R** – 209, 280.

*Brachionus plicatilis* Müller, 1786 [*B. mülleri*]; **D** – 29, 33, 49, 58, 62, 64, *68, *69, 76, 77, 78, *79, 84, *86, 88; **Z** – ham, (k); **E** – M-B, 6-20‰, eh, p; **R** – 105, 106, 190, 207, 374, 389.

*Brachionus quadridentatus* Hermann, 1783 [*B. bakeri*]; **D** – 58, 59, 60, *68, 77, *86, 92, 96; **Z** – ham, (k); **E** – L, 3-16‰, l, p; **R** – 209, 278, 280, 374, 389.

*Brachionus rubens* Ehrenberg, 1838; **D** – 69, 96; **Z** – ? kclm, (sk); **E** – L, p; **R** – 160, 209, 373.

*Brachionus urceolaris* Müller, 1773 [*B. urceus*]; **D** – 58, 60, *68, 77, 78, 79, 80, 96; **Z** – 58, 60, *68, 77-80, 96; **E** – L, 4.3-15‰, eh, p; **R** – 190, 209, 293, 353a, 374, 389.

*Kellicottia longispina* (Kellicott, 1879); **D** – 60; **Z** – ace; **E** – L, p, sw; **R** – 279, 280.

*Keratella cochlearis* (Gosse, 1851) [*Anuraea*]; **D** – 7, 58, 59, 60, *68, *69, 77, 78, 79, 80; **Z** – anamnp, (sk); **E** – L-B, 10-16‰, eh, p, sw; **R** – 193, 209, 278, 280, 293, 317, 319, 353a, 389.

*Keratella cruciformis* (Thompson, 1892); **D** – 29, 96; **Z** – clp, (h); **E** – B-M, p; **R** – 209, 212.

*Keratella hiemalis* Carlin, 1943; **D** – 58, 60, 96; **Z** – (hnat); **E** – L, sw; **R** – 209, 280.

*Keratella irregularis* (Lauterborn, 1898); **D** – 60, 96; **Z** – (hnat, ? h); **E** – L, sw; **R** – 209, 280.

*Keratella quadrata* (Müller, 1786) [*Anuraea aculeata*]; **D** – 58, 59, 60, *68, 71, 77, 78, 79, 80, 92; **Z** – ace, (k); **E** – L, 0.5-6‰, p, sw; **R** – 190, 252, 278, 280, 353a, 374, 389.

*Keratella tecta* (Gosse, 1851); **D** – 59, 78; **Z** – cpwp, (k); **E** – L, p, sw; **R** – 209, 293, 353a.

*Keratella testudo* (Ehrenberg, 1832); **D** – 60, 96; **Z** – lm, (hat); **E** – L, p, sw; **R** – 209, 280.

*Keratella tropica* (Apstein, 1907); **D** – 58, 59, 60, 78; **Z** – lm, (k), i; **E** – L-B, is, p, sw; **R** – 209, 210, 280, 353a.

*Keratella valga* (Ehrenberg, 1834); **D** – 58, 59, 77, 79, 80, 96; **Z** – lm, (k); **E** – L-B, p, sw; **R** – 193, 209, 278.

*Notholca acuminata* (Ehrenberg, 1832) [*N. bipalium* var. *acuminata*]; **D** – 60, *69, 77, 78, *79, 85, *86, 88; **Z** – neamj, (pat); **E** – L-B, 0.1-18‰, p, sw; **R** – 209, 258, 280, 374.

*Notholca labis* Gosse, 1887; **D** – 77; **Z** – neamj, (dp); **E** – L, 25-54‰, p, eh; **R** – 258.

*Notholca squamula* (Müller, 1786) [*Brachionus*]; **D** – 58, 60; **Z** – lmnz, (? k); **E** – L-B, 2‰, p, sw; **R** – 209, 252, 278, 280.

*Notholca striata* (Müller, 1786) [*Brachionus*]; **D** – 33, *69, *78; **Z** – neamwp, (hna); **E** – M-B-L, 10-13‰, p; **R** – 105, 106, 374, 389105, 106, 374, 389.

## Dicranophoridae

*Dicranophorus bulgaricus* Althaus, 1957; **D** – 16, 35; **Z** – ● p; **E** – M, bt, ps, gw, if; **R** – 2, 209, 249, 392.

*Dicranophorus forcipatus* (Müller, 1786) [*Cercaria, Notommata*]; **D** – 60, 96; **Z** – bam, (k); **E** – L-B, bt, l, ph; **R** – 209, 280.

*Dicranophorus rostratus* (Dixon-Nuttall & Freeman, 1902) [*Diglena coenura*]; **D** – *69; **Z** – bam, (h); **E** – L, 3‰, l; **R** – 374, 389.

*Encentrum arenarium* Althaus, 1957; **D** – 16, 35; **Z** – ● p; **E** – M, bt, l, ps, gw, if; **R** – 2, 209, 249, 392.

*Encentrum marinum* (Dujardin, 1841) [*Furcularia*]; **D** – 13, 32, 33, 35, 78; **Z** – cacpnz, (h); **E** – M-B-L, 30‰, l, sp, if; **R** – 106, 188, 209, 249, 353a, 389, 392.

*Encentrum psammophilum* Althaus, 1957; **D** – 7, 35; **Z** – clp; **E** – M, bt, l, ps, gw, if; **R** – 2, 209, 249, 392.

*Encentrum striatum* Althaus, 1957; **D** – 7, 34, 35; **Z** – calp; **E** – M, bt, l, ps, gw, if; **R** – 2, 209, 249, 392.

*Encentrum valkanovi* Althaus, 1957; **D** – 7, 35; **Z** – ● p; **E** – M, bt, l, ps, gw, if; **R** – 2, 111, 209, 249, 392.

## Euchlanidae

*Euchlanis dilatata* Ehrenberg, 1832; **D** – 59, 60, *68, *79; **Z** – namep, (k); **E** – L, 2.5‰, p, ph, sw; **R** – 209, 374, 389.

*Euchlanis lyra* Hudson, 1886; **D** – 60; **Z** – (? hnata); **E** – L, 0.8‰, p, sw; **R** – 280.

*Euchlanis oropha* Gosse, 1887; **D** – 59; **Z** – (e, ? k); **E** – L, 0.8‰, p, sw; **R** – 280.

*Euchlanis pyriformis* Gosse, 1851; **D** – 77; **Z** – (e, ? sk); **E** – L, p, sw; **R** – 258.

*Tripleuchlanis plicata* (Levander, 1894); **D** – 59; **Z** – namnep, (sk); **E** – B-L, 0.8‰, p, sw; **R** – 280.

## Lecanidae

*Lecane althausi* Rudescu, 1960; **D** – 35; **Z** – (? sk); **E** – L-B, l, ps, gw, if; **R** – 209, 249.

*Lecane closterocerca* (Schmarda, 1859); **D** – 58, 59, 60; **Z** – cpnz, (sk); **E** – L-B, 0.8-2‰, p, sw; **R** – 209, 280.

*Lecane copeis* (Harring & Myers, 1926) [*Monostyla*]; **D** – 96; **Z** – (hn); **E** – L, p, sw; **R** – 209, 372, 373.

*Lecane cornuta* (Müller, 1786) [*Monostyla*]; **D** – *68, *69; **Z** – (hn); **E** – L, 3‰, l, ph, sw; **R** – 249, 374, 389.

*Lecane furcata* (Murray, 1913) [*Monostyla*]; **D** – 58, 59, 60; **Z** – (sk); **E** – L, p, sw; **R** – 209, 280.

*Lecane lamellata* (Daday, 1893) [*Monostyla*]; **D** – 76, *86, 88; **Z** – (h); **E** – L-B, 1-12‰, p; **R** – 374, 389.

*Lecane luna* (Müller, 1776) [*Cathypna*]; **D** – 60, *68, *79; **Z** – namnz, (? k); **E** – L-B, 2‰, p; **R** – 209, 280, 374, 389.

*Lecane lunaris* (Ehrenberg, 1832); **D** – 59; **Z** – (k); **E** – L-B, 0.5-2‰, p; **R** – 280.

## Lepadellidae

*Lepadella ovalis* (O. F. Müller, 1786); **D** – 33; **Z** – nam, (k); **E** – L-B-M, 12‰, l; **R** – 106, 209, 389.

*Lepadella patella* (O. F. Müller, 1773) [*Metopidia oblonga, Squamella*]; **D** – 59, 60, *68, 76, 96; **Z** – antamp, ? sk; **E** – L-B, 2‰, p; **R** – 209, 280, 374, 389.

*Lepadella pontica* Althaus, 1957; **D** – 7, 35; **Z** – • p; **E** – M, bt, ps, sls; **R** – 2, 249.

*Colurella adriatica* Ehrenberg, 1831 [*Colurus leptus*]; **D** – 13, 33, 58, 59, 60, *68, *69, 78, *79, *86, 88, 90, 93, 94; **Z** – namswp, ? SK, (k); **E** – M-B-L, 15‰, p, l; **R** – 106, 188, 209, 280, 353a, 374, 389.

*Colurella colurus* (Ehrenberg, 1831) [*Monura loncheres, Colurus*]; **D** – 13, 33, 35, *69, 76, *86, 88, 96; **Z** – namswp, ? K, SK, (k); **E** – M-B-L, 5-15‰, bt-p, l, gw, if; **R** – 104, 106, 188, 209, 374, 389.

*Colurella marinovi* Althaus, 1957; **D** – 7, 16, 34; **Z** – • p; **E** – M, bt, 12‰, l, ps; **R** – 2, 209, 249, 389.

? *Colurella monodactilos* Althaus, 1957 [? undetermined ciliate]; **D** – 7, 25; **Z** – • p; **E** – M, bt, 12‰, l, ps; **R** – 2, 249, 392.

*Colurella obtusa* (Gosse, 1886); **D** – 96; **Z** – clmnz, (k); **E** – M-B-L, p, sw; **R** – 209.

*Colurella uncinata* (Müller, 1773) [*Colurus bicuspidatus, Brachionus*]; **D** – ?; **Z** – namnz, (k); **E** – M-L, 9‰, p, l, ph; **R** – 209.

*Squatinella longispinata* (Tatem, 1867) [*Stephanops*]; **D** – 96; **Z** – (hoa); **E** – L-B, p; **R** – 209, 373.

## Lindiidae

*Lindia tecusa* Harring & Myers, 1922 [*Halodigma, Halolindia*]; **D** – 49, *69; **Z** – clp; **E** – M, 12-13‰, bt, l; **R** – 111, 209, 249, 374.

# Microcodidae

*Microcodon clavus* Ehrenberg, 1830; **D** – 96; **Z** – (h, ? k); **E** – L, p, sw; **R** – 209, 373, 374, 389.

# Mytilinidae

*Mytilina ventralis* (Ehrenberg, 1830) [*Salpina*]; **D** – 59, 60, 96; **Z** – (hno); **E** – L-B, p, 0.8‰, sw, l; **R** – 209, 280.

# Notommatidae

*Monommata aequalis* (Ehrenberg, 1830) [*Notommata*]; **D** – *68, 96; **Z** – (hnoa); **E** – L, 1‰, p, sw; **R** – 209, 372, 373, 389.

*Cephalodella auriculata* (O. F. Müller, 1773) [*Diaschiza lacinulata*]; **D** – *68, 96; **Z** – (hnoa); **E** – L-B, bt, l, ph, sw; **R** – 209, 249, 374.

*Cephalodella catellina* (O. F. Müller, 1786); **D** – 33, 59, 60, *69, 96; **Z** – K, (k); **E** – L-B, 12‰, eh, bt, l; **R** – 105, 107, 209, 249.

*Cephalodella gibba* (Ehrenberg, 1830) [Furcularia]; **D** – 96; **Z** – lm, (k); **E** – L, p; **R** – 209, 389.

*Cephalodella hoodii* (Gosse, 1886) [*Diaschiza*]; **D** – 96; **Z** – (ho); **E** – L, p, ph; **R** – 209, 372.

*Cephalodella reimanni* Donner, 1950; **D** – 79, 96; **Z** – (hop); **E** – L; **R** – 209, 373, 374.

*Cephalodella ventripes* (Dixon-Nuttall, 1901); **D** – 59, 60; **Z** – (hna); **E** – L, p, sw; **R** – 209, 280.

# Proalidae

*Proales commutata* Althaus, 1957; **D** – 7, 35; **Z** – ● p; **E** – M, bt, ps; **R** – 2, 209, 249.

*Proales halophila* Remane, 1929 [*P. globulifera halophilus*]; **D** – 7, 35; **Z** – clp; **E** – M-B, bt, l, ps, gw, if; **R** – 104, 209, 249.

*Proales reinhardti* (Ehrenberg, 1834) [*Furcularia*]; **D** – 13, 35; **Z** – clp; **E** – M, bt-p, ps, ph, gw, if; **E** – 188, 209, 212, 314.

*Proales similis* de Beauchamp, 1907; **D** – 32, 33; **Z** – p, (hpta); **E** – M-B, 10‰, bt, l; **R** – 106, 209, 249, 389.

## Synchaetidae

*Synchaeta baltica* Ehrenberg, 1834; **D** – *69; **Z** – namnz; **E** – M-B, 12‰, p; **R** – 317, 389.

*Synchaeta cecilia* Rousselet, 1902; **D** – 59, 60, *69; **Z** – bamswp; **E** – M-B, 0.5-12‰, p; **R** – 105, 209, 280, 317, 374, 389.

*Synchaeta gyrina* Hood, 1887; **D** – 33; **Z** – clp; **E** – M-B, 12‰, p; **R** – 106, 389.

*Synchaeta neapolitana* Rousselet, 1902; **D** – 29; **Z** – clmnz; **E** – M, p; **R** – 109, 194, 201, 209.

*Synchaeta oblonga* Ehrenberg, 1831; **D** – 77; **Z** – bam, (k); **E** – L, 1.8‰, p; **R** – 258.

*Synchaeta pectinata* Ehrenberg, 1832; **D** – 7, 59, 60, 76, 77, 80; **Z** – amnz, (sk); **E** – M-B-L, eh, p; **R** – 209, 280, 319.

*Synchaeta pontica* Rodewald-Rudescu, 1960 ?; **D** – 29; **Z** – • p; **E** – M, 14‰, p; **R** – 109, 111, 201, 209.

*Synchaeta stylata* Wierzejski, 1893; **D** – *68, *69; **Z** – namnz, (k); **E** – B-L, 10‰, p, l; **R** – 105, 209, 374, 389.

*Synchaeta tavina* Hood, 1893; **D** – 33; **Z** – clp, (ha); **E** – M-B, 12‰, p, l; **R** – 106, 389.

*Synchaeta tremula* (O. F. Müller, 1786); **D** – *69; **Z** – namnz, (k); **E** – M-B-L, p, eh, l; **R** – 105, 209, 389.

*Synchaeta vorax* Rousselet, 1902; **D** – 7, 13, 29, *69, 76, 77, 78, 80; **Z** – clm, (h); **E** – M-B, 12‰, p; **R** – 105, 190, 191, 249, 289, 317, 389, 398.

*Polyarthra dolichoptera* Idelson, 1925 [*P. platyptera*]; **D** – 59, 60, *68, 71, 77, *79, 78, 88, 96; **Z** – clm, (k); **E** – L-B, 0.8‰, p, β; **R** – 209, 280, 293, 353a, 374.

*Polyarthra longiremis* Carlin, 1943; **D** – 58, 60; **Z** – (sk); **E** – L, 0.8‰, p; **R** – 209, 278, 280.

*Polyarthra major* Burckhardt, 1900; **D** – 78; **Z** – (sk); **E** – L, p; **R** – 209, 293.

*Polyarthra remata* Skorikov, 1896; **D** – 59, 78, 79, 80; **Z** – clp, (sk); **E** – L-B, 16.8‰, p, β; **R** – 192, 193, 252, 280, 353a.

*Polyarthra vulgaris* Carlin, 1943; **D** – 58, 78, 79, 80; **Z** – clm, (k); **E** – L-B, 0.5‰, p, β; **R** – 192, 209, 278, 293, 353a.

*Ploesoma hudsoni* (Imhof, 1891); **D** – 77; **Z** – clm, (e); **E** – B-L, 24-58‰, p, eh; **R** – 258.

## Trichocercidae

*Trichocerca agnatha* Wulfert, 1939; **D** – 78; **Z** – (pata); **E** – L, p; **R** – 209, 293.

*Trichocerca capucina* (Wierzejski & Zacharias, 1893); **D** – 78; **Z** – nam, (k); **E** – B-L, p, eu, o-β; **R** – 353a.

*Trichocerca mucosa* (Stokes, 1896) [*Mastigocerca*]; **D** – 78, 96; **Z** – (h); **E** – L, p; **R** – 209, 293.

*Trichocerca musculus* (Hauer, 1937) [*Diurella*]; **D** – 60; **Z** – (hata); **E** – L, p, sw; **R** – 209, 280.

*Trichocerca porcellus* (Gosse, 1851); **D** – 96; **Z** – namnz, (k); **E** – B-L, p; **R** – 209.

*Trichocerca pusilla* (Jennings, 1903) [*Rattulus*]; **D** – 58, 78; **Z** – namnz, (k); **E** – B-L, 2‰, p; **R** – 209, 278, 293, 353a.

*Trichocerca similis* (Wierzejski, 1893); **D** – 59; **Z** – neamnz, (k); **E** – L, 0.5-0.8‰, p; **R** – 209, 280.

*Trichocerca stylata* (Gosse, 1851)[*Monocerca*]; **D** – 58; **Z** – K, (k); **E** – L, 2‰, p; **R** – 209.

*Trichocerca tenuior* (Gosse, 1886) [*Diurella*]; **D** – *68; **Z** – (k); **E** – L, 1‰, p; **R** – 209, 374, 389.

## Trichotriidae

*Trichotria pocillum* (O. F. Müller 1776); **D** – 77; **Z** – (h); **E** – L, p; **R** – 155a.

*Trichotria tetractis* (Ehrenberg, 1830) [*Dinocharis*]; **D** – 96; **Z** – namnz, (k); **E** – M-B-L, p; **R** – 209.

*Macrochaetus subquadratus* (Perty, 1850) [*Dinocharis, Polychaetus*]; **D** – 96; **Z** – (k); **E** – L-B, p; **R** – 209, 373.

FLOSCULARIIDA

## Conochilidae

*Conochilus hippocrepis* (Schrank, 1803) [*Linza*]; **D** – 96; **Z** – (k); **E** – L-B, p; **R** – 209.

## Trichosphaeridae

*Filinia longiseta* (Ehrenberg, 1834) [*Triarthra*]; **D** – 58, 59, 60, 77, 79, 80; **Z** – clm, (k); **E** – L-B, 0.8-15‰, p, sw; **R** – 193, 252, 278, 280.

*Filinia passa* (O. F.Muller 1786); **D** – 78; **Z** – ? (k), i; **E** – L-B, p, sw; **R** – 353a.

*Filinia terminalis* (Plate, 1886) [*Triarthra*]; **D** – 59, 60, 78; **Z** – clm, (k); **E** – L-B, 0.8‰, p, sw; **R** – 209, 280, 353a.

## Flosculariidae

*Beauchampia crucigera* (Dutrochet, 1812) [*Cephalosiphon candidus, Rotifer*]; **D** – 79; **Z** – (e, ? k); **E** – L-B, p; **R** – 209, 373.

*Ptygura melicerta* Ehrenberg 1832; **D** – 78; **Z** – (k); **E** – L, sw, ph; **R** – 353a.

## Hexarthridae

*Hexarthra fennica* (Levander, 1892) [*Pedalia*]; **D** – 33, *69, *74, *78, 88; **Z** – amswp, (k); **E** – M-B, 5-50‰, p, eh; **R** – 105, 106, 317, 389.

*Hexarthra mira* (Hudson, 1841) [*Pedalion*]; **D** – 59, 60, 77, 80, 96; **Z** – namnz, (k); **E** – M-B-L, p, eh; **R** – 209, 280.

*Hexarthra oxyuris* (Sernov, 1903) [*Pedalia*]; **D** – 7, *69; **Z** – (k); **E** – L-B, 10-15‰, p, eh; **R** – 317, 319, 389.

## Testudinellidae

*Testudinella clypeata* (Müller, 1786) [*Brachionus, Pterodina clipeata*]; **D** – 29, 32, 33, *78, *79; **Z** – clm, (h); **E** – M-B, 7-19.5‰, bt, p; **R** – 106, 201, 374, 389.

*Testudinella emarginula* (Stenroos, 1898); **D** – 78; **Z** – (k); **E** – L-B, bt, eh, et, eu, ph, ps; **R** – 353a.

*Testudinella obscura* Althaus, 1957; **D** – 7, 35; **Z** – calm; **E** – M, bt, ps, gw, if; **R** – 2, 209, 249.

*Testudinella parva* (Ternetz, 1892); **D** – 78; **Z** – (k); **E** – L, bt-p, eu, x-β, **R** – 353a.

*Testudinella patina* (Hermann, 1783); **D** – 59, 77; **Z** – amnz, (k); **E** – M-L, 0.8‰, p-bt, l, s; **R** – 280.

*Pompholyx complanata* Gosse, 1851; **D** – 59, 60, 78; **Z** – (e, ? k); **E** – L, 0.5-0.8‰, p; **R** – 280, 293, 353a.

# ANNELIDA

## POLYCHAETA

### PHYLLODOCIDA

## Phyllodocidae

*Phyllodoce maculata* (Linnaeus, 1767); **D** – 25, 27, 28; **V** – 0-100; **Z** – abambp; **E** – M, bt, eb, pe, ms, phs; **R** – 245, 249.

*Phyllodoce mucosa* Örsted, 1843; **D** – 3, 4, 5, 6, 7, 8, 13, 25, 28, 69; **V** – 15-100; **Z** – anamnep; **E** – M, bt, eb, ps-pe, ms; **R** – 219, 245, 249, 389.

*Genetyllis tuberculata* (Bobretzky, 1868) [*Phyllodoce rubiginosa*]; **D** – 2, 5, 7, 11, 12, 13, 16, 24, 27; **V** – 0-70; **Z** – lm, ? em, ? p; **E** – M, bt, eh, mb, slc, ms; **R** – 67, 91, 84, 219, 389.

*Nereiphylla rubiginosa* (Saint-Joseph, 1888) [*Phyllodoce rubiginosa*]; **D** – 69; **V** – 2.5-10; **Z** – clm; **E** – M, bt, ps-pe; **R** – 358.

*Eulalia viridis* (Johnston, 1829) [*Phyllodoce*]; **D** – 11, 13, 24; **V** – 0-30; **Z** – anamip; **E** – M, bt, ep, slc, slr, mc; **R** – 91, 245, 249.

*Eumida sanguinea* (Örsted, 1843) [*Eulalia*]; **D** – 6, 7, 24, 25, 69; **V** – 1-25; **Z** – amip; **E** – M, bt, eh, ep, ps-s-ph; **R** – 84, 219, 245, 249, 358, 374.

*Sige macroceros* (Grube, 1860) [*Eulalia, Pterocirrus*]; **D** – 11, 12, 24; **V** – 0-23; **Z** – namnp; **E** – M, bt, ep, ph, slr, ps; **R** – 91, 245, 249, 389.

*Mysta picta* (Quatrefages, 1865) [*Eteone armata*]; **D** – 13, *69, 25; **V** – 1-5, 25; **Z** – clm; **E** – M, bt, sep, ps-pe-mc; **R** – 84, 91, 219, 235, 245, 249, 389.

*Pseudomystides limbata* (Saint-Joseph, 1888) [*Mystides*]; **D** – 6, 25; **V** – 1; **Z** – anclm; **E** – M, bt, sep, ps, ph, r; **R** – 219, 245, 249, 389.

*Hesionura coineaui* (Laubier, 1962) [*Eteonides, Mystides*]; **D** – 25, 35; **V** – -15; **Z** – namnep; **E** – M, bt, sep, ps, gw; **R** – 180, 226, 245, 249.

**Polynoidae**

*Harmothoe extenuata* (Grube, 1840) [*Lagisca*]; **D** – 25, 26; **V** – 0-35, 50; **Z** – anamnep; **E** – M, bt, mb, ps-pe, phc; **R** – 245, 249.

*Harmothoe imbricata* (Linnaeus, 1767); **D** – 24, 25, 69, 76; **V** – 0-25, 70; **Z** – anaminp, ? K; **E** – M, bt, mb, slr, ps, s; **R** – 84, 180, 245, 358.

*Harmothoe impar* (Johnston, 1839); **D** – 7, 69; **V** – 2.5-18; **Z** – anam; **E** – M, bt, ep; **R** – 358.

*Harmothoe reticulata* (Claparède, 1870); **D** – 7, 12, 24, 25, 27; **V** – 0-80; **Z** – clm; **E** – M, bt, mb, ph-pe-ps; **R** – 84, 91, 219, 245.

*Malmgreniella lunulata* (Delle Chiaje, 1830) [*Harmothoe*]; **D** – 76; **Z** – amnep; **E** – M, bt, eh, s; **R** – 119, 396.

*Polynoe scolopendrina* Savigny, 1822; **D** – 11, 12, 13, 23, 24; **V** – 0-10; **Z** – clm; **E** – M, bt, ep, slr, slz, r; **R** – 91, 245.

## Pholoidae

*Pholoe inornata* Johnston, 1839 [*Ph. synophthalmica*]; **D** – 7, 24, 25, 26; **V** – 5-30, 130; **Z** – ace; **E** – M, bt, eb, slr, sls, s; **R** – 84, 180, 245, 249.

## Sigalionidae

*Sthenelais boa* (Johnston, 1833) [*Sigalion*]; **D** – 13, 23, 24, 25; **V** – 0-3; **Z** – amiwp; **E** – M, bt, sep, sls, slz; **R** – 245, 249.

## Pisionidae

*Pisione remota* (Southern, 1914) [*Praegeria*]; **D** – 7, 17, 19, 35, 51; **V** – 0-0.50; **Z** – namrsnep; **E** – M, bt, ls, ps, gw; **R** – 226, 245, 249.

## Hesionidae

*Hesionides arenaria* Friedrich, 1937; **D** – 7, 11, 25, 35, 51; **V** – 0-10; **Z** – aminwp; **E** – M, bt, sep, sls, ps, gw, ■; **R** – 114, 219, 245, 249, 360, 387, 389.

*Microphthalmus fragilis* Bobretzky, 1870; **D** – 7, 13, 25, 35, 51; **V** – 0-10; **Z** – vclm; **E** – M, bt, sep, ls, sls; **R** – 219, 245, 249, 389.

*Microphthalmus sczelkowii* Metschnikow, 1865; **D** – 7, 25, 27; **V** – 10-30, 70; **Z** – namnp; **E** – M, bt, ep, sls, ms; **R** – 234, 245, 252.

*Microphthalmus similis* Bobretzky, 1870; **D** – 13, 25, 51; **V** – 0-10; **Z** – nam; **E** – M, bt, sep, l, sls, slr; **R** – 219, 245, 249, 389.

## Syllidae

*Haplosyllis spongicola* (Grube, 1855) [*Syllis*]; **D** – 17, 24, 50; **Z** – K; **E** – M, bt, ep, ph, l, sl, r; **R** – 221, 245, 249, 392.

*Syllis gracilis* Grube, 1840; **D** – 7, 24, 25, 50; **V** – 1-17; **Z** – K; **E** – M, bt, sep, ph, lt, sls; **R** – 219, 245, 249, 389.

*Syllis hyalina* Grube, 1863; **D** – 7, 24, 25; **V** – -30; **Z** – K; **E** – M, bt, ep, ps, ph, r; **R** – 234, 245, 249, 392.

*Syllis prolifera* Crohn, 1852; **D** – 7, 11, 13, 24, 25; **V** – 0-12; **Z** – K; **E** – M, bt, ep, slr, mc, sls; **R** – 91, 219, 245, 249.

*Amblyosyllis formosa* (Claparède, 1863) [*Pterosyllis*]; **D** – 69, 84; **Z** – aminwp, ? K; **E** – M, bt; **R** – 102, 247, 249, 252.

*Streptosyllis varians* Webster & Benedict, 1887; **D** – 11, 25; **V** – 5-10; **Z** – bap; **E** – M, bt, sep, sls, ps; **R** – 245, 249, 360a, 392.

*Syllides longocirratus* (Örsted, 1845); **D** – 7, 20, 49; **V** – 0.5-17; **Z** – namip; **E** – M, bt, sep, slr, r; **R** – 221, 245, 249, 392.

*Nudisyllis pulligera* (Krohn, 1852) [*Pionosyllis*]; **D** – 11, 24, 25, 26, 27; **V** – 0-60; **Z** – clm; **E** – M, bt, mb, ph, ps, s; **R** – 221, 245, 392.

*Neopetitia amphophthalma* (Siewing, 1956) [*Petitia*]; **D** – 7, 16, 17, 35, 51; **Z** – cst; **E** – M, bt, 4-7‰, ls, gw; **R** – 238, 245, 249, 252.

*Salvatoria clavata* (Claparède, 1863) [*Grubea*]; **D** – 2, 3, 5, 7, 8, 24, 69; **V** – 0-17; **Z** – amrs, ? K; **E** – M, bt, ep, ph, slr; **R** – 84, 219, 245, 249.

*Salvatoria limbata* (Claparède, 1868) [*Grubea*]; **D** – 7, 25, 68; **Z** – clm, ? clmrs; **E** – M, bt, ps-pe, ph; **R** – 219, 245, 389.

*Salvatoria tenuicirrata* (Claparède, 1864) [*Grubea*]; **D** – 13, 24; **V** – 1-30; **Z** – klm; **E** – M, bt, ep, slr-ph; **R** – 91, 245, 389.

*Sphaerosyllis bulbosa* Southern, 1914; **D** – 1, 2, 3, 4, 5, 6, 7, 8, 9, 10, 11, 12, 13, 14, 15, 16, 17, 18, 19, 20, 21, 22, 25, 26, 27, 28; **V** – 5-105; **Z** – nam; **E** – M, bt, eb, ps-pe; **R** – 221, 245, 249, 392.

*Sphaerosyllis hystrix* Claparède, 1863; **D** – 10, 13, 17; **V** – 17-47; **Z** – SK; **E** – M, bt, eb, ps, pe, et; **R** – 356.

*Exogone naidina* Örsted, 1845 [*E. gemmifera, Paedophylax claviger*]; **D** – 12, 13, 25, 26, 27; **V** – 0-70, 213; **Z** – SK; **E** – M, bt, eb, ps-s, sg, ph; **R** – 91, 221, 245, 389.

## Nereididae

*Nereis pelagica* Linnaeus, 1758; **D** – 7, 24; **Z** – aaminp; **E** – M, bt, eb, slr, ph; **R** – 84, 245, 249, 389.

*Nereis rava* Ehlers, 1864; **D** – 7, 24; **V** – 8; **Z** – lm; **E** – M, bt, ep, slr, ph; **R** – 84, 245, 249, 389.

*Nereis zonata* Malmgren, 1867; **D** – 2, 7, 24, 25, 26, 27, 69; **V** – 0-60; **Z** – anamnp, ? K; **E** – M, bt, mb, mc-ph-ps; **R** – 84, 245, 249, 358.

*Hediste diversicolor* (O. F. Müller, 1776) [*Nereis*]; **D** – 2, 3, 5, 6, 7, 10, 11, 12, 13, *68, *69, *74, 77, *78, *79, 84, *86, 88, 92, 93, 94; **V** – 0-90; **Z** – anam; **E** – M, bt, eh-0.5-36‰, eb, s, pe-ps, eu; **R** – 67, 84, 91, 219, 245, 358, 374, 389.

*Alitta succinea* (Frey & Leuckart, 1847) [*Neanthes, Nereis*]; **D** – 7, 13, 16, 24, 27, 68, 69, 77; **V** – 0-10, 30; **Z** – amip; **E** – M, bt, eh, ep, mc-s-ps; **R** – 84, 219, 245, 374.

*Perinereis cultifera* (Grube, 1840); **D** – 7, 16, 23, 24, 25; **V** – 0-30; **Z** – amip; **E** – M, bt, ep, lt, ps, slz; **R** – 84, 91, 245, 249.

*Platynereis dumerili* (Audouin & Milne-Edwards, 1834); **D** – 3, 7, 13, 16, 24, 25, 27, 69; **V** – 0.5-50; **Z** – amip; **E** – M, bt, mb, ph, lt, s; **R** – 67, 84, 91, 245, 358.

*Nemanereis pontica* (Borbretzky, 1872) [*N. quadriticeps, Lycastopsis*]; **D** – 46, 51; **V** – 0-0.5, 90; **Z** – nam; **E** – M, bt, sps-zc, s-ps; **R** – 234, 245, 249.

## Nephthyidae

*Micronephtys stammeri* (Augener, 1932); **D** – 1, 2, 3, 4, 5, 6, 7, 8, 9, 10, 11, 12, 13, 14, 15, 16, 17, 18, 19, 20, 21, 22, 25, 26; **V** – 0-40, 80; **Z** – adp; **E** – M, bt, mb, ps, ps-pe; **R** – 226, 245, 249, 392.

*Nephtys ciliata* (Müller, 1788); **V** – 0-50; **Z** – anamnp, i; **E** – M, bt, ep-mb, eu; **R** – 360a.

*Nephtys cirrosa* (Ehlers, 1868); **D** – 7, 9, 25, 76; **V** – 0-28, 100; **Z** – clm; **E** – M, bt, ep, sls, ps; **R** – 84, 245, 249, 358.

*Nephtys hombergii* Savigny in Lamarck, 1818; **D** – 2, 3, 4, 5, 6, 7, 8, 13, 17, *69, 76; **V** – 0-184; **Z** – clm; **E** – M, bt, eh-6‰, eb, eu; **R** – 67, 84, 91, 245, 358.

## Glyceridae

*Glycera alba* (O. F. Müller, 1776); **D** – 2, 3, 5, 7, 8, 25, 26, *69; **V** – 0-40; **Z** – namiwp; **E** – M, bt, mb, sls, ps-pe; **R** – 84, 245, 249, 389.

*Glycera capitata* Örsted, 1842; **V** – 0-100; **Z** – aamip, K, i; **E** – M, bt, eb, eu; R – 360a.

*Glycera convoluta* Keferstein, 1862; **D** – 5, 6, 7, 8, 9, 10, 11, 12, 13, 14, 15, 16, 17, 18, 19, 20, 21, 22, 25, 26, 27, 69; **V** – 0-40; **Z** – eamip; **E** – M, bt, mb, sls, ps-pe; **R** – 221, 245, 249, 358.

*Glycera gigantea* Quatrefages, 1866 [*G. decorata*]; **D** – 7, 25, *69; **V** – 0-30; **Z** – clmi; **E** – M, bt, ep, sls, ps; **R** – 84, 221, 245, 249.

*Glycera tesselata* Grube, 1840; **D** – 7, 11, 12, 13; **Z** – amip; **E** – M, bt; **R** – 247, 252, 255, 288.

*Glycera unicornis* Savigny in Lamarck, 1818 [*G. rouxii*]; **D** – 7, 11, 12, 13; **Z** – eamip; **E** – M, bt; **R** – 247, 249, 252, 288.

## Goniadidae

*Goniadella bobrezkii* (Annenkova, 1929) [*Goniada*]; **D** – 11, 25; **V** – 8, 13-26; **Z** – clm; **E** – M, bt, ep, sls, ps; **R** – 221, 245, 248, 392.

EUNICIDA

## Eunicidae

*Eunice vittata* (Delle Chiaje, 1828); **D** – 11, 12, 13, 23, 24, 25; **V** – 1-25; **Z** – amip; **E** – M, bt, ep, ro, ps, zc, s; **R** – 91, 245, 249, 389.

*Lysidice ninetta* Audouin & Milne-Edwards, 1833; **D** – 6, 7, 11, 12, 13, 24; **V** – 0-30, 40; **Z** – amip; **E** – M, bt, ep-mb, slr; **R** – 84, 91, 245, 389.

## Dorvilleidae

*Protodorvillea kefersteini* (McIntosh, 1869) [*Staurocephalus*]; **D** – 7, 8, 13, 23, 24, 25; **V** – 1.5-20; **Z** – anam; **E** – M, bt, sep, zc, ro, ps-s; **R** – 219, 245, 249, 389.

*Dorvillea rubrovittata* (Grube, 1855) [*Staurocephalus*]; **D** – 11, 12, 13, 24, 25; **V** – 0-5, 40; **Z** – anamrs; **E** – M, bt, mb, sls, ro, r; **R** – 91, 245, 249, 389.

*Schistomeringos rudolphi* (Delle Chiaje, 1828) [*Staurocephalus*]; **D** – 23, 24, 25; **V** – 3-25; **Z** – amp; **E** – M, bt, ep, sls, zc, sg; **R** – 234, 245, 249.

ORBINIIDA

## Orbiniidae

*Orbinia latreillii* (Audouin & Milne-Edwards, 1833) [*Aricia*]; **D** – 11, 25; **Z** – clm; **E** – M, bt, ep, sls, r; **R** – 221, 245, 249, 392.

## Paraonidae

*Aricidea claudiae* Laubier, 1967 [*A. jeffreysii*]; **D** – 1, 2, 3, 5,7, 13, 17, 25, 27, 28, 69; **V** – 5-92, 200; **Z** – klm; **E** – M, bt, eb, s, ps-pe, sg; **R** – 178, 180, 219, 226, 230, 245, 358, 389.

*Paradoneis harpagonea* (Storch, 1967) [*Cirrophorus, Paraonis fulgens*]; **D** – 11, 25, 27; **V** – 15-44; **Z** – ep, ? lm; **E** – M, bt, mb, ps, s; **R** – 178, 180, 221, 230, 245, 249, 392.

SPIONIIDA

## Spionidae

*Scolelepis ciliata* (Keferstein, 1862); **D** – 7, 16, 17, 51; **V** – 0.5; **Z** – clm; **E** – M, bt, sep, ps, sls, r; **R** – 221, 245, 392.

*Scolelepis squamata* (Müller, 1806) [*Nerine cirratulus*]; **D** – 7, 25, 51; **V** – 0-20; **Z** – amip; **E** – M, bt, sep, ps, ls, sls; **R** – 319, 245, 249, 389.

*Pseudomalacoceros tridentata* (Southern, 1914) [*Nerinides*]; **D** – 7, 25; **V** – 3-; **Z** – klm; **E** – M, bt, sep, ps, sls; **R** – 84, 219, 245, 389.

72

*Aonides paucibranchiata* Southern, 1914 [*A. ornatus*]; **D** – 4, 7, 8, 9, 10, 11, 12, 13, 25, 27, 28; **V** – 5-125; **Z** – ham; **E** – M, bt, eb, ps, sg, s; **R** – 180, 219, 245, 358.

*Spio filicornis* (O. F. Müller, 1776); **D** – 7, 25, 26, 68, 69; **V** – 0-30; **Z** – abambp; **E** – M, bt, ep, ps-pe; **R** – 84, 245, 358, 389.

*Pygospio elegans* Claparède, 1863; **D** – 7, 25, 26, 27, 69; **V** – 0-70; **Z** – anamnp; **E** – M, bt, eb, ps-pe; **R** – 219, 245, 358, 389.

*Polydora ciliata* (Johnston, 1838); **D** – 7, 68, 69, 76, 84; **V** – 27; **Z** – aanamip; **E** – M, bt, ep, ro, mc; **R** – 219, 245, 249, 389.

*Polydora cornuta* Bosc, 1802; **D** – 7, 13, 68, 69; **V** – 0.1-32; **Z** – amp, i; **E** – M, bt, ep, s, ps, sg, is; **R** – 358, 360a

*Dipolydora quadrilobata* (Jacobi, 1883); **D** – 24, 49; **Z** – bapbp, i; **E** – M, bt, is; **R** – 355, 360a.

*Prionospio cirrifera* Wirén, 1883; **D** – 7, 9, 13, 25, 26, 27, 69; **V** – 5-84; **Z** – aamip; **E** – M, bt, eb, sg, s; **R** – 219, 222, 226, 249.

*Prionospio malmgreni* Claparède, 1868; **D** – 6, 23, 25; **V** – 4-20; **Z** – aamip; **E** – M, bt, eb, ps, sg, zc, r; **R** – 180, 219, 226, 245.

*Streblospio shrubsolii* (Buchanan, 1890); **D** – 7, 68, 69, *78, *79; **V** – 12; **Z** – clm; **E** – M-B, bt, eh, s, ps-pe; **R** – 219, 245, 358, 360a, 389.

## Magelonidae

*Magelona papillicornis* F. Müller, 1858; **D** – 7, 11, 18, 25, 69; **V** – -23; **Z** – amip; **E** – M, bt, ep, ps; **R** – 221, 226, 245, 249.

*Magelona rosea* Moore, 1907; **D** – 7, 11, 18, 25; **V** – 4-28; **Z** – nam; **E** – M, bt, ep, ps, ps-s; **R** – 219, 180, 245, 392.

## Cirratulidae

*Cirriformia tentaculata* (Montagu, 1808) [*Audouinia*]; **D** – 23, 25; **V** – 0-20; **Z** – amiswp; **E** – M, bt, ep, ps, sg, zc, r; **R** – 245.

*Caulleriella bioculata* (Keferstein, 1862) [*Cirratulus viridis, Heterocirrus*]; **D** – 7, 13, 25; **V** – 5-25; **Z** – namp; **E** – M, bt, ep, ps-sg; **R** – 91, 226, 245, 389.

OPHELIIDA

## Opheliidae

*Ophelia bicornis* Savigny, 1918; **D** – 2, 6, 7, 19, 21, 25, 51; **V** – 0.5-1.5; **Z** – bam; **E** – M, bt, sep, l-sl, ps, ■; **R** – 236, 245, 249, 389.

*Ophelia limacina* (Rathke, 1843); **D** – 7, 11, 12, 25; **V** – 10-37; **Z** – abapnep; **E** – M, bt, ep, sl, ps; **R** – 221, 236, 245, 249.

*Polyophthalmus pictus* (Dujardin, 1839); **D** – 13, 24, 50; **V** – 0-10; **Z** – amip; **E** – M, bt-p, sep, ph-lt; **R** – 91, 222, 245, 389.

CAPITELLIDA

## Capitellidae

*Notomastus latericeus* M. Sars, 1851; **D** – ?; **Z** – aanamip; **E** – M, bt, eu; **R** – 245, 249.

*Notomastus profundus* (Eisig, 1887); **D** – 1, 2, 3, 4, 5, 6, 7, 8, 28; **V** – 65-; **Z** – clm, ? clmrs; **E** – M, bt, shb, phs; **R** – 219, 245, 249, 389.

*Heteromastus filiformis* (Claparède, 1864); **D** – 1, 2, 3, 4, 5, 6, 7, 8, 13, 17, 25, 27, 28, 59, 69; **V** – 0-16-200; **Z** – aamip; **E** – M, bt, eb, pe, ms, phs; **R** – 84, 245, 249, 358.

*Capitella capitata* (Fabricius, 1780); **D** – 4, 5, 6, 7, 25, 26, *69; **V** – 0-30, 80; **Z** – aamip; **E** – M, bt, ep, ps-pe; **R** – 84, 245, 249, 358.

*Capitella minima* Langerhans, 1880 [*Capitomastus*]; **D** – 7, 13, 25, 26, 69; **V** – 5-110; **Z** – aam; **E** – M, bt, eb, ps, pe, sg; **R** – 245 349, 358, 389.

*Capitellides giardi* Mesnil, 1897; **D** – 7, 25; **Z** – bam; **E** – M, bt, sep, ps; **R** – 177, 219, 245, 249.

? *Capitellethus dispar* (Ehlers, 1907); **Z** – miwp, i; **E** – M-L, bt; R – 360a.

74

## Arenicolidae

*Arenicola marina* Lamarck, 1801; **D** – 7, 12, 25, 51, 84; **Z** – bamnep; **E** – M, bt, ep, ps, ls; **R** – 219, 226, 245, 249.

*Arenicolides branchialis* (Audouin & Milne-Edwards, 1833) [*Arenicola grubii*]; **D** – 25; **V** – 1.5-12; **Z** – lm; **E** – M, bt, sep, sg; **R** – 245, 249.

## Maldanidae

*Clymene collaris* (Claparède, 1868) [*Praxylla*]; **D** – 13, 23, 25; **V** – 2-30; **Z** – lm; **E** – M, bt, ep, ps-s, sg, zc; **R** – 91, 226, 245, 249.

*Leiochone leiopygos* (Grube, 1860) [*Clymenura clypeata*]; **D** – 7, 25, 26, 69; **V** – 0-35; **Z** – clm; **E** – M, bt, ep, ps, ps-pe; **R** – 84, 219, 226, 245, 249, 358, 289.

CTENODRILIDA

## Ctenodrilidae

*Zeppelina dentata* Monticelli, 1897 [nomen dubium ?]; **D** – 13; **Z** – lm; **E** – M, bt; **R** – 252.

## Parergodrilidae

*Stygocapitella subterranea* Knöllner, 1934; **D** – 7, 35; **Z** – bamswp; **E** – M, ls, gw, r; **R** – 221, 245, 249, 387.

TEREBELLIDA

## Sabellariidae

*Sabellaria taurica* (Rathke, 1837) [*S. spinulosa, Centrocorone*]; **D** – 7, 12, 13, 24; **V** – -30; **Z** – lm, ? clm; **E** – M, bt, ep, sl, sg; **R** – 84, 91, 219, 222, 245, 249, 389.

## Pectinariidae

*Lagis koreni* Malmgren, 1866 [*Pectinaria*]; **D** – 7, 11, 13, 25, 26, 27, 69; **V** – 0-30, 50; **Z** – clm; **E** – M, bt, mb, ps-pe, ms, sg, s; **R** – 84, 91, 219, 226, 245, 249, 358, 389.

## Ampharetidae

*Melinna palmata* Grube, 1870; **D** – 2, 3, 4, 5, 6, 7, 13, 17, 26, 27, 28, *69; **V** – 15-200; **Z** – clmnwi; **E** – M, bt, eh, eb, s, ps-pe, sg; **R** – 67, 84, 178, 219, 245, 249, 358.

*Hypania invalida* (Grube, 1860); **D** – 59, 60; **V** – 2-5, 40; **Z** – cp, ? pc, Rc; **E** – M-B, bt, 0.5-0.8‰, s, sg; **R** – 206, 245, 249, 252, 320, 386.

## Trichobranchiidae

*Terebellides stroemii* Sars, 1835; **D** – 2, 3, 7, 10, 11, 13, 16, 17; **V** – 36-200; **Z** – aanamip, ? K; **E** – M, bt, eb-hb, s, phs; **R** – 67, 84, 91, 178, 245.

## Terebellidae

*Polycirrus jubatus* Bobretzky, 1869; **D** – 7, 12, 16, 25, 51; **V** – 0-30; **Z** – ● p; **E** – M, bt, ep, sg, ps-pe; **R** – 91, 226, 245, 389.

SABELLIDA

## Sabellidae

*Fabricia sabella* (Ehrenberg, 1836); **D** – 7, 25, 34, 69; **V** – 12; **Z** – aannam; **E** – M, bt, 3-38‰, ep, ps; **R** – 219, 245, 249, 389.

*Fabricia stellaris* (O. F. Müller, 1774)  [? *F. sabella*]; **D** – 7, 68, 69; **Z** – bam; **E** – M, bt, ep; **R** – 358.

? *Manayunkia caspica* Annenkova, 1928; **D** – ?; **Z** – pc, cpc, Rc; **E** – M-B-L, bt; **R** – 245, 249, 320.

*Oriopsis armandi* (Claparède, 1864) [*Oridia*]; **D** – 1, 2, 3, 4, 5, 6, 7, 8, 9, 10, 11, 12, 13, 14, 15, 16, 17, 18, 19, 20, 21, 22, 27, 28; **V** – -130, 200; **Z** – kclm, ? clmrs; **E** – M, bt, hb, ms, phs; **R** – 234, 245, 249, 358.

## Serpulidae

*Hydroides norvegicus* Gunnerus, 1768; **D** – 7, 24; **V** – 0-30; **Z** – nami; **E** – M, bt, ep, lt, r; **R** – 84, 245, 389.

*Ficopomatus enigmaticus* (Fauvel, 1923) [*Mercierella*]; **D** – 5, 7, 68, 69, 77, *79, 84, 87; **V** – 0-15; **Z** – amip, i; **E** – M, bt, 0-55‰, ep, is; **R** – 219, 245, 249, 389.

*Vermiliopsis infundibulum* (Philippi, 1844); **D** – 3, 7, 12, 13; **V** – 10-15; **Z** – amrsp; **E** – M, bt, ep, lt, ro; **R** – 67, 84, 91, 245, 249.

*Spirobranchus triqueter* (Linnaeus, 1758) [*Pomatoceros, P. triquetroides*]; **D** – 3, 7, 24, 25, 27; **V** – 0-70; **Z** – clm; **E** – M, bt, mb, lt, ms, sg; **R** – 67, 84, 219, 226, 245, 389.

*Ditrupa arietina* (O. F. Müller, 1776); **D** – 7, 26; **Z** – namip; **E** – M, bt, ep, pe; **R** – 84, 245, 249, 389.

*Pileolaria militaris* Claparède, 1870 [*Spirorbis*]; **D** – 24; **V** – 0-30; **Z** – neaminz; **E** – M, bt, ep-mb, slc, phc; **R** – 230, 245, 249.

### Spirorbidae

*Spirorbis pusilla* Rathke, 1837; **D** – 1, 2, 3, 4, 5, 6, 7, 8, 9, 10, 11, 12, 13, 14, 15, 16, 17, 18, 19, 20, 21, 22, 24, 50; **V** – 0-30, 75; **Z** – neamnp; **E** – M, bt, mb, lt, ro, ph, phc, slc, sg, zc; **R** – 91, 230, 245, 249, 389.

Archiannelida

### Polygordiidae

*Polygordius lacteus* Schneider, 1868 [*P. neapolitanus ponticus, P. ponticus*]; **D** – 25; **V** – 0-25; **Z** – clm; **E** – M, bt, ep, ps, sg, sls; **R** – 180, 226, 245, 249.

### Protodrilidae

*Protodrilus flavocapitatus* (Uljanin, 1877); **D** – 7, 25, 27, 35, 51; **V** – 0-80; **Z** – lm, ? clm; **E** – M, bt, eb, l-sl, ps, pe; **R** – 220, 245, 249, 358.

### Saccocirridae

*Saccocirrus papillocercus* Bobretzky, 1872; **D** – 7, 51; **V** – 0.5-0.7; **Z** – lm, ? clm; **E** – M, bt, sep, l-sl, ps; **R** – 220, 245, 249, 389.

## Nerillidae

*Nerilla antennata* Schmidt, 1848; **D** – 7, 25, 34, 35, 51, *69; **V** – 80; **Z** – naminz; **E** – M, bt, ep, l-sl, ps, gw; **R** – 220, 239, 245, 386.

## Dinophilidae

*Dinophilus gyrociliatus* O. Schmidt, 1857; **D** – 25, 34, 51; **Z** – naminz; **E** – M, bt, sep, l-sl, ps; **R** – 220, 245, 249, 389.

*Trilobodrilus heideri* Remane, 1925; **D** – 7, 35; **V** – 0-1; **Z** – clm; **E** – M, bt, sep, ps, if, r; **R** – 220, 245, 249, 389.

### APHANONEURA

#### AEOLOSOMATIDA

## Aeolosomatidae

*Aeolosoma hemprichi* Ehrenberg, 1831; **D** – 58, *68; **Z** – (hnata, ? sk); **E** – L, bt, cr, po, sw; **R** – 88, 249, 365, 389.

### OLIGOCHAETA

#### OPISTOPHORA

## Criodrilidae

*Criodrilus lacuum* Hoffmeister, 1845; **D** – 71, 72; **Z** – (eca, ? h); **E** – L, bt; **R** – 365.

## Lumbricidae

*Eiseniella tetraedra* (Savigny, 1876); **D** – 71, 77; **Z** – (h, ? k); **E** – L-TL, bt, tx; **R** – 365.

#### TUBIFICIDA

## Naididae

*Stylaria lacustris* (Linnaeus, 1767); **D** – 1, 2, 3, 4, 5, 6, 7, 8, 9, 10, 11, 12, 13, 14, 15, 16, 17, 18, 19, 20, 21, 22, 35, 58, 77, 78, 84; **Z** – nam, (hno); **E** – M, bt, 7‰, ps, gw; **R** – 42, 84, 88, 249, 353a, 365.

*Pristina rosea* (Piguet, 1906); **D** – 78; **Z** – pno, ? sk; **E** – L, bt, pe-ps; **R** – 353a.

*Slavina appendiculata* (Udekem, 1855); **D** – 58; **Z** – amnz, (k); **E** – M, bt, 1-2‰; **R** – 365.

*Dero digitata* (O. F. Müller, 1773); **D** – 78; **Z** – (k); **E** – L, bt; **R** – 353a, 365.

*Dero obtusa* Udekem, 1855; **D** – 58, 78; **Z** – (k); **E** – L, bt, 1-2‰, α; **R** – 353a, 365.

*Nais barbata* O. F. Müller, 1773 ; **D** – 58, 71, 77, 78; **Z** – (hoa); **E** – L, bt; **R** – 353a, 365.

*Nais bretscheri* Michaelsen, 1899; **D** – 58; **Z** – (hop, ? hno); **E** – L, bt, 1-2‰, o-β; **R** – 265.

*Nais communis* Piguet, 1906 ; **D** – 88, 90 ; **Z** – ham, (k) ; **E** – M-L, bt, eh-16‰, eu; **R** – 88, 249, 365, 374.

*Nais elinguis* O. F. Müller, 1774 ; **D** – 54, 58, *68, *69, 71, 74, 77, *78, 82, 84, *86, 87, 88, 90, 92; **Z** – aminz, (k); **E** – M-B-L, bt, 18‰, eh, eu, sw-po; **R** – 84, 87, 88, 249, 365, 374, 389.

*Nais pardalis* Piguet, 1906; **D** – 58, *69, 78, 84; **Z** – (hno); **E** – L, bt, 10‰, ph, ps-pe; **R** – 88, 249, 353a, 365, 389.

*Nais variabilis* Piguet, 1906; **D** – 58, 74, 79, 82; **Z** – aminz, (k); **E** – M-B, bt, sw-po; **R** – 365.

*Ophidonais serpentina* O. F. Müller, 1774; **D** – 58, 79, 84, 92; **Z** – bap, (hno); **E** – L, bt, sw-po; **R** – 365.

*Homochaeta naidina* Bretscher, 1896 [*Paranais*]; **D** – 7, 58; **Z** – (po); **E** – L-M, bt, 15‰; **R** – 84, 249, 365, 389.

*Paranais frici* Hrabe, 1941; **D** – 84; **Z** – amip, (hno); **E** – M-B, bt, po; **R** – 365.

*Paranais litoralis* (O. F. Müller, 1780); **D** – 7, 58, 87, 88; **Z** – namp, (hn); **E** – M-B, bt, 16‰; **R** – 84, 88, 249, 365.

*Amphichaeta leydigi* Tauber, 1879; **D** – 58, 87; **Z** – (h, ? hno); **E** – L, 2‰, sw; **R** – 365.

*Chaetogaster cristallinus* Vejdovsky, 1883; **D** – *68, 84; **Z** – (hpt, ? hptn); **E** – L-M, bt, 10‰, sw; **R** – 84, 88, 365, 374.

*Chaetogaster diaphanus* (Gruithuisen, 1828); **D** – 58; **Z** – SK, (sk); **E** – M-L, bt, sw, po; **R** – 88, 365, 372.

## Tubificidae

*Tubifex nerthus* Michaelsen, 1908; **D** – 84, 96; **Z** – clp, ? bap; **E** – M-B, bt, sw; **R** – 107, 249, 364, 365.

*Tubifex tubifex* (O. F. Müller, 1774); **D** – 58, 71, 72, 74, 77, 78, 80, 84, 87, 92; **Z** – (k); **E** – M-B-L, bt, eu; **R** – 353a, 365.

*Rhyacodrilus coccineus* (Vejdovsky 1876); **D** – 78; **Z** – (sk); **E** – L-B, bt,o-β; **R** – 353a.

*Branchiura sowerbyi* Beddard, 1892; **D** – 59; **Z** – (h); **E** – L, 0.78‰, bt, ro; **R** – 365.

*Aulodrilus pluriseta* (Piguet, 1906); **D** – 58, 96; **Z** – (hoa); **E** – L, bt, eu; **R** – 365.

*Limnodrilus claparedeanus* Ratzel, 1868; **D** – 71, 78, 80, 92; **Z** – (k); **E** – L, bt, eu, sw, po; **R** – 363a, 365.

*Limnodrilus hoffmeisteri* Claparède, 1862; **D** – 58, *68, 71, 74, 78, 87, 92; **V** – 0.5-120; **Z** – SK, (k); **E** – M-L, bt, sw, po, α, s; **R** – 84, 249, 353a, 365, 389.

*Limnodrilus profundicola* (Verrill, 1871); **D** – 78; **Z** – nam, (ho); **E** – L-B, bt, sw, po; **R** – 353a, 365.

*Limnodrilus udekemianus* Claparède, 1862; **D** – 58, 65, *68, 74, 79, 80, 84, 87; **Z** – SK, (k); **E** – L-B, 1‰, bt, eu, α; **R** – 84, 249, 365, 389.

*Psammoryctides albicola* (Michaelsen, 1901) [*Psammoryctes, Tubifex*]; **D** – 58, *68; **Z** – clm, (h); **E** – L-B, 2‰, bt, pe, α; **R** – 365.

*Psammoryctides barbatus* (Grube, 1861); **D** – 58; **Z** – clm, (h); **E** – M-L, 2‰, bt, ps, α-β; **R** – 365.

*Psammoryctides moravicus* (Hrabe, 1934); **D** – 58; **Z** – (wpat); **E** – L, 2‰, bt, po, sw, r; **R** – 365.

*Potamothrix bavaricus* (Oeschmann, 1913); **D** – 58; **Z** – clmnz, (hna); **E** – M-L, 2‰, bt, sw, r; **R** – 365.

*Potamothrix bedoti* (Piguet, 1913); **D** – 58; **Z** – (ho); **E** – L, 2‰, bt, po, sw; **R** – 365.

*Potamothrix hammoniensis* (Michaelsen, 1901); **D** – 58, 78, 84, 88; **Z** – clm, (ho); **E** – M-L, bt, sw, po, pe; **R** – 205, 365.

*Potamothrix vejdovskyi* (Hrabě, 1941); **D** – 58, 78; **Z** – pc, (eit, ? h); **E** – M-B-L, bt, po, sw; **D** – 353a, 365.

*Clitellio arenarius* (Müller, 1776); **D** – 84, 96; **Z** – bap; **E** – M, bt, l-sl, s, ro, ps; **R** – 249, 364, 365.

*Aktedrilus monospermathecus* Knöllner, 1935; **D** – 1, 2, 3, 4, 5, 6, 7, 8, 9, 10, 11, 12, 13, 14, 15, 16, 17, 18, 19, 20, 21, 22, 35, 49; **Z** – clm, ? clmnei; **E** – M, bt, l-sl, gw, ps; **R** – 150, 244, 249, 365.

*Heterochaeta costata* Claparède, 1863 [*Tubifex*]; **D** – 11, 58, 84, 88, 96; **Z** – clp; **E** – M-B, bt, l, ps, pe-ps; **R** – 249, 364, 365.

*Tubificoides benedii* (Udekem, 1855) [*Peloscolex*]; **D** – 1, 2, 3, 4, 5, 6, 7, 8, 9, 10, 11, 12, 13, 14, 15, 16, 17, 18, 19, 20, 21, 22, 35; **Z** – bap; **E** – M-B, bt, l, ps, gw; **R** – 43, 249.

*Tubificoides euxinicus* (Hrabě, 1966) [*Tubifex euxinus* ?]; **D** – 1, 2, 3, 4, 5, 6, 7, 8, 9, 10, 11, 12, 13, 14, 15, 16, 17, 18, 19, 20, 21, 22, 35; **Z** – ? bap; **E** – M, bt, l, ps, gw; **R** – 244, 249, 252.

*Tubificoides swirencowi* Jaroschenko, 1948 [*T. heterochaetus, Pristina papillosa*]; **D** – 13, 84; **Z** – pc, cp; **E** – M, bt, 15‰, pe-ps; **R** – 88, 249, 365, 389.

*Monopylephorus rubroniveus* Levinsen, 1884 [*Rhizodrilus ponticus*, *M. rubroniveus ponticus*]; **D** – 1, 2, 3, 4, 5, 6, 7, 8, 9, 10, 11, 12, 13, 14, 15, 16, 17, 18, 19, 20, 21, 22, 35; **Z** – ? namnp; **E** – M, bt, l, pe-ps, gw, ph; **R** – 43, 249, 252.

## Enchytraeidae

*Enchytraeus albidus* Henle, 1837; **D** – 11, 35; **Z** – napnei, (hna); **E** – M-L-TL, l, ps, gw; **R** – 84, 88, 249, 365.

*Marionina achaeta* (Hagen, 1954); **D** – 1, 2, 3, 4, 5, 6, 7, 8, 9, 10, 11, 12, 13, 14, 15, 16, 17, 18, 19, 20, 21, 22, 35; **Z** – cp; **E** – M, 15‰, l, ps, gw; **R** – 244, 249, 252.

*Marionina elongata* Lasserre, 1964; **D** – 1, 2, 3, 4, 5, 6, 7, 8, 9, 10, 11, 12, 13, 14, 15, 16, 17, 18, 19, 20, 21, 22, 35; **Z** – ? nam; **E** – M, bt, 15‰, l, ps, gw; **R** – 244, 249, 252.

*Marionina spicula* (Leuckart, 1847); **D** – 1, 2, 3, 4, 5, 6, 7, 8, 9, 10, 11, 12, 13, 14, 15, 16, 17, 18, 19, 20, 21, 22, 35; **Z** – bap; **E** – M, bt, 15‰, l, ps, gw; **R** – 244, 249, 252.

*Marionina subterranea* (Knöllner, 1935); **D** – 1, 2, 3, 4, 5, 6, 7, 8, 9, 10, 11, 12, 13, 14, 15, 16, 17, 18, 19, 20, 21, 22, 35, 54; **Z** – bapbp, (h); **E** – M-L, bt, l, ps, gw, cr; **R** – 244, 249, 252, 365.

## BRANCHIOBDELLEA

BRANCHIOBDELLIDA

## Branchiobdellidae

*Branchiobdella astaci* Odier, 1823; **D** – 60, ? 61; **Z** – (csee, ? cse); **E** – L, epi, 0.1-2‰; **R** – 365, 389.

## HIRUDINEA

RHYNCHOBDELLIDA

## Glossiphoniidae

*Glossiphonia complanata* (Linnaeus, 1758); **D** – 58, *68, 94; **Z** – neamj, (h); **E** – L, bt, ph, 2‰; **R** – 374, 389.

*Helobdella stagnalis* (Linnaeus, 1758); **D** – 58, *68, 77; **Z** – (h); **E** – L, bt, ph, 1‰; **R** – 155a, 374, 389.

*Hemiclepsis marginata* (O. F. Müller, 1774); **D** – 58, 59; **Z** – (po); **E** – L, bt, ph, 0.3‰; **R** – 374, 389

*Placobdella costata* (Fr. Müller, 1846); **D** – 58; **Z** – (wp); **E** – L, bt, ph, 0.3‰.

*Theromyzon tessulatum* (O. F. Müller, 1774) [*Protoclepsis*]; **D** – *67; **Z** – (? h); **E** – L, bt; **R** – 372, 389.

### Piscicolidae

*Piscicola geometra* (Linnaeus, 1758); **D** – *68; **Z** – (hn, ? hnat); **E** – L-B-M, bt, 1‰; **R** – 84, 389.

ARHYNCHOBDELLIDA

### Hirudinidae

*Hirudo verbana* Carena, 1820 [*H. medicinalis*]; **D** – 58, *68, *69, *74; **Z** – (cse); **E** – L, bt, ph, 10‰, ▲; **R** – 374, 389.

### Haemopidae

*Haemopis sanguisuga* (Linnaeus, 1758) [*Aulastoma gulo*]; **D** – 58, *68, *74; **Z** – (ena); **E** – L, bt, ph, 10‰; **R** – 374, 389.

### Erpobdellidae

*Erpobdella ococulata* (Linnaeus, 1758); **D** – 58, 77; **Z** – (po); **E** – L, bt, ph, lt, 5-7‰; **R** – 258.

# TARDIGRADA

## HETEROTARDIGRADA

### ARTHROTARDIGRADA

### Batillipedidae

*Batillipes mirus* Richters, 1909; **D** – 16, 18, 25, 51; **V** – 1-8; **Z** – nami, ? K; **E** – M, bt, l-sl, ps, gw, if; **R** – 249, 388, 389.

### Halechiniscidae

*Halechiniscus guiteli* Richters, 1908; **D** – 16, 35, 51; **Z** – lm, ? clm; **E** – M, bt, l-sl, ps, s, gw; **R** – 249, 388, 389.

### Stygarctidae

*Stygarctus bradypus* Schulz, 1951; **D** – 35; **Z** – cpnei, ? cg; **E** – M, bt, l, ps, gw; **R** – 249, 392.

### ECHINISCOIDEA

### Echiniscoididae

*Echiniscoides sigismundi* Plate, 1889; **D** – 3, 4, 5, 6, 7, 8, 9, 10, 11, 12, 13, 14, 15, 16, 17, 18, 19, 20, 21, 22, 50; **Z** – anamp; **E** – M, bt, eh, eu, l, ph; **R** – 84, 249, 382, 388.

### EUTARDIGRADA

### PARACHETA

### Hypsibiidae

*Halobiotus stenostomus* (Richters, 1908) [*Hypsibius*]; **D** – 24, 25, 49; **V** – -25; **Z** – clm; **E** – M, bt, sl, ps, ph; **R** – 249, 316.

# ARTHROPODA

## ARACHNIDA

### ACARINA

## Hydrozetidae

*Hydrozetes lacustris* (Michael, 1882); **D** – 70, *86, 94; **Z** – (csee, ? e); **E** – L, 4‰; **R** – 374, 389.

## Limnozetidae

*Limnozetes rugosus* (Sellnick, 1923); **D** – 94; **Z** – (e); **E** – L, 6‰; **R** – 374, 389.

## Halacaridae

*Rhombognathus magnirostris* Trouessart, 1889; **D** – 7, 11, 12, 13, 16, 24; **Z** – clm; **E** – M-B, bt, slc; **R** – 84, 89, 275, 389.

*Rhombognathus notops* (Gosse, 1855); **D** – *69; **Z** – abap; **E** – M, bt, 12‰, l-sl; **R** – 249, 389, 402.

*Rhombognathides pascens* (Lohmann, 1889) [*Rhombognathus*]; **D** – 11, 12, 13, 16; **Z** – nam; **E** – M-B, bt, sl; **R** – 89, 249, 389.

*Halacarellus capuzinus* (Lohmann, 1893) [*Halacarus*]; **D** – 35; **Z** – bap; **E** – M-B, bt, l-sl, ps, gw; **R** – 249, 252, 307.

*Halacarellus phreaticus* Petrova, 1972; **D** – 7, 16; **Z** – hom; **E** – M-B-L, bt; **R** – 249, 252, 306.

*Halacarellus procerus* (Viets, 1927); **D** – 35; **Z** – bam; **E** – M-B, bt, 15‰, l-sl, ps, gw, ■; **R** – 114, 249, 252, 307.

*Halacarellus subterraneus* Schulz, 1933; **D** – 3, 5, 84; **Z** – abam, ? anam; **E** – M-B, eh, bt, l, gw; **R** – 249, 252, 306.

*Anomalohalacarus marcandrei* Monniot, 1967 [*Halacarus, Halacarellus*]; **D** – 8, 18; **Z** – clm, ? clmi; **E** – M, bt, sl; **R** – 249, 252, 307.

*Thalassarachna affinis* (Trouessart, 1896) [*Halacarus basteri* var. *affinis*]; **D** – 11, 12, 13, 16, 24, 27, 28; **V** – -150; **Z** – hom; **E** – M-B, bt, l-sl, pe, slc; **R** – 89, 249, 389.

? *Thalassarachna basteri* (Johnston, 1836) ? [*Halacarellus*]; **D** – ?; **V** – -46; **Z** – abam, ? anam; **E** – M-B, eh, bt, l-sl; **R** – 249.

*Thalassarachna hexacantha* (Viets, 1927) [*Halacarus, Halacarellus*]; **D** – 7; **V** – 0.5-28; **Z** – cp; **E** – M, bt, l-sl; **R** – 84, 249, 389.

*Copidognathus brachystomus* Viets, 1940; **D** – *69; **Z** – clm; **E** – M-B, bt, l; **R** – 249, 389, 402.

*Copidognathus brevirostris* Viets, 1927; **D** – 7, 24; **V** – 0-10; **Z** – clm; **E** – M-B, bt, l-sl, ph, lt; **R** – 84, 249, 389.

*Copidognathus extensus* Viets, 1940; **D** – 7; **V** – 12; **Z** – hom; **E** – M, bt, l-sl; **R** – 84, 249, 389.

*Copidognathus fabricii* (Lohmann, 1889); **D** – 11, 12, 13, 16, *79; **Z** – clmi; **E** – M-B, bt, l-sl; **R** – 89, 249, 374, 389.

*Copidognathus magnipalpus* (Police, 1909) [*C. magnipalpus ponticus, C. magnipalpus serratiseta*]; **D** – *86; **Z** – lmwi, ?R; **E** – M-B, bt, 1-20‰, l; **R** – 249, 389, 401.

*Copidognathus oculatus* (Hodge, 1863) [*Copidognathopsis, Halacarus*]; **D** – 11, 12, 13, 16, *69; **Z** – clm; **E** – M-B, bt, 12‰, sl; **R** – 89, 249, 374, 389.

*Copidognathus quadricostatus* (Trouessart, 1894) [*Copidognathopsis*]; **D** – 11, 12, 16; **Z** – clm; **E** – M, bt, sl; **R** – 89, 249, 275, 389.

*Copidognathus rhodostigma* (Gosse, 1855); **D** – 11, 12, 16, 24; **Z** – clm; **E** – M-B, bt, l-sl, ph, sg; **R** – 89, 249, 275, 389.

*Copidognathus tabellio* (Trouessart, 1894); **D** – 11, 12, 13, 16; **V** – 9-28; **Z** – lm; **E** – M-B, bt, sl; **R** – 89, 249, 275, 389.

*Copidognathus tectiporus* (Viets, 1935) [*Copidognathopsis*]; **D** – *67, *69; **Z** – lm, ?R; **E** – B-L-(M), bt, eu; **R** – 249, 389, 400, 402.

*Agaue chevreuxi* (Trouessart, 1889) [*Halacarus*]; **D** – 11, 12, 13, 16, 24; **V** – -20; **Z** – lm, ? clm; **E** – M-B, bt, l-sl, slc, phc; **R** – 89, 249, 389.

*Agauopsis brevipalpus* (Trouessart, 1889) [*A. pontica, Agaue brevipalpus* var. *ponticus*]; **D** – 11, 12, 13, 16, 25, 27; **V** – 9-28; **Z** – namnei; **E** – M-B, bt, sg, ms, phc; **R** – 89, 249, 389.

*Agauopsis marinovi* Petrova, 1976; **D** – 7; **Z** – ham; **E** – M-B, bt, l; **R** – 249, 252, 308.

*Actacarus pygmaeus* Schulz, 1937; **D** – 25, 35; **V** – -5; **Z** – lmnei; **E** – M-B, bt, gw-if, ps, sl; **R** – 229, 249, 389.

*Arhodeoporus gracilipes* (Trouessart, 1889) [*Copidognathopsis*]; **D** – 12, 16, 25, 28; **V** – -150; **Z** – clm; **E** – M-B, bt, eb, sg, phs; **R** – 89, 249, 275, 389.

*Acarochelopodia delamarei* Angelier, 1954; **D** – 7, 35; **Z** – hom; **E** – M-B, bt, ps, gw; **R** – 229, 249, 392.

*Lohmannella falcata* (Hodge, 1863); **D** – 12, 27, 28; **V** – 45-100; **Z** – bam; **E** – M-B, bt, eb, ms, phs; **R** – 89, 249, 389.

*Caspihalacarus hyrcanus* Viets, 1928 [*C. hyrcanus danubialis*]; **D** – *69; **V** – 1-2; **Z** – pc, clm, Rc; **E** – B-L, bt, ep, ro, s; **R** – 252, 389, 402.

### Hydrachnidae

*Hydrachna cruenta* O. F. Müller, 1776; **D** – 42; **Z** – (h); **E** – L, bt, l; **R** – 84, 389.

### Eylaidae

*Eylais infundibulifera* Koenike, 1897; **D** – 42; **Z** – (h); **E** – L, bt, l; **R** – 84, 389.

*Eylais rimosa* Piersig, 1899; **D** – 42; **Z** – (h); **E** – L, bt, l; **R** – 84, 389.

### Hydryphantidae

*Hydryphantes crassipalpis* Koenike, 1914; **D** – *68; **Z** – (wesa); **E** – L, bt, 1‰, l; **R** – 389, 400.

## Hydrodromidae

*Hydrodroma despiciens* (O. F. Müller, 1776); **D** – 88; **Z** – (sk, ? k); **E** – L, bt, 1‰; **R** – 389, 402.

## Pontarachnidae

*Pontarachna valkanovi* Petrova, 1978; **D** – 11, 12, 25, 35, 51, 92; **Z** – ● p; **E** – M, bt, l-sl, ps, gw; **R** – 249, 252, 309, 392.

## Limnesiidae

*Limnesia koenikei* Piersig, 1894; **D** – 90; **Z** – (h); **E** – L, bt; **R** – 389, 402.

*Limnesia undulata* (O. F. Müller, 1776); **D** – *68; **Z** – (h); **E** – L, bt, l; **R** – 389, 400.

## Aturidae

*Brachypoda versicolor* (O. F. Müller, 1776); **D** – *67; **Z** – (? tp); **E** – L, bt; **R** – 389, 400.

## Arrenuridae

*Arrenurus bruzelii* Koenike, 1885; **D** – 93; **Z** – (wp, ? tp); **E** – L, bt, 3‰; **R** – 389, 402.

### PYCNOGONIDA

#### PANTOPODA

## Ammotheidae

*Tanystylum conirostre* (Dohrn, 1881); **D** – 7, 24; **V** – 1-2; **Z** – kclm; **E** – M, bt, ep, sl, ph; **R** – 39, 249, 389.

## Callipallenidae

*Callipallene phantoma* (Dohrn, 1881); **D** – 1, 2, 3, 4, 5, 6, 7, 8, 9, 10, 11, 12, 13, 14, 15, 16, 17, 18, 19, 20, 21, 22, 28; **V** – 35-80; **Z** – anam; **E** – M, bt, shb, phs; **R** – 247, 249, 252.

CRUSTACEA

SARSOSTRACA: ANOSTRACA

## Artemiidae

*? Artemia urmiana* Günther, 1899 [**?** = *Artemia parthenogenetica*]; **D** – 76, 77; **Z** – (omcaa, ? mwca); **E** – B, 20-100-340‰, p; **R** – 119, 155a, 359.

*Artemia salina* (Linnaeus, 1758); **D** – *64, 76, 77; **Z** – (sp, ? sk); **E** – B, 20-80-340‰, p; **R** – 86, 91, 100, 155a, 372, 389.

*Artemia* is a problematic genus with an unclear taxonomic status of the part of the species. Some authors impugn the existence of *A. parthenogenetica* (Abatzopoulos et al., 2002; Asem et al., 2010). They accept that these are parthenogenetic populations of different species of the genus *Artemia*. These populations (known as *A. parthenogenetica*) have been established in other continents as well. After DNA analysis Munoz et al. (2010) bring closer the parthenogenetic *Artemia* from the Atanasovsko Lake to *Artemia urmiana* Günther, 1899 (endemic from the Urmia Lake in Iran).

## Thamnocephalidae

*Branchinella spinosa* (H. Milne Edwards, 1840); **D** – *64; **Z** – lm, (atm); **E** – B-M, 30‰, p, ■; **R** – 100, 114, 389.

PHYLLOPODA: CLADOCERA

## Sididae

*Penilia avirostris* Dana, 1849 [*P. schmackeri*]; **D** – 7, 29, *69, 77; **V** – 0-50; **Z** – amiwp; **E** – M-B-L, p, eh, epp; **R** – 198, 202, 319, 389.

*Diaphanosoma brachyurum* (Liévin, 1848); **D** – 58, *68, *79, 80; **Z** – amwp, (hat); **E** – M-B-L, p, eh, 6-8‰; **R** – 97, 278, 374, 389.

*Diaphanosoma lacustris* Korinek, 1981; **D** – 59, 60; **Z** – (pat); **E** – L, p, 0.5-0.8‰; **R** – 279, 280.

89

## Macrothricidae

*Macrothrix laticornis* (Jurine, 1820); **D** – *68; **Z** – (sk); **E** – L, 1‰, p-bt, s; **R** – 97, 279, 389.

*Ilyocryptus sordidus* (Liévin, 1848); **D** – *68; **Z** – amwp, (pat); **E** – B-L, 1‰, p; **R** – 279, 374, 389.

## Bosminidae

*Bosmina coregoni* Baird, 1857; **D** – 59, 60; **Z** – apswp, (h); **E** – M-B-L, 0.5-0.8‰, p; **R** – 279, 280.

*Bosmina longirostris* (O. F. Müller, 1776) [*B. longirostris similis*]; **D** – 58, 59, 60, *68, 77, 78, 80; **Z** – anam, (k); **E** – M-B-L, 1-2‰, p, epp; **R** – 278, 279, 280, 353a, 389.

## Chydoridae (Eurycercidae)

*Acroperus harpae* (Baird, 1834); **D** – 92; **Z** – (hnat); **E** – M-B-L, 0.5‰, p; **R** – 279, 374, 389.

*Alona guttata* Sars, 1862; **D** – 78; **Z** – (sk); **E** – L, 0-20‰, p; **R** – 252, 278, 279.

*Alona quadrangularis* (O. F. Müller, 1776); **D** – 58, *68; **Z** – namp, (sk); **E** – M-B-L, 0.2-2‰, p; **R** – 278, 279, 374, 389.

*Alona rectangula* Sars, 1861; **D** – 58, 59, 60, *68, *69, 76, 77, *79, *86, 88, 92, 93; **Z** – (sk); **E** – B-L, 0.5-6‰, p; **R** – 277, 279, 280, 374, 389.

*Alonella excisa* (Fischer, 1854); **D** – 59, 60; **Z** – (k); **E** – L, 0.5-0.8‰, p; **R** – 279, 280.

*Alonella exigua* (Lilljeborg, 1853); **D** – 42; **Z** – (h); **E** – L, p; **R** – 252, 278, 279.

*Chydorus piger* Sars, 1862; **D** – 60; **Z** – (e); **E** – L, p, 0.5-0.8‰; **R** – 279, 280.

*Chydorus sphaericus* (O. F. Müller, 1785); **D** – 58, 59, 60, *68, *69, 71, 77, *79, 80, 88, 92; **Z** – amswp, (k); **E** – M-B-L, 3‰, p, epp; **R** – 155a, 278, 279, 280, 374, 389.

*Disparalona rostrata* (Koch, 1841) [*Phrixura, Rhynchotalona*]; **D** – 42, 86; **Z** – (h, ? sk); **E** – B-L, p; **R** – 252, 277, 279.

*Graptoleberis testudinaria* (Fischer, 1848); **D** – 60; **Z** – (k); **E** – M-B-L, p, 0.5-0.8‰; **R** – 279, 280.

*Leydigia acanthocercoides* (Fischer, 1854); **D** – 95; **Z** – (sk); **E** – B-L, p, 0.5‰; **R** – 279, 374, 389.

*Leydigia leydigi* (Schoedler, 1863); **D** – *68; **Z** – (sk); **E** – B-L, p, 1‰; **R** – 97, 279, 389.

*Monospilus dispar* Sars, 1862; **D** – *69; **Z** – (hata); **E** – B-L, p; **R** – 279, 374, 389.

*Oxyurella tenuicaudis* (Sars, 1862); **D** – 59, 60, *69; **Z** – (hata); **E** – L, p, 0.5-0.8‰; **R** – 279, 289, 375, 389.

*Pleuroxus trigonellus* (O. F. Müller, 1776); **D** – 59, 60, *68, *86, 88; **Z** – (hat); **E** – L, p, 1.5‰; **R** – 279, 280, 374, 389.

*Pleuroxus uncinatus* Baird, 1850; **D** – 58; **Z** – (hop, ? h); **E** – L, p, 0.5-2‰; **R** – 252, 278, 279.

### Moinidae

*Moina hartwigi* Welter, 1898; **D** – 81; **Z** – (? atm); **E** – L, p, 1‰, r; **R** – 279, 374, 389.

*Moina macrocopa* (Straus, 1819); **D** – 58; **Z** – (h, ? hpt); **E** – L, p, 2‰; **R** – 252, 278, 279.

*Moina micrura* Kurz, 1874 [*M. micrura dubia, M. propinqua*]; **D** – 58, 59, 60, 78, 79; **Z** – ? SK, (sk); **E** – B-L, p, 0.5-20‰; **R** – 190, 193, 252, 278, 280, 353a, 374, 389.

*Moina rectirostris* (Leydig, 1860); **D** – *69; **Z** – (hpt); **E** – B-L, p, 4‰; **R** – 374, 389.

# Daphniidae

*Daphnia cucullata* Sars, 1862 [*D. cucullata apicata, D. cucullata breviensis*]; **D** – 58, 59, 60, 78, 80; **Z** – (h); **E** – M-B-L, p, 0.8‰; **R** – 252, 277, 278, 280, 353a.

*Daphnia curvirostris* Eylmann, 1887 [? complex]; **D** – 60, 78, 82, 88, 92; **Z** – (hat); **E** – L-B, p, 0.5-15‰; **R** – 252, 277, 279, 280, 353a.

*Daphnia galeata* Sars, 1863; **D** – 59, 60, 78; **Z** – (h); **E** – B-L, p, 0.8‰; **R** – 279, 280, 353a.

*Daphnia hyalina* Leydig, 1860; **D** – 68, 79; **Z** – (hes); **E** – M-B-L, p, 1-12‰; **R** – 252, 277, 278, 279.

*Daphnia longispina* (O. F. Müller, 1776); **D** – 13, 71, 78; **Z** – (hnat); **E** – B-L, p; **R** – 279, 353a, 374, 389.

*Daphnia magna* Straus, 1820; **D** – 58, 78, 79, 80; **Z** – (hat,? hpt); **E** – B-L, p, 0.5-2‰; **R** – 192, 193, 252, 279, 353a.

*Daphnia pulex* (Leydig, 1860): **D** – 74, 77, 78, 80; **Z** – (k); **E** – B-L, p, 5-17‰; **R** – 258, 279, 353a, 374, 389.

*Ceriodaphnia affinis* Lilljeborg, 1900; **D** – 78, 79, 80; **Z** – (h); **E** – L-B, p; **R** – 190, 192, 193, 252.

*Ceriodaphnia dubia* Richard, 1894 [*Moina*]; **D** – 58, 79; **Z** – (sk); **E** – L, p, 0.5-8‰; **R** – 252, 278, 279, 374.

*Ceriodaphnia pulchella* Sars, 1862; **D** – 58; **Z** – (sk); **E** – L, p, 0.5-8‰; **R** – 252, 278, 279.

*Ceriodaphnia quadrangula* (O. F. Müller, 1785); **D** – 58, 59, 60, 78; **Z** – (sk); **E** – B-L, p, 0.5-2‰; **R** – 277, 278, 279, 280.

*Ceriodaphnia reticulata* (Jurine, 1820); **D** – 58, 84; **Z** – (sk); **E** – M-B-L, p, 0.5-2‰; **R** – 252, 277, 278, 279.

*Scapholeberis mucronata* (O. F. Müller, 1776); **D** – 58, 59, 60, 80, 88; **Z** – (sk); **E** – B-L, p, 0.5-2‰; **R** – 278, 280, 374, 389.

*Simocephalus vetulus* (O. F. Müller, 1776); **D** – 58, 59, 60, *68, 77, 80; **Z** – (sk); **E** – B-L, p, 0.5-2‰; **R** – 97, 278, 279, 280.

## Podonidae

*Evadne nordmanni* Lovén, 1836; **D** – 29, 69; **V** – 0-25; **Z** – namp; **E** – M, p, eh, epp, et; **R** – 201, 317, 319, 389.

*Evadne spinifera* P. E. Müller, 1867; **D** – 7, 11, 12, 29; **Z** – amwp; **E** – M, p, epp; **R** – 91, 198, 319, 389.

*Pseudevadne tergestina* (Claus, 1877) [*Evadne, Pleopis*]; **D** – 7, 11, 12, 16, 29; **V** – 0-20; **Z** – amip; **E** – M, p; **R** – 91, 198, 319, 389.

*Podon intermedius* Lilljeborg, 1853; **D** – 29; **Z** – bamswp; **E** – M, p, epp; **R** – 201, 202.

*Podon leuckartii* G. O. Sars, 1862; **D** – 7, 29, *69; **Z** – abapbp; **E** – M, p, epp, 10‰; **R** – 201, 319, 374, 389.

*Pleopis polyphaemoides* (Leucart, 1859) [*Podon*]; **D** – 7, 13, 29, 69, 72, 77; **V** – 0-50; **Z** – amnp; **E** – M-B, p, eh; **R** – 198, 201, 304, 317.

*Podonevadne trigona* (G. O. Sars, 1897) [*P. trigona ovum, Podon ovum*]; **D** – *69; **Z** – pc, Rc; **E** – M-B-L, p, 5‰; **R** – 317, 389.

## Cercopagidae

*Cercopagis pengoi* (Ostroumov, 1891); **D** – *68, *69; **Z** – pc, Rc; **E** – M-B-L, p, 3‰; **R** – 317, 385, 389.

OSTRACODA: PODOCOPIDA

### Darwinulidae

*Darwinula stevensoni* (Brady et Robertson, 1870); **D** – 93; **Z** – (k); **E** – L-B, bt, 1‰; **R** – 185, 249, 389.

### Candonidae

*Paracypris polita* G. O. Sars, 1866; **D** – 11, 25, 27; **V** – 8, 100; **Z** – clm; **E** – M, bt, eb, ps; **R** – 233, 249, 252.

*Candona neglecta* G. O. Sars, 1887; **D** – *68; **Z** – (wp); **E** – L-B, bt, 1-5‰; **R** – 185, 249, 389.

*Candonopsis kingsleii* (Brady et Robertson, 1870); **D** – 93; **Z** – (h); **E** – L-B, bt, 1‰; **R** – 185, 249, 389.

*Fabaeformiscandona levanderi* (Hirschmann, 1912) [*Candona*]; **D** – *68; **Z** – (e); **E** – L-B, bt, 1‰; **R** – 185, 249, 389.

*Physocypria kraepelini* G. W. Müller, 1903 [*Phycocypria kliei*]; **D** – *68; **Z** – (k); **E** – L, bt, 1‰; **R** – 97, 389.

**Cyprididae**

*Cyprinotus salinus* (Brady, 1868) [*Heterocypris fretensis*]; **D** – *68, *69, *74, *86, 88, 89, 90, 93, 94; **Z** – (ean); **E** – M-B-L, bt, 0-20‰; **R** – 185, 249, 374, 389.

*Eucypris inflata* (G. O. Sars, 1903); **D** – 61, 64, 76, 77, *79; **Z** – (mwca); **E** – B, bt, eh-150‰; **R** – 185, 249, 389.

*Sarscypridopsis aculeata* (Costa, 1847) [*Cypridopsis*]; **D** – 96; **Z** – (hnat); **E** – L-B, bt; **R** – 185, 249, 389.

*Cypridopsis vidua* (O. F. Müller, 1776); **D** – 93; **Z** – (sk); **E** – L-B, bt, 8‰; **R** – 185, 249, 389.

*Plesiocypridopsis newtoni* (Brady et Robertson, 1870) [*Cypridopsis*]; **D** – 93, 95; **Z** – (wp); **E** – L-B, bt, 8‰; **R** – 185, 249, 389.

*Potamocypris steueri* Klie, 1935; **D** – 94; **Z** – (hom); **E** – L-B, bt, 10‰; **R** – 185, 249, 389.

*Heterocypris maura* (Masi, 1932); **D** – *68; **Z** – (pm); **E** – L-B, bt, 1‰; **R** – 185, 389.

*Heterocypris incongruens* (Ramdohr, 1808); **D** – 42; **Z** – (k); **E** – L, bt; **R** – 84, 389.

*Trajancypris serrata* G. W. Müller, 1900 [*Eucypris*]; **D** – *68; **Z** – (eca); **E** – L, bt, 1‰; **R** – 97, 389.

## Ilyocyprididae

*Ilyocypris biplicata* (Koch, 1838); **D** – *68; **Z** – (h); **E** – L, bt, 1‰; **R** – 97, 389.

## Cushmanideidae

*Pontocythere bacescoi* (Caraion, 1960) [*Cyteridea*]; **D** – 11, 12, 25, 71; **V** – 0.7-10; **Z** – ● p; **E** – M, bt, ep, ps; **R** – 228, 249, 392.

*Pontocythere tchernijawskii* Dubowsky, 1939; **D** – 26, 27; **V** – 10-43; **Z** – ep; **E** – M, bt, mb, pe, s-ps; **R** – 249.

## Cytherideidae

*Cyprideis torosa* (Jones, 1850) [*C. littoralis*]; **D** – *68, *69, 76, *78, *79, 84, 93, 96; **Z** – clm, (hat); **E** – M-B, bt, ep, 0-30‰, eh, ps, pe, ph; **R** – 84, 249, 374, 389.

*Cytheridea acuminata* (Bosquet, 1952); **D** – 25, 26, 27, 28; **V** – 20-100; **Z** – m, ? em; **E** – M, bt, eb, pe, s-ps; **R** – 249.

## Leptocytheridae (Cytheridae)

*Amnicythere quinquetuberculata* (Schweyer, 1949) [*Leptocythere*]; **D** – 3; **Z** – pc, Sf, Rc; **E** – B-M, bt; **R** – 237, 249, 252.

*Amnicythere striatocostata* (Schweyer, 1949) [*Leptocythere*]; **D** – 3; **Z** – pc, Sf, Rc; **E** – B-M, bt; **R** – 237, 249, 252.

*Euxinocythere lopatici* (Schornikov, 1964) [*Leptocythere*]; **D** – 3; **Z** – pc, Sf, Rc; **E** – B-M, bt; **R** – 237, 249, 252.

*Leptocythere devexa* Schornikov, 1966; **D** – 3, 7, 10, 25, 26, 27, 28; **V** – 1-100; **Z** – hom; **E** – M, bt, eb, pe, sg; **R** – 249, 327, 252.

*Leptocythere diffusa* G. W. Müller, 1894) [*Callistocythere*]; **D** – 1, 2, 3, 4, 5, 6, 7, 8, 9, 10, 11, 12, 13, 14, 15, 16, 17, 18, 19, 20, 21, 22, 26, 27; **V** – 13-70; **Z** – lm; **E** – M, bt, eb, pe; **R** – 237, 249, 252.

*Leptocythere macallana* (Brady et Robertson, 1869); **D** – 1, 2, 3, 4, 5, 6, 7, 8, 9, 10, 11, 12, 13, 14, 15, 16, 17, 18, 19, 20, 21, 22, 25; **V** – 8-12; **Z** – ce; **E** – M, bt, ep, pe, ps, sg; **R** – 237, 249, 252.

*Leptocythere mediterranea* (G. W. Müller, 1894) [*Callistocythere*]; **D** – 1, 2, 3, 4, 5, 6, 7, 8, 9, 10, 11, 12, 13, 14, 15, 16, 17, 18, 19, 20, 21, 22, 24, 50; **V** – 0-10; **Z** – mj; **E** – M, bt, ep, pe, ps, ph; **R** – 237, 249, 252.

*Leptocythere multipunctata* (Seguenza, 1942); **D** – 3, 7, 10, 25, 27, 28, 35; **V** – -100; **Z** – mj; **E** – M, bt, hb, pe, s-ps, sg; **R** – 237, 249, 252.

*Leptocythere nitida* Schornikov, 1966; **D** – 1, 2, 3, 4, 5, 6, 7, 8, 9, 10, 11, 12, 13, 14, 15, 16, 17, 18, 19, 20, 21, 22, 25, 26; **V** – 8-15; **Z** – ● p; **E** – M, bt, ep, ps, pe; **R** – 237, 249, 252.

*Leptocythere relicta* Schornikov, 1964; **D** – 3; **Z** – pc, Sf, Rc; **E** – M-B, bt; **R** – 237, 249, 252.

## Trachyleberididae

*Carinocythereis carinata* (Roemer, 1838) [*C. antiquata, Cythereis*]; **D** – 1, 2, 3, 4, 5, 6, 7, 8, 9, 10, 11, 12, 13, 14, 15, 16, 17, 18, 19, 20, 21, 22, 25, 26; **V** – 20-80; **Z** – clm; **E** – M, bt, eb-hb, pe, s-ps; **R** – 225, 249, 392.

*Hiltermannicythere rubra* (G. W. Müller, 1894) [*H. rubra pontica, Carinocythereis*]; **D** – 1, 2, 3, 4, 5, 6, 7, 8, 9, 10, 11, 12, 13, 14, 15, 16, 17, 18, 19, 20, 21, 22, 25, 26, 27, 28; **V** – 2-100; **Z** – m, ? em; **E** – M, bt, eb, pe, s, ps; **R** – 225, 249, 392.

## Hemicytheridae

*Aurila dubowskyi* Schornikov, 1969; **D** – 24; **V** – 0.4-10; **Z** – ● p; **E** – M, bt, ep, ph, ro; **R** – 249.

*Hemicythere sicula* (Brady, 1902) [*Cythere*]; **D** – *68, *69; **Z** – pca; **E** – M-B, bt, 1-12‰; **R** – 185, 249, 374, 389.

## Limnocytheridae

*Limnocythere inopinata* (Baird, 1843); **D** – *68; **Z** – (h); **E** – B-L, bt, 1‰; **R** – 185, 249, 389.

## Cytheromatidae (Cytheromidae)

*Cytheroma karadagiensis* Dubowsky, 1939; **D** – 7, 24, 25, 26, 27; **V** – 5-80; **Z** – m; **E** – M, bt, eb, ps, ph, pe; **R** – 225, 249, 392.

*Cytheroma marinovi* Schornikov, 1969; **D** – 24, 25, 26; **V** – 15-30; **Z** – • p; **E** – M, bt, ep, ps, ph, pe; **R** – 249.

*Cytheroma variabilis* G. W. Müller, 1894; **D** – 26, 27; **V** – 20-80; **Z** – adep; **E** – M, bt, eb, pe; **R** – 249.

## Cytheridae

*Pontocytheroma arenaria* Marinov, 1963; **D** – 7, 11, 25; **V** – 10-25; **Z** – • p; **E** – M, bt, ep, ps; **R** – 227, 249, 392.

*Paracytheridea paulii* Dubowsky, 1939; **D** – 8, 11, 12, 25; **V** – 2-20; **Z** – • p; **E** – M, bt, ep, ps; **R** – 228, 249, 392.

*Parvocythere hartmanni* Marinov, 1962; **D** – 25, 35; **Z** – • p; **E** – M, bt, ps, gw; **R** – 225, 249, 392.

## Loxoconchidae

*Loxoconcha aestuarii* Marinov, 1963; **D** – 86; **Z** – • p; **E** – M-B, bt, pe, s-ps; **R** – 228, 249, 392.

*Loxoconcha bulgarica* Caraion, 1961; **D** – 16, 23, 26; **V** – -5; **Z** – • p; **E** – M, bt, ep, pe, zc; **R** – 80, 249, 392.

*Loxoconcha elliptica* Brady, 1868 [*L. gauthieri*]; **D** – *68, 84, *86, 88; **Z** – clm; **E** – M-B, bt, eh; **R** – 185, 249, 374, 389.

*Loxoconcha granulata* G. O. Sars, 1866; **D** – 1, 2, 3, 4, 5, 6, 7, 8, 9, 10, 11, 12, 13, 14, 15, 16, 17, 18, 19, 20, 21, 22, 26, 27, 28; **V** – 20-100; **Z** – clm; **E** – M, bt, eb, pe; **R** – 225, 249, 392.

*Loxoconcha impressa* (Baird, 1850); **D** – 7; **Z** – bam; **E** – M, bt, sl; **R** – 225, 392.

*Loxoconcha littoralis* G. W. Müller, 1894 [*L. pennatus, Sagmatocythere*]; **D** – 8, 25; **Z** – m, ? hom; **E** – M, bt, ps; **R** – 228, 249, 392.

*Loxoconcha nana* Marinov, 1962 [*Tuberoloxoconcha*]; **D** – 7, 25, 33; **Z** – ● p; **E** – M, bt, ps, if; **R** – 225, 249, 392.

*Loxoconcha pontica* Klie, 1937; **D** – 7, 16, 24, 77, *86; **V** – 0-5; **Z** – clm; **E** – M, bt, sep, ph; **R** – 84, 185, 249, 389.

*Loxoconcha rhomboidea* (Fischer, 1855): **D** – 3, 7, 10, 17, 24, 26, 27; **V** – 5-50; **Z** – clm, ? bam; **E** – M, bt, mb, eh, s-ph-ro; **R** – 247, 249, 252.

*Microloxoconcha marinovi* Schornikov, 1969; **D** – 7, 21, 22, 25; **V** – 2-3; **Z** – ● p, Ebg; **E** – M, bt, sep, ps; **R** – 233, 249, 252.

## Cytheruridae

*Eucytherura bulgarica* Klie, 1937 [*Hemicytherura*]; **D** – 16, 24, 89; **V** – 0-15; **Z** – ● p; **E** – M, bt, ep, ph; **R** – 185, 249, 389.

*Pseudocytherura pontica* Dubowsky, 1939; **D** – 8, 11, 12, 25; **V** – 5-25; **Z** – ● p; **E** – M, bt, ep, ps; **R** – 228, 249, 392.

*Semicytherura calamitica* Schornikov, 1969; **D** – 3, 7, 10, 25; **V** – -25; **Z** – ● p; **E** – M, bt, ep, ps; **R** – 249, 252.

*Semicytherura euxinica* (Caraion, 1967); **D** – 3, 7, 10, 17, 25; **V** – -20; **Z** – ● p; **E** – M, bt, ep, ps; **R** – 249, 252.

*Semicytherura pontica* (Marinov, 1962) [*Cytherura, Levocytherura*]; **D** – 7, 11, 25; **V** – -7; **Z** – ● p, Ebg; **E** – M, bt, ep, ps; **R** – 225, 249, 252.

*Semicytherura virgata* Schornikov, 1969; **D** – 2, 3, 5, 7, 10, 25; **V** – -25; **Z** – ● p; **E** – M, bt, ep, ps; **R** – 249, 252.

*Levocytherura remanei* (Marinov, 1962) [*Cytherura, Semicytherura*]; **D** – 8, 11, 12, 25; **V** – 5-10; **Z** – ● p; **E** – M, bt, ep, ps; **R** – 228, 249, 392.

*Microcytherura fulvoides* Dubowsky, 1939; **D** – 11, 25; **Z** – ● p; **E** – M, bt, ep, ps; **R** – 225, 249, 392.

*Microcytherura nigrescens* G. W. Müller, 1894; **D** – 3, 4, 7, 25; **V** – 1-30; **Z** – m, ? hom; **E** – M, bt, ep-me, ps; **R** – 247, 249, 252.

## Xestoleberididae

*Xestoleberis aurantia* (Baird, 1838); **D** – 24*, 69; **V** – -5; **Z** – namni; **E** – M, bt, ep, ph; **R** – 185, 249, 389

*Xestoleberis cornelii* Caraion, 1963; **D** – 24, 25, 27; **V** – 1.5-90; **Z** – m; **E** – M, bt, eb, ph, pe, ps-s; **R** – 249.

*Xestoleberis decipiens* (G. W. Müller, 1894); **D** – 24, *86; **V** – 0-25; **Z** – hom; **E** – M, bt, eb, ph; **R** – 185, 249, 389.

## Microcytheridae

*Microcythere longiantennata* Marinov, 1962; **D** – 7, 11, 12, 13, 14, 15, 16, 17, 18, 19, 20, 21, 22, 25, 51; **Z** – ● p, Ebg; **E** – M, bt, eb, ps; **R** – 225, 249, 392.

*Microcythere varnensis* Marinov, 1962; **D** – 25, 35, 51; **Z** – ● p, Ebg; **E** – M, bt, eb, l, ps, gw; **R** – 225, 249, 389.

## Bythocytheridae

*Bythocythere turgida* G. O. Sars, 1866; **D** – 3, 7, 10, 21, 22, 24, 28; **V** – -70, 100; **Z** – abam; **E** – M, bt, hb, ph, phs; **R** – 247, 249, 252.

*Sclerochilus gewemuelleri* Dubowsky, 1939 [*S. gewemuelleri dubowskyi*]; **D** – 7, 23, 24, 27, 28; **V** – 3-100; **Z** – clm; **E** – M, bt, eb, ph-zc-phs; **R** – 225, 249, 392.

## Paradoxostomatidae, Cytheridae

*Cytherois carcinitica* Marinov, 1964; **D** – 24, 25; **V** – 10-20; **Z** – ● p; **E** – M, bt, ep, ps, ph, ro; **R** – 249.

*Cytherois cepa* Klie, 1937; **D** – 23, 24, *79, *86; **V** – -2 ; **Z** – ● p; **E** – M, bt, seb, eh, ph; **R** – 185, 249, 389.

*Cytherois messambriensis* Marinov, 1964 [*C. pseudovitrea messambriensis*]; **D** – 11, 25; **Z** – ● p, Ebg; **E** – M, bt, ep, ps; **R** – 228, 249, 392.

*Cytherois pontica* Marinov, 1966; **D** – 11, 25; **V** – 10-15; **Z** – ● p, Ebg; **E** – M, bt, ep, ps; **R** – 234, 249, 392.

*Cytherois pseudovitrea* Dubowsky, 1939 [*C. pseudovitrea pseudovitrea*]; **D** – 24, 25; **Z** – • p; **E** – M, bt, ep, ps, ph; **R** – 249.

*Cytherois valkanovi* Klie, 1937; **D** – 23, 24, *69; **V** – -25; **Z** – • p; **E** – M, bt, ep, ph, zc; **R** – 186, 249, 389.

*Paradoxostoma abbreviata* G. O. Sars, 1866; **D** – 7, 20; **Z** – clm; **E** – M, bt; **R** – 228, 392.

*Paradoxostoma convexum* Schornikov, 1965; **D** – 23, 24; **V** – -4; **Z** – • p; **E** – M, bt, sep, ph, zc; **R** – 249.

*Paradoxostoma guttatum* Schornikov, 1965; **D** – 24; **V** – -30; **Z** – • p; **E** – M, bt, ep-mb, ph, ro; **R** – 249.

*Paradoxostoma intermedium* G. W. Müller, 1894; **D** – 23, 24,*69, *86; **V** – -30; **Z** – m, ? hom; **E** – M, bt, ep, ph, slc, zc; **R** – 185, 249, 389.

*Paradoxostoma ponticum* Klie et Whittaker, 1942; **D** – 24, *86; **V** – -10; **Z** – lm; **E** – M, bt, ep, ph, lt; **R** – 185, 186, 249, 392.

*Paradoxostoma simile* G. W. Müller, 1894; **D** – 20, 24, 27, 28; **V** – 15-83; **Z** – m, hom; **E** – M, bt, eb, phc-ms-phs; **R** – 228, 249, 392.

Copepoda: Calanoida

### Calanidae

*Calanus euxinus* Hulsemann, 1991 [? *C. helgolandicus*]; **D** – 7, 10, 11, 12, 13, 14, 15, 16, 17, 29; **Z** – • p; **E** – M, p; **R** – 160, 163, 269, 334.

? *Calanus helgolandicus* (Claus, 1863); **D** – 3, 7, 10, 17, 29; **V** – 0-150; **Z** – amp; **E** – M, p, et; **R** – 120, 198, 319, 389.

### Paracalanidae

*Paracalanus parvus* (Claus, 1863); **D** – 7, 10, 11, 12, 13, 14, 15, 16, 17, 29, 77; **V** – 0-150; **Z** – K; **E** – M, p, pp, et; **R** – 91, 201, 304, 389.

## Clausocalanidae (Pseudocalanidae)

*Pseudocalanus elongatus* (Boeck, 1865); **D** – 7, 10, 11, 12, 13, 14, 15, 16, 17, 29, 69; **V** – 0-200; **Z** – anamnp; **E** – M, p; **R** – 198, 317, 319, 389.

## Temoridae

*Eurytemora affinis* (Poppe, 1880); **D** – 7, 29, 58, 80; **Z** – anapnep, (h); **E** – M-B, p, epp, eh; **R** – 110, 252, 278, 279.

*Eurytemora lacustris* (Poppe, 1887); **D** – 59; **Z** – cp; **E** – M, p, eh, 0.5-0.8‰; **R** – 280.

*Eurytemora velox* (Lilljeborg, 1853): **D** – 59, 60, 71, 77, 80, 88, 93; **Z** – cpc, (tp); **E** – M-B, p, eh, 0-10‰; **R** – 280, 294, 374, 389.

## Centropagidae

? *Centropages kroyeri* Giesbrecht, 1893; **D** – 3, 7, 10, 11, 12, 13, 14, 15, 16, 17, 76, 77; **Z** – amip; **E** – M, p; **R** – 91, 304, 319, 374.

*Centropages ponticus* Karavaev, 1895 [*C. kroyeri pontica*, ? *C. kroyeri*]; **D** – 7, 11, 12, 29, *69, 76, 77; **Z** – mrs; **E** – M, p, ■; **R** – 114, 198.

## Diaptomidae

*Arctodiaptomus byzantius* Mann, 1940 [*A. byzantinus*]; **D** – 42; **Z** – (seea); **E** – L; **R** – 252, 277, 279.

*Arctodiaptomus salinus* (Daday, 1885) [*Diaptomus*]; **D** – ?; **Z** – (po); **E** – L, p, 10‰; **R** – 91, 279, 389.

*Neolovenula alluaudi* (Guerne et Richard, 1890) [*Lovenula, Paradiaptomus*]; **D** – 78; **Z** – (hom); **E** – L, p; **R** – 192, 252, 279.

## Pseudodiaptomidae

*Calanipeda aquaedulcis* Kritchagin, 1873; **D** – 7, 13, 29, 59, 60, *68, 69, 70, *74, 76, 77, 78, 79, 80, 83, 85, 88, 90, 92, 93, 94, 95; **Z** – lm, (hom); **E** – B-L, p, eh, et; **R** – 198, 201, 279, 280, 317, 319, 374, 389.

## Pontellidae

*Anomalocera patersoni* Templeton, 1837; **D** – 7, 29; **Z** – amip; **E** – M, p, et, ■; **R** – 198, 201, 319, 389.

*Labidocera brunescens* (Czerniavsky, 1868); **D** – 7, 29; **Z** – lmm; **E** – M, p, eh, th, ■; **R** – 198, 201, 319, 389.

*Pontella mediterranea* (Claus, 1863); **D** – 7, 12, 29, *69; **Z** – lmm; **E** – M, p, ■; **R** – 91, 198, 319, 389.

## Acartiidae

*Acartia clausi* Giesbrecht, 1889; **D** – 7, 10, 11, 12, 13, 14, 15,16, 17, 29, 68, 69, 76, 77, 80, 84; **V** – 0-50; **Z** – K, aamip; **E** – M, p, 10‰, et; **R** – 91, 198, 201, 304, 317, 319, 374, 389.

*Acartia tonsa* Dana, 184; **D** – 29; **Z** – antamip, K, i; **E** – M, p, th; **R** – 160, 204, 355, 360a.

*Paracartia latisetosa* (Kritchagin, 1873) [*Acartia*]; **D** – 7, *69, *79, 84; **Z** – lmmwi; **E** – M, p, 10‰, th; **R** – 279, 317, 319, 374.

COPEPODA: MONSTRILLOIDA

## Monstrillidae

*Monstrilla grandis* Giesbrecht, 1891; **D** – 7, 11, 29; **Z** – amrsp; **E** – M, p; **R** – 91, 198, 319, 360a, 389.

COPEPODA: CYCLOPOIDA

## Oithonidae

*Oithona brevicornis* Giesbrecht, 1891; **D** – 29; **Z** – amip; **E** – M-B-L, p; **R** – 324.

*Oithona davisae* Ferrari F.D. & Orsi, 1984; **D** – 2, 3, 4, 5, 6, 7, 8, 9, 10, 11, 12, 13, 14, 15, 16, 17; **Z** – SK, i; **E** – M-B-L, p, eh, is; **R** – 269, 355, 360a.

*Oithona minuta* (Krichagin, 1877; Scott, 1894) [? *O. nana, Dioithona*]; **D** – 7, 11, 12, 29, *68, *79, 84, *86, 88; **V** – 20; **Z** – amip; **E** – M-B-L, p, eh, ■; **R** – 84, 114, 198, 201, 317, 319, 374, 389.

*Oithona nana* Giesbrecht, 1893 [? = *O. minuta*]; **D** – 10, 11, 12, 13, 14, 15, 16, 17, 29; **Z** – K; **E** – M-B-L, p, eh; **R** – 160, 324.

*Oithona similis* Claus, 1866; **D** – 7, 10, 11, 12, 13, 14, 15, 16, 17, 29; **V** – 20-; **Z** – K; **E** – M-B-L, p, eh, epp; **R** – 198, 201, 319, 389.

## Cyclopidae

*Halicyclops rotundipes* Kiefer, 1935 [*H. rotundipes rotundipes*]; **D** – *69, 84, 88; **Z** – (nem); **E** – B, p, 0-10‰; **R** – 279, 374, 389.

*Macrocyclops albidus* (Jurine, 1820); **D** – 59, 60, 77; **Z** – (k); **E** – L, p, 0.5-0.8‰; **R** – 155a, 279, 280.

*Macrocyclops fuscus* (Jurine, 1820); **D** – 59, 60; **Z** – (hn, ? sk); **E** – L, p, 0.5-0.8‰; **R** – 279, 280.

*Eucyclops macruroides* (Lilljeborg, 1901) [*E. macruroides macruroides*]; **D** – 85; **Z** – (pat); **E** – L, p, 0-30‰; **R** – 252, 277.

*Eucyclops serrulatus* (Fischer, 1851) [*Cyclops*]; **D** – 58, 59, 60, *68, 76, 77, *79, 80, 85, *86, 88, 92, 93; **Z** – (sk, ? k); **E** – L, p, 0.5-8‰; **R** – 252, 277, 278, 280, 374, 389.

*Eucyclops speratus* (Lilljeborg, 1901); **D** – 60; **Z** – (hptn, ? sk); **E** – L, p, 0.5-0.8‰; **R** – 279, 280.

*Euryte longicauda* Philippi, 1843; **D** – 3; **Z** – nam, ? namnz; **E** – M, p, r; **R** – 160, 355, 360a.

*Paracyclops affinis* (G. O. Sars, 1863); **D** – 60; **Z** – (hpt); **E** – L, p, 0.5-0.8‰; **R** – 279, 280.

*Paracyclops fimbriatus* (Fischer, 1853); **D** – 90; **Z** – (sk); **E** – L, p, 0-30‰; **R** – 252, 277.

*Ectocyclops phaleratus* (Koch, 1838); **D** – 60; **Z** – (sk); **E** – L, p, 0.5-0.8‰; **R** – 279, 280.

*Cyclops strenuus* Fischer, 1851; **D** – 58, 60, 78, 79, 80, 84, *86, 92; **Z** – (h); **E** – L, p, eh; **R** – 192, 193, 252, 277, 278, 280.

*Cyclops vicinus* Uljanin, 1875 [*C. vicinus vicinus*]; **D** – 7, 13, 29, 58, 59, 60, 78, 79, 84, 92; **Z** – (ho); **E** – L, p, eh; **R** – 110, 192, 193, 198, 177, 201, 278, 280, 353a.

*Megacyclops latipes* (Lowndes, 1927) [*Acanthocyclops*]; **D** – 92; **Z** – (h); **E** – L, p; **R** – 252, 277.

*Megacyclops viridis* (Jurine, 1820) [*M. v. viridis, Cyclops*]; **D** – 58, 59, 60, *68, *69, *86, 88; **Z** – (hptn, ? sk); **E** – L, p, 6‰; **R** – 278, 280, 374, 389.

*Acanthocyclops robustus* (G. O. Sars, 1863) [? *A. americanus*]; **D** – 58, 59, 60, 77, 80, *86; **Z** – (hna, ? sk), i; **E** – L, p, is; **R** – 252, 277, 278, 280.

*Acanthocyclops vernalis* (Fischer, 1853); **D** – 78, 79, 85, 92; **Z** – (sk, ? k); **E** – L, p, epp; **R** – 192, 193, 252, 277, 353a.

*Diacyclops bicuspidatus* (Claus, 1857)[*D. bicuspidatus odessanus, Acanthocyclops, Cyclops*]; **D** – 59, 60, 79, 84, 85, *86, 92; **Z** – (h, ? hn); **E** – L, p, 20‰, eh; **R** – 252, 277, 280, 374.

*Diacyclops bisetosus* (Rehberg, 1880) [*Acanthocyclops, Cyclops*]; **D** – 33; **Z** – (ha); **E** – L, p, 10‰; **R** – 252, 277, 374, 389.

*Metacyclops planus* (Gurney, 1909) [*Microcyclops*]; **D** – 42; **Z** – (wp); **E** – L, p; **R** – 252, 278.

*Microcyclops minutus* Claus, 1863 [? *Metacyclops*]; **D** – 69; **Z** – (? pat); **E** – L, p; **R** – 279, 334.

*Microcyclops varicans* (G. O. Sars, 1863); **D** – 76; **Z** – (sk, ? k); **E** – L, p, eh; **R** – 119, 259.

*Mesocyclops leuckarti* (Claus, 1857); **D** – 59, 60, 76, 92; **Z** – (tp); **E** – L, p, 0.5-0.8‰; **R** – 252, 277, 279, 280.

*Thermocyclops crassus* (Fischer, 1853) [*Mesocyclops*]; **D** – 79, 80; **Z** – (hptn, ? sk); **E** – L, p, 0.5‰; **R** – 193, 252.

*Thermocyclops dybowskii* (Landé, 1890); **D** – 60; **Z** – (wcp, ? hop); **E** – L, p, 0.5-0.8‰; **R** – 279, 280.

*Thermocyclops oithonoides* (G. O. Sars, 1863); **D** – 60, 78; **Z** – (hop, ? h); **E** – L, p, 0.5-0.8‰; **R** – 279, 280, 353a.

## Lernaeidae

*Lernaea cyprinacea* Linnaeus, 1758; **D** – 60; **Z** – (h); **E** – L, ec; **R** – 389.

Copepoda: Harpacticoida

## Longipediidae

*Longipedia minor* T. Scott et A. Scott, 1893 [*L. pontica*]; **D** – 7, 10, 12, 24, 25, 27; **V** – 10-100; **Z** – cp; **E** – M, bt-p, eb, ph-sg-s; **R** – 13, 17, 31, 249, 252.

## Canuellidae

*Canuella perplexa* T. Scott et A. Scott, 1893; **D** – 7, 24, 25, 27, 61, 76, 78, 88; **V** – -65; **Z** – nam; **E** – M, bt-p, mb, eh, 12-50‰, ph-ps-s; **R** – 185, 249, 252, 304, 389.

*Canuella furcigera* G. O. Sars, 1903; **D** – 1, 2, 3, 4, 5, 6, 7, 8, 9, 10, 11, 12, 13, 14, 15, 16, 17, 18, 19, 20, 21, 22, 24, 25, 26, 76, 96; **V** – -50; **Z** – abam; **E** – M, bt, mb, ps, ps-s; **R** – 17, 18, 31, 43, 240.

*Canuella pontica* Apostolov, 1971; **D** – 25, 35, 51, 86, 87; **Z** – ● p; **E** – M, bt, l, ps, gw; **R** – 16, 18, 31, 249, 252.

*Sunaristes paguri* Hesse, 1867; **D** – ?; **Z** – cp, ? amip; **E** – M, bt, co, ep; **R** – 17, 31.

## Ectinosomatidae

*Ectinosoma melaniceps* Boeck, 1865; **D** – 16, 24, 25, 26, 27, *69, 76, *86; **V** – 1-100; **Z** – amip; **E** – M, bt-p, eb, ph, ps, s; **R** – 31, 84, 185, 249.

*Ectinosoma normani* T. Scott et A. Scott, 1894; **D** – 24, 25, 27; **V** – 1-100; **Z** – clm; **E** – M, bt, eb, ph, ps, s-ps; **R** – 17, 31, 249.

105

*Ectinosoma soyeri* Apostolov, 1975; **D** – 19, 20, 35, 51; **Z** – ● p; **E** – M, bt, l, ps, gw; **R** – 24, 31, 249, 252.

*Halectinosoma abrau* (Kritchagin, 1877) [*Ectinosoma*]; **D** – 35, 51; **Z** – abam, (wcp); **E** – M-B, bt, eh, ps, gw; **R** – 16, 31, 249, 252.

*Halectinosoma brevirostre* (G. O. Sars, 1904) [*Ectinosoma*]; **D** – 7, 12, 23, 25, 27, 29; **V** – 1.5-50; **Z** – bap; **E** – M, bt-p, mb, zc, ps, s; **R** – 17, 31, 249, 252.

*Halectinosoma curticorne* (Boeck, 1872) [*Ectinosoma*]; **D** – 25, 35, 84, 96; **V** – 0-15; **Z** – abam, (h); **E** – M, bt, ep, ps, ps-s, gw; **R** – 31, 249, 252, 263.

*Halectinosoma elongatum* (G. O. Sars, 1904) [*Ectinosoma*]; **D** – 7, 24, 25, 26; **V** – -30; **Z** – bap; **E** – M, bt, ep, ph, ps, ps-s; **R** – 17, 31, 43, 249, 252.

*Halectinosoma herdmani* (T. Scott et A. Scott, 1894) [*Ectinosoma*]; **D** – 1, 2, 3, 4, 5, 6, 7, 8, 9, 10, 11, 12, 13, 14, 15, 16, 17, 18, 19, 20, 21, 22, 24, 25, 26, 29; **V** – 5-6, 30; **Z** – cp, ? clm; **E** – M, bt-p, eu, ph, ps, s; **R** – 13, 17, 31, 249, 252.

*Pseudobradya beduina* Monard, 1935; **D** – 7, 17, 24, 25; **V** – 2-12; **Z** – cp, ? clm; **E** – M, bt, ep, ps, ph, ps-s; **R** – 26, 31, 249, 252.

*Pseudobradya minor* (T. Scott et A. Scott, 1896); **D** – 3, 7, 10, 13, 25, 26; **V** – 5-25; **Z** – acp; **E** – M, bt, ep, ps, ps-s; **R** – 31, 246, 249, 252.

*Glabrotelson bodini* Apostolov, 1974 [*Hastigerella*]; **D** – 2, 35, 51; **Z** – ● p; **E** – M, bt, ls, ps, gw; **R** – 23, 31, 249, 252.

*Noodtiella wellsi* Apostolov, 1974; **D** – 2, 35; **V** – 0.80; **Z** – ● p; **E** – M, bt, ls, ps, gw; **R** – 23, 31, 249, 252.

## Darcythompsoniidae

*Leptocaris brevicornis* (van Douwe, 1905) [*Horsiella*]; **D** – 25, *78, *79; **Z** – nam, (wcp); **E** – M, bt, ep, l, eh, ph, ps; **R** – 31, 249, 386, 389.

## Tachidiidae

*Microarthridion littorale* (Poppe, 1881); **D** – 7, 25; **Z** – abap, (h); **E** – M-B, bt, l, ep, ps, if; **R** – 31, 240, 246, 249.

# Harpacticidae

*Harpacticus flexus* Brady et D. Robertson, 1873; **D** – 1, 2, 3, 4, 5, 6, 7, 8, 9, 10, 11, 12, 13, 14, 15, 16, 17, 18, 19, 20, 21, 22, 24, 25; **V** – 1-2; **Z** – clm, ? cp; **E** – M, bt, ep, ph, ps, s, eu; **R** – 14, 15, 31, 43, 182.

*Harpacticus gracilis* Claus, 1863 [*H. nicaeensis* var. *pontica*]; **D** – 7, 24, 25, 29, 35; **Z** – namni; **E** – M, bt, ep, ph, (p, ps); **R** – 31, 249, 304, 319.

*Harpacticus littoralis* G. O. Sars, 1910; **D** – 13, 16, 24, 25, 29, 76; **Z** – namni; **E** – M, ep, ph, (p, ps), co; **R** – 14, 15, 31, 249, 252.

*Harpacticus nicaeensis* Claus, 1866; **D** – 1, 2, 3, 4, 5, 11, 12, 13, 24, 25; **Z** – nam, ? bap; **E** – M, bt, ep, ph, (ps); **R** – 31, 91, 157, 249.

*Harpacticus obscurus* T. Scott, 1895; **D** – 7, 24, 29; **Z** – clp, ? clm; **E** – M, bt, ep, ph, (p); **R** – 31, 84, 249, 389.

*Harpacticus uniremis* Kröyer, 1842; **D** – 20, 24, 35, 51; **Z** – anap, (h); **E** – M, bt, eh, ph, gw, s, r; **R** – 15, 17, 31, 249, 252.

*Tigriopus fulvus* (Fischer, 1860) [*Harpacticus*]; **D** – 32, 33; **Z** – neamnp; **E** – M, 10-60‰, spr; **R** – 57, 392.

# Tisbidae

*Tisbe dilatata* Klie, 1949; **D** – 20, 21, 22, 24; **V** – 2-25; **Z** – cp; **E** – M, bt, ep, sl, ph-ro; **R** – 31, 84, 249, 389.

*Tisbe furcata* (Baird, 1837) [*Idya*]; **D** – 7, 13, 24, 29, 76; **Z** – anaminp, ? K; **E** – M, bt, p, sl, ph, eu; **R** – 31, 84, 249, 304.

*Scutellidium arthuri* Poppe, 1884; **D** – 24; **Z** – abap; **E** – M, bt, ep, ph; **R** – 17, 31, 249.

*Scutellidium longicauda* (Philippi, 1840) [*Machairopus, Psammathe*]; **D** – 7, 11, 12, 24; **Z** – am; **E** – M, bt, ep, ph; **R** – 31, 91, 157, 249.

# Porcellidiidae

*Porcellidium viride* (Philippi, 1840); **D** – 7, 24; **Z** – clmwi; **E** – M, bt, ep, ph; **R** – 17, 31, 249, 252.

## Peltidiidae

*Alteutha typica* Czerniavski, 1868; **D** – 1, 2, 3, 4, 5, 6, 7, 8, 9, 10, 11, 12, 13, 14, 15, 16, 17, 18, 19, 20, 21, 22, 24; **Z** – • p; **E** – M, bt, ep, ph; **R** – 20, 31, 243, 249.

## Tegastidae

*Tegastes longimanus* (Claus, 1863); **D** – 1, 2, 3, 4, 5, 6, 7, 8, 9, 10, 11, 12, 13, 14, 15, 16, 17, 18, 19, 20, 21, 22, 24, 35; **Z** – clp, ? clm; **E** – M, bt, ep, ph, (gw); **R** – 17, 25, 31, 249, 252.

## Thalestridae

*Thalestris longimana* Claus, 1863; **D** – 13, 24, 25, 29; **V** – -35, 100; **Z** – bam, ? bap; **E** – M, bt (p), eb, ph, ps, s; **R** – 16, 31, 249, 252.

*Thalestris rufoviolascens* Claus, 1866; **D** – 13, 24, 25; **V** – -30; **Z** – clm; **E** – M, bt, ep, ph, ps-sg, s; **R** – 15, 16, 31, 249, 252.

*Parathalestris clausi* (Norman, 1868); **D** – 20, 24, 25, 26, 27; **V** – 25-50; **Z** – clp; **E** – M, bt, mb, ph, ps-s, et; **R** – 17, 31, 249, 252.

*Parathalestris dovi* Marcus, 1966; **D** – 13, 24, 25; **V** – -12; **Z** – lp, ? clp; **E** – M, bt, ep, sl, ph-lt; **R** – 17, 31, 182, 249.

*Parathalestris harpactoides* (Claus, 1863);  **D** – 7, 24, 25, 35; **V** – 0-50; **Z** – cp, ? clm; **E** – M, mb, ph, ps, s, co; **R** – 31, 84, 249, 389.

*Phyllothalestris mysis* (Claus, 1863) [*Phyllopodopsillus*]; **D** – 11, 12, 24, 25, 27; **V** – -90; **Z** – clmi; **E** – M, bt, eb, ps, ph, s, et; **R** – 31, 91, 249, 389.

## Rhynchothalestridae

*Ambunguipes rufocincta* (Brady, 1880) [*Rhynchothalestris*]; **D** – 3, 24, 25; **Z** – nami, ? bami; **E** – M, bt, eb, ph, ps-pe; **R** – 17, 31, 249, 252.

## Dactylopusiidae

*Diarthrodes assimilis* (G. O. Sars, 1906) ; **D** – 24; **Z** – cpnei; **E** – M, bt, ep, sl, ph; **R** – 17, 31, 249.

*Diarthrodes minutus* (Claus, 1863); **D** – 24, *69; **Z** – bam; **E** – M, bt, ep, sl, ph, eh; **R** – 31, 185, 249, 389.

*Diarthrodes nobilis* (Baird, 1845); **D** – 7, 16, 20, 24, 29, 66; **V** – -25; **Z** – amswp; **E** – M, bt (p), ep, slc, ps-s; **R** – 17, 20, 31, 243, 249.

*Diarthrodes ponticus* (Kritchagin, 1873) [*D. ponticus orientalis*]; **D** – 7, 13, 16, 24, 25; **V** – -25; **Z** – clmwi; **E** – M, bt, ep, ph, ps, sg; **R** – 17, 24, 31, 182, 263.

*Diarthrodes pygmaeus* (T. Scott et A. Scott, 1895); **D** – 16, 24; **Z** – ami; **E** – M, bt, ep, ph, r; **R** – 17, 31, 249, 252.

*Dactylopusia tisboides* (Claus, 1863) [*Dactylopodia*]; **D** – 7, 11, 12, 24, 25, 35; **V** – -30, 50; **Z** – aminz, ? SK; **E** – M, ph, gw, (sg, s), co; **R** – 31, 91, 185, 249.

*Dactylopusia vulgaris* G. O. Sars, 1905 [*Dactylopodia*]; **D** – 20, 24, 25, 29, 35; **V** – -100; **Z** – abam; **E** – M, bt (p), ph, gw (ps); **R** – 13, 14, 17, 31, 249.

*Paradactylopodia brevicornis* (Claus, 1866) [*Dactylopusia*]; **D** – 1, 11, 12, 24, 25, 35; **V** – -40, 100; **Z** – naminz, ? SK; **E** – M, bt, ph, ps, ro, gw; **R** – 31, 91, 249, 389.

*Paradactylopodia latipes* (Boeck, 1865); **D** – 7, 20, 24, 25, 35; **V** – -20, 100; **Z** – namni; **E** – M, bt, ph, gw, sg, s, r; **R** – 14, 15, 17, 20, 31, 182, 249, 252.

### Pseudotachidiidae

*Dactylopodella flava* (Claus, 1866); **D** – 7, 24, 25, 26, 27; **V** – 8-110; **Z** – acmnz; **E** – M, bt, eb, ph, ps, s; **R** – 31, 243, 249, 252.

### Parastenheliidae

*Parastenhelia hornelli* I. C. Thompson et A. Scott, 1903; **D** – 16, 20, 24; **Z** – anpip; **E** – M, bt, ep, ph, r; **R** – 20, 31, 182, 249.

*Parastenhelia reducta* Apostolov, 1975; **D** – 20, 21, 25; **V** – 5; **Z** – ● p; **E** – M, bt, ep, ps, r; **R** – 24, 31, 249, 252.

*Parastenhelia spinosa* (Fischer, 1860) [*P. spinosa bulgarica, P. forficula litoralis*]; **D** – 7, 16, 24; **V** – -25; **Z** – naminz, ? SK; **E** – M, bt, ep, ph, ps; **R** – 31, 84, 249, 389.

## Miraciidae, Diosaccidae

*Stenhelia elisabethae* Por, 1960 [*Delavalia*]; **D** – 1, 2, 3, 4, 5, 6, 7, 8, 9, 10, 11, 12, 13, 14, 15, 16, 17, 18, 19, 20, 21, 22, 25, 26; **Z** – cp, ? clp; **E** – M, bt, sl, pe, co; **R** – 31, 240, 249, 252.

*Stenhelia normani* (T. Scott, 1905) [*Delavalia*]; **D** – 7, 13, 24, 25, 27, 28; **V** – 6-100; **Z** – clm; **E** – M, bt, eb, ps, pe, ph; **R** – 16, 18, 31, 43, 249.

*Stenhelia palustris* (Brady, 1868) [*Delavalia*]; **D** – 7, 25, 26; **V** – 10; **Z** – bam, ? bap; **E** – M, bt, ep, eh, ps, pe ; **R** – 17, 31, 43, 249, 252.

*Stenhelia proxima* G. O. Sars, 1907; **D** – 7, 25, 26, 27; **V** – 10-30; **Z** – cp; **E** – M, bt, ep, ps, pe, ph; **R** – 17, 31, 43, 249.

*Stenhelia reflexa* Brady et D. Robertson, 1880 [*Delavalia*]; **D** – 10, 25 ; **V** – -25; **Z** – anam; **E** – M, bt, sl, ep, ps; **R** – 17, 31, 246, 249.

*Stenhelia tethysensis* Monard, 1928 [*Delavalia*]; **D** – 7, 12, 25, 27; **V** – 10-100; **Z** – m, ? nm; **E** – M, bt, eb, pe, ps-s; **R** – 31, 84, 91, 249, 389.

*Diosaccus tenuicornis* (Claus, 1863); **D** – 7, 17, 18, 24, 25, 35; **V** – 15-30; **Z** – nam; **E** – M, bt, ep, ph, ps, gw; **R** – 20, 31, 249, 252.

*Diosaccus varicolor* (Farran, 1913) [*D. varicolor biarticulatus*]; **D** – 24; **Z** – lm, ? clm; **E** – M, bt, ph; **R** – 26, 31, 249.

*Robertsonia knoxi* (I. C. Thompson et A. Scott, 1903); **D** – 7, 16, 20, 25, 27, 28; **V** – 15-100; **Z** – nami; **E** – M, bt, ps-s, ms, phs; **R** – 10, 31, 248, 249.

*Robertsonia monardi* (Klie, 1937) [*Varnaia*]; **D** – 24, 25, 69; **V** – 0.3-3; **Z** – cp; **E** – M, bt, sep, ps-s, ph; **R** – 31, 185, 249, 389.

*Sarsamphiascus caudaespinosus* (Brian, 1927) [*Amphiascus*]; **D** – 24, 27, 28; **V** – -150; **Z** – m, ? nm; **E** – M, bt, hb, pe, ph; **R** – 31, 42.

*Sarsamphiascus gracilis* Lang, 1936 [*Amphiascus*]; **D** – 24; **Z** – ap; **E** – M, bt, ph, lt; **R** – 15, 17, 31.

*Sarsamphiascus propinquus* (G. O. Sars, 1906) [*Amphiascus, A. angustipes*]; **D** – 7, 18, 24, 25, 35; **V** – 0-20; **Z** – ami; **E** – M, bt, ep, ps, ph, gw; **R** – 31, 242, 249.

*Sarsamphiascus sinuatus* (G. O. Sars, 1906) [*Amphiascus*]; **D** – 16, 24, 25, 27, 28, 29; **V** – 30-150; **Z** – nam, ? bam; **E** – M, bt (p), ph, ps, pe; **R** – 14, 15, 31, 249, 252.

*Amphiascopsis cinctus* (Claus, 1866); **D** – 13, 24; **V** – 6; **Z** – amip, ? SK; **E** – M, bt, sep, ph, lt; **R** – 17, 26, 31, 249, 252.

*Amphiascopsis minutus* (Claus, 1863) [*Amphiascus*]; **D** – 11, 13, 20, 24, 25, 35; **V** – 0-10; **Z** – amnep, SK; **E** – M, bt, ep, ph, ps, gw; **R** – 14, 15, 31, 243, 249.

*Amphiascopsis thalestroides* (G. O. Sars, 1911) [*Amphiascus latilobus*]; **D** – 7, 24, 25; **Z** – clm; **E** – M, bt, ps, ph; **R** – 31, 84, 249, 389.

*Amonardia normani* (Brady, 1872); **D** – 13, 20, 24, 25, 27, 35; **V** – -85; **Z** – ap; **E** – M, bt, eb, ph, ps, pe; **R** – 15, 17, 31, 243, 249.

*Amonardia similis* (Claus, 1866); **D** – 24, 25, 26; **V** – -30; **Z** – lm; **E** – M, bt, ep, ph, ps, s; **R** – 13, 31, 249.

*Pseudamphiascopsis attenuatus* (G. O. Sars, 1906); **D** – 13, 18, 25; **V** – -20; **Z** – clm, ? nam; **E** – M, bt, ep, ps; **R** – 26, 31, 249, 252.

*Amphiascus longirostris* (Claus, 1863) [*Paramphiascopsis*]; **D** – 1, 2, 3, 4, 5, 6, 7, 8, 9, 10, 11, 12, 13, 14, 15, 16, 17, 18, 19, 20, 21, 22, 25, 27, 28; **V** – 5-98; **Z** – ham; **E** – M, bt (p), eb, eh, et, eu, ph, ps, s; **R** – 14, 15, 17, 31, 240, 246, 249, 252.

*Bulbamphiascus imus* (Brady, 1872); **D** – 1, 2, 3, 4, 5, 6, 7, 8, 9, 10, 11, 12, 13, 14, 15, 16, 17, 18, 19, 20, 21, 22, 25, 26, 27, 28; **V** – 10-200; **Z** – naminz; **E** – M, bt, eb, eu, pe, sg, s, ms, phs, ph; **R** – 17, 31, 243, 246, 248, 249, 252.

*Robertgurneya oligochaeta* Noodt, 1955; **D** – 1, 2, 3, 4, 5, 6, 7, 8, 9, 10, 11, 12, 13, 14, 15, 16, 17, 18, 19, 20, 21, 22, 25, 35; **V** – 0-20; **Z** – lm; **E** – M, bt, ep, ps, ps-s, gw; **R** – 17, 18, 27, 31, 240, 249, 252.

*Robertgurneya rostrata* (Gurney, 1927); **D** – 25; **Z** – namnei; **E** – M, bt, ps-pe; **R** – 249.

*Robertgurneya similis* (A. Scott, 1896); **D** – 7, 24, 25; **V** – 8-40; **Z** – lm; **E** – M, bt, mb, ps, sg, ph; **R** – 18, 31, 246, 249.

*Robertgurneya soyeri* Apostolov, 1974; **D** – 20, 21, 22, 25; **V** – 10; **Z** – ●p; **E** – M, bt, ep, ps; **R** – 23, 31, 249, 252.

*Robertgurneya spinulosa* (G. O. Sars, 1911); **D** – 13, 14, 25, 27; **V** – -40; **Z** – cp, ? clm; **E** – M, bt, mb, ps, pe; **R** – 27, 31, 249, 252.

*Typhlamphiascus confusus* (T. Scott, 1902); **D** – 16, 25, 27; **V** – -80; **Z** – clm; **E** – M, bt, eb, ps, pe, rh; **R** – 31, 246, 249, 252.

*Typhlamphiascus typhlops* (G. O. Sars, 1906); **D** – 17, 25, 26, 39; **V** – 10-15; **Z** – bam; **E** – M, bt, ep, ps, ps-s, s; **R** – 17, 19, 31, 249, 252.

*Amphiascoides brevifurca* (Czerniavsky, 1868) [*A. speciosus*]; **D** – 24, 25, 26, *86; **Z** – lm; **E** – M, bt, ep, ph, pe, ps; **R** – 17, 31, 185, 249.

*Amphiascoides debilis* (Giesbrecht, 1881) [*Amphiascella*]; **D** – 7, 24, 25, 26; **Z** – anam; **E** – M, bt, ph, ps, s; **R** – 31, 84, 249, 389.

*Amphiascoides neglecta* (Norman et T. Scott, 1905) [*Amphiascella*]; **D** – 24, 25, 26; **Z** – cp, ? clm; **E** – M, bt, ps, ps-s, s; **R** – 249.

*Amphiascoides subdebilis* (Willey, 1935) [*Amphiascella*]; **D** – 7, 24, 25, 26; **V** – 8; **Z** – antami; **E** – M, bt, ep, ph, ps, s; **R** – 12, 17, 31, 243, 249.

*Mesamphiascus junodi* (Monard, 1935) [*Haloschizopera*]; **D** – 18, 25, 26; **V** – 30; **Z** – nam; **E** – M, bt, ep, ps, pe; **R** – 19, 31, 249, 252.

*Haloschizopera pontarchis* Por, 1959; **D** – 1, 2, 3, 4, 5, 6, 7, 8, 9, 10, 11, 12, 13, 14, 15, 16, 17, 18, 19, 20, 21, 22, 27, 28; **V** – 25-200; **Z** – adp; **E** – M, bt, hb, ms, phs, s; **R** – 31, 240, 249, 252.

*Schizopera brusinae* Petkovski, 1954; **D** – 7, 22, 25, 35, 51, 95; **Z** – m, ? nm; **E** – M, bt, l-sl, ps, gw; **R** – 22, 30, 31, 242, 249.

*Schizopera chaetosa* Petkovski, 1954; **D** – 25, 71, 92; **Z** – m, adp; **E** – M, bt, ps, if; **R** – 15, 18, 30, 31, 249.

*Schizopera clandestina* (Klie, 1924); **D** – 25, *86, 88, 96; **Z** – neamnz, (pa); **E** – M, bt, 3‰, ps, eh; **R** – 14, 31, 185, 249.

*Schizopera compacta* De Lint, 1922; **D** – 7, 25, 69; **Z** – clm, (cse); **E** – M, bt, ep, ps, eh; **R** – 15, 22, 31, 249, 252.

*Schizopera jugurtha* (Blanchard et Richard, 1891); **D** – 7, 25, 35; **V** – 0-5; **Z** – hom, (atm); **E** – M, bt, l-sl, ep, ps, if; **R** – 16, 17, 22, 31, 249.

*Schizopera kunzi* Apostolov, 1967; **D** – 7, 25, 35; **Z** – • p; **E** – M, bt, ps, if; **R** – 9, 13, 14, 31, 249.

*Schizopera langi* Petkovski, 1954; **D** – 7, 25, 35; **Z** – m, adp; **E** – M, bt, ep, ps, gw, r; **R** – 16, 22, 30, 31, 249.

*Schizopera meridionalis* Petkovski, 1954; **D** – 7, 13, 25, 35; **Z** – cm, (e); **E** – M, bt, l, ps, gw; **R** – 22, 30, 31, 240, 249.

*Schizopera neglecta* Akatova, 1935; **D** – 25, 94; **Z** – pc, (? wcp); **E** – M, bt, eh, ps; **R** – 12, 18, 22, 31, 249.

*Schizopera petkovskii* Apostolov, 1971; **D** – 25, 35, 51, 92; **Z** – • p; **E** – M, bt, l, ps, gw, r; **R** – 15, 17, 31, 249, 252.

*Schizopera pontica* Chappuis et Serban, 1953; **D** – 7, 25, 35; **Z** – • p; **E** – M, bt, l, ps, gw, if; **R** – 17, 31, 243, 249.

*Eoschizopera gligici* (Petkovski, 1957); **D** – 7, 11, 26, 35, 84; **Z** – m, adp; **E** – M, bt, l, ps, if, s, r; **R** – 17, 22, 31, 249, 261.

*Schizoperopsis arenicola* (Chappuis et Serban, 1953) [*Schizopera*]; **D** – 7, 25, 35; **Z** – • p; **E** – M, bt, l, ps, if; **R** – 17, 22, 30, 31, 249.

*Schizoperopsis varnensis* (Apostolov, 1967); **D** – 7, 11, 25, 35; **Z** – • p; **E** – M, bt, l, ps, if; **R** – 9, 13, 14, 17, 30, 31.

**Metidae**

*Metis ignea* Philippi, 1843; **D** – 7, 16, 24, 25, 35; **V** – 0-10; **Z** – namwi, (hat); **E** – M, bt, ep, ph, ps, if, s; **R** – 17, 19, 31, 182, 249.

**Ameiridae**

*Ameira divagans* Nicholls, 1939 [*A. divagans pontica*]; **D** – 7, 24, 25; **V** – 8; **Z** – bap; **E** – M, bt, ep, ph, ps, r; **R** – 31, 241, 246, 249.

*Ameira longipes* Boeck, 1865; **D** – 20, 24, 25; **V** – 5-72; **Z** – amnei; **E** – M, bt, eb, ph, ps, et; **R** – 15, 31, 249, 252.

*Ameira parvula* (Claus, 1866) [*A. parvula tenuiseta, A. tau*]; **D** – 7, 11, 12, 13, 24, 25, 76, *86; **V** – 0-12; **Z** – nami; **E** – M, bt-p, sep, ph, ps, ps-s, s, co; **R** – 249, 389.

*Ameira scotti* G. O. Sars, 1911 [*A. scotti brevicornis*]; **D** – 7, 20, 24, 25, 27; **V** – 0-70; **Z** – ham; **E** – M, bt, eb, ph, ps, if, s; **R** – 15, 31, 243, 249.

*Filexilia attenuata* (Thompson, 1893) [*Ameira tenella*]; **D** – 20, 25, 35; **Z** – cp; **E** – M, bt, ph, ps, s (gw); **R** – 15, 17, 31, 249, 252.

*Filexilia brevipes* (Kunz, 1954) [*Ameira, A. brevipes pontica*]; **D** – 1, 7, 12, 13, 16, 25, 35; **V** – 0-10; **Z** – cp; **E** – M, bt, ep, ph, ps, if, r; **R** – 12, 15, 17, 26, 31.

*Filexilia pestae* (Petkovski, 1955) [*Ameira, A. brevipes pestae*]; **D** – 1, 7, 12, 22, 24, 25, 27; **V** – 10-70; **Z** – m, adp; **E** – M, bt, eb, ph, pe, ps; **R** – 19, 31, 240, 249.

*Nitokra affinis* Gurney, 1927 [*N. affinis californica*]; **D** – 7, 13, 25, 35, 95; **V** – 0-10; **Z** – amip, (m); **E** – L-eh, bt, ps, gw-if, ph; **R** – 18, 28, 31, 244, 249.

*Nitokra divaricata* Chappuis, 1923; **D** – 25, 35, 79; **Z** – adpc, (cseea); **E** – L-B, bt, gw, ps-s, co; **R** – 17, 31,  249, 250.

*Nitokra fallaciosa* Klie, 1937; **D** – 35, *69, 76; **Z** – clm, (wp); **E** – L-B, 60‰, bt, pe, gw; **R** – 31, 86, 185, 389.

*Nitokra fragilis* G. O. Sars, 1905; **D** – 24, 69; **Z** – amiswp, (ha); **E** – B-L, bt, sep, ph; **R** – 17, 31, 249.

*Nitokra hibernica* (Brady, 1880) [*N. hibernica bulgarica, N. hibernica hialina, N. inuber*]; **D** – 2, 24, 35, 41, 71, 92; **Z** – clm, (et); **E** – B-L, bt, ph, ps-if, co; **R** – 25, 31, 185, 389.

*Nitokra lacustris* (Shmankevich, 1875); **D** – 16, 18, 20, 24, 25, 35, 76, 77, 95; **Z** – SK, (pat); **E** – L-B, eh-0-60‰, bt, ps, gw, pe, cr, ph, et; **R** – 14, 15, 21, 31, 243, 249, 252, 263.

*Nitokra mediterranea* Brian, 1928 [*N. mediterranea pontica*]; **D** – 25, 95; **V** – 10; **Z** – ep; **E** – L, eh, bt, ep, ps; **R** – 28, 31, 249, 252.

*Nitokra pontica* Jakubisiak, 1938 [*N. typica* var. *pontica*]; **D** – 4, 25; **Z** – ●
p; **E** – L, eh, bt, ep, ps, r; **R** – 31, 157, 249, 389.

*Nitokra pusilla* G. O. Sars, 1911; **D** – 7, 20, 25, 35; **Z** – cp, ? nap; **E** – L,
eh, bt, ps, gw; **R** – 31, 84, 249, 389.

*Nitokra spinipes* Boeck, 1865; **D** – 7, 20, 22, 24, 25, 35, *69, 76; **Z** – nami,
? K, (h); **E** – L, eh, bt (p), ph, ps, if; **R** – 31, 185, 249, 389.

*Nitokra stygia* (Apostolov, 1976); **D** – 20, 25; **Z** – ● p, (Ebg); **E** – L-B, eh,
ps, cr; **R** – 25, 31, 249, 252.

*Nitokra typica* Boeck, 1865 [*N. typica adriatica*]; **D** – 7, 16, 20, 25, 35; **V**
– 0-35; **Z** – anam, (h); **E** – L, eh, bt, mb, ps, gw; **R** – 14, 15, 31, 84, 249.

*Ameiropsis reducta* Apostolov, 1973; **D** – 11, 17, 24, 25, 85; **V** – 10; **Z** – ●
p; **E** – M, bt, ep, ph, ps; **R** – 20, 31, 182, 249.

*Sicameira intermedia* Marinov, 1973; **D** – 10, 17, 25; **V** – 26; **Z** – ● p; **E** –
M, bt, ep, ps; **R** – 31, 241, 249, 252.

*Pseudoleptomesochrella halophila* (Noodt, 1952); **D** – 16, 17, 19, 22, 25,
35; **V** – 0.8; **Z** – cp; **E** – M, bt, l, ps, sg, gw-if; **R** – 12, 14, 31, 249, 263.

*Leptomesochra africana* Kunz, 1951 [*Paraleptomesochra*]; **D** – 7, 25, 35;
**V** – 0-25; **Z** – aminp; **E** – M, bt, ep, ps, sg, if; **R** – 31, 241, 249, 252.

*Parevansula wellsi* (Marinov, 1973) [*Philoleptomesochra*]; **D** – 10, 11, 17,
25; **V** – -20; **Z** – ● p; **E** – M, bt, ep, ps, ps-s; **R** – 31, 241, 244, 249.

**Paramesochridae**

*Paramesochra helgolandica* Kunz, 1936; **D** – 7, 11, 18, 25; **V** – 10-30; **Z** –
namnei; **E** – M, bt, ep, ps, ps-s; **R** – 18, 31, 240, 249.

*Paramesochra similis* Kunz, 1936; **D** – 7, 11, 25, 35; **Z** – cp; **E** – M, bt, ep,
l-sl, ps, gw; **R** – 14, 17, 31, 249, 252.

*Kliopsyllus constrictus* (Nicholls, 1935); **D** – 17, 20, 25, 35, 94; **V** – 0-10;
**Z** – cclm, ? bam; **E** – M, bt, ep, l-sl, ps, gw; **R** – 14, 17, 31, 240, 249.

*Kliopsyllus holsaticus* (Klie, 1929); **D** – 25; **Z** – amnei; **E** – M, bt, ps, r; **R** – 31, 249.

*Scottopsyllus herdmani* (I. C. Thompson et A. Scott, 1899) [*Paramesochra*]; **D** – 1, 2, 3, 4, 5, 6, 7, 8, 9, 10, 11, 12, 13, 14, 15, 16, 17, 18, 19, 20, 21, 22, 25, 35; **V** – 0-20; **Z** – bap; **E** – M, bt, ep, ps, ps-s, if; **R** – 31, 240, 249, 252.

*Scottopsyllus intermedius* (T. Scott et A. Scott, 1895); **D** – 7, 25, 26, 35; **Z** – cp; **E** – M, bt, ep, ps, pe, gw; **R** – 19, 31, 252.

*Scottopsyllus minor* (T. Scott et A. Scott, 1895); **D** – 20, 25, 35; **Z** – bap; **E** – M, bt, ep, l-sl, ps, gw; **R** – 12, 14, 17, 31, 249.

*Scottopsyllus robertsoni* (T. Scott et A. Scott, 1895); **D** – 11, 22; **V** – 0.5-10; **Z** – cp; **E** – M, bt, ep, ps; **R** – 18, 31, 240, 249.

## Tetragonicipitidae

*Phyllopodopsyllus briani* Petkovski, 1955; **D** – 7, 22, 25, 35, 95; **Z** – adp; **E** – M, bt, ep, l-sl, ps, gw; **R** – 13, 17, 31, 244, 249.

*Phyllopodopsyllus pauli* Crisafi, 1960 [*Ph. ponticus*]; **D** – 7, 22, 25, 35, 95; **V** – 8-12; **Z** – hom; **E** – M, bt, ep, l-sl, ps, gw; **R** – 11, 19, 31, 249, 252.

*Phyllopodopsyllus thiebaudi* Petkovski, 1955; **D** – 7, 22, 25, 35, 95; **Z** – lmwp; **E** – M, bt, ep, l-sl, ps, if; **R** – 17, 18, 31, 249, 252.

*Diagoniceps kunzi* Marinov, 1973; **D** – 17, 18, 25; **V** – 20; **Z** – M, bt, ep, ps, r; **E** – M, bt, ep, ps, r; **R** – 31, 242, 249, 252.

## Canthocamptidae (Orthopsyllidae, Cylindropsyllidae part)

*Mesochra aestuarii* Gurney, 1921 [*M. apostolovi, M. pontica*]; **D** – 25, *68, *69, *74, 84, 88, 90, 94, 96; **Z** – neamj, (tp); **E** – M-B-L, eh, 1-16‰, bt, ps, ps-s; **R** – 17, 31, 185, 249, 252, 389.

*Mesochra heldti* Monard, 1935; **D** – 16, 24, 25, 49, 76; **Z** – lm, (hom); **E** – B-M, bt, l-sl, ph, ps, s; **R** – 31, 84, 249, 389.

*Mesochra lilljeborgi* Boeck, 1864; **D** – 24, 25, 35, *86; **Z** – ham, (h); **E** – B-M, bt (p), ph, ps-s; **R** – 31, 185, 249, 389.

*Mesochra pestai* Lang, 1948; **D** – 24, 25; **Z** – ● p; **E** – B, bt, ph, ps; **R** – 31, 249, 262, 276.

*Mesochra pygmaea* (Claus, 1863); **D** – 12, 24, 25, 26, *69, 76; **V** – 8-120; **Z** – namip; **E** – B, bt-p, eu, ph, ps, pe; **R** – 31, 91, 185, 249.

*Mesochra rapiens* (Schmeil, 1894); **D** – 24, 25, *69, 76; **Z** – cpj, (h); **E** – B, bt, ph, ps-pe; **R** – 31, 185, 249, 389.

*Mesochra xenopoda* Monard, 1935; **D** – 7, 24, 26, 35, *69; **Z** – ham, (atm); **E** – B, bt, l-sl, ph, pe, gw; **R** – 31, 185, 249, 389.

*Orthopsyllus linearis* (Claus, 1866); **D** – 12, 25, 27, 28; **V** – 35-200; **Z** – ami; **E** – L, bt, eu, ph, ps, s, et; **R** – 26, 31, 249, 252.

*Itunella intermedia* Apostolov, 1975; **D** – 25, 92; **Z** – ● p, (Ebg); **E** – B-L, bt, l, ps, if; **R** – 24, 31, 249, 252.

*Itunella muelleri* (Gagern, 1923); **D** – 7, 25, 35; **Z** – clm, (e); **E** – B-L, l, ps-gw, if; **R** – 17, 18, 31, 249, 252.

*Nannomesochra arupinensis* (Brian, 1925); **D** – 13, 24, 25, 28; **V** – -100; **Z** – namp; **E** – M, bt, eb, ph, ps, phs; **R** – 26, 31, 249, 252.

*Stenocaris minor* (T. Scott, 1892); **D** – 7, 11, 12, 25, 35, 92; **V** – 0-20; **Z** – clm, ? bap; **E** – L-M, bt, ep, ps, gw; **R** – 14, 17, 31, 249, 252.

*Vermicaris pontica* (Chappuis et Serban, 1953) [*Stenocaris*]; **D** – 1, 2, 3, 4, 5, 6, 7, 8, 9, 10, 11, 12, 13, 14, 15, 16, 17, 18, 19, 20, 21, 22, 25, 35; **Z** – cp; **E** – L-M, l, ps, gw, if; **R** – 17, 22, 31, 240, 244, 249, 252.

*Stenocaris valkanovi* Marinov, 1974 [*Stenocaropsis*]; **D** – 7, 18, 25, 35; **V** – 0-25; **Z** – ● p; **E** – L-M, bt, ep, ps, (gw); **R** – 19, 31, 242, 249.

## Leptastacidae (Cylindropsyllidae part)

*Leptastacus laticaudatus* Nicholls, 1935 [*L. laticaudatus intermedius*]; **D** – 11, 25; **V** – 10; **Z** – cp; **E** – M, bt, ep, ps, r; **R** – 18, 31, 249, 252.

*Leptastacus macronyx* (T. Scott, 1892); **D** – 10, 11, 21, 25; **V** – 8-50; **Z** – nap; **E** – M, bt, mb, ps, sg; **R** – 31, 240, 246, 249.

*Leptastacus taurica* Marinov, 1973; **D** – 10, 21, 25; **V** – -18; **Z** – cp; **E** – M, bt, ep, ps, sg; **R** – 31, 242, 246, 249.

*Paraleptastacus holsaticus* Kunz, 1937; **D** – 7, 20, 25, 35, 95; **Z** – bap, ? nap; **E** – M, bt, eh, ps, gw, if; **R** – 14, 15, 17, 31, 249.

*Paraleptastacus spinicaudus* (T. Scott et A. Scott, 1895); **D** – 7, 11, 25, 35; **Z** – clp; **E** – M, bt, eh, ps, gw, if; **R** – 13, 14, 17, 18, 31.

*Psammastacus confluens* Nicholls, 1935; **D** – 18, 25, 35; **Z** – cp, ? nap; **E** – M, bt, ph, ps, gw, r; **R** – 26, 31, 249, 252.

### Leptopontiidae (Arenopontiidae)

*Leptopontia curvicauda* T. Scott, 1892; **D** – 7, 11, 16, 25; **Z** – cp; **E** – M, bt, ep, ps; **R** – 15, 19, 31, 249, 252.

*Arenopontia subterranea* Kunz, 1937 [*A. pontica*]; **D** – 1, 2, 3, 4, 5, 6, 7, 8, 9, 10, 11, 12, 13, 14, 15, 16, 17, 18, 19, 20, 21, 22, 23, 24, 25, 26, 27, 28, 29, 30, 31, 32, 33, 35; **Z** – clmi; **E** – M, bt, ep, ps, if, gw; **R** – 13, 14, 31, 240, 249.

*Psammoleptastacus stygius* (Noodt, 1955) [*Arenopontia*]; **D** – 17, 25, 35, 84; **Z** – nap, ? calp; **E** – M, bt, ep, ps, (gw), r; **R** – 31, 240, 249, 252.

### Cletodidae

*Cletodes limnicola* Brady, 1872; **D** – 1, 2, 3, 4, 5, 6, 7, 8, 9, 10, 11, 12, 13, 14, 15, 16, 17, 18, 19, 20, 21, 22, 25, 28; **V** – 20-100; **Z** – clm, ? clmi; **E** – M, bt, hb, pe, s-ps, sg; **R** – 31, 246, 249, 252.

*Cletodes longicaudatus* (Boeck, 1872); **D** – 1, 2, 3, 4, 5, 6, 7, 8, 9, 10, 11, 12, 13, 14, 15, 16, 17, 18, 19, 20, 21, 22, 25, 28; **V** – 20-100; **Z** – bam, ? nam; **E** – M, bt, hb, pe, s-ps, sg; **R** – 31, 246, 249, 252.

*Cletodes tenuipes* T. Scott, 1896; **D** – 3, 10, 17, 25, 27, 28; **V** – 20-250; **Z** – nam; **E** – M, bt, hb, s-ps, ms, s; **R** – 31, 246, 249, 252.

*Enhydrosoma caeni* Raibaut, 1965; **D** – 7, 25, 26, 76; **Z** – lm; **E** – M, bt, pe, s-ps, (ps); **R** – 16, 31, 249, 252.

*Enhydrosoma gariensis* Gurney, 1930 [*E. gariene*]; **D** – 7, 25; **Z** – cp; **E** – M, bt, eh, ps-s, ps; **R** – 16, 31, 249, 252.

*Enhydrosoma longifurcatum* G. O. Sars, 1909; **D** – 25, 26, 27, 76; **V** – -50; **Z** – nam; **E** – M, bt, mb, pe, ps-s; **R** – 31, 249.

*Enhydrosoma propinquum* (Brady et D. Robertson, 1880); **D** – 1, 2, 3, 4, 5, 6, 7, 8, 9, 10, 11, 12, 13, 14, 15, 16, 17, 18, 19, 20, 21, 22, 25, 27, 28; **V** – 4-90; **Z** – nam; **E** – M, bt, eb, pe, s-ps, ps, et, eu; **R** – 18, 31, 43, 240, 246, 249, 252.

*Enhydrosoma sordidum* Monard, 1926; **D** – 1, 2, 3, 4, 5, 6, 7, 8, 9, 10, 11, 12, 13, 14, 15, 16, 17, 18, 19, 20, 21, 22, 25, 27, 28; **V** – 15-150; **Z** – ham; **E** – M, bt, hb, pe, s-ps, (ps); **R** – 16, 19, 31, 43, 240, 249, 252.

*Cletocamptus confluens* (Schmeil, 1894); **D** – 25, 26, 27, 28, 35, *66, *69; **Z** – SK, (sk); **E** – M-L, 60‰, ps, s, gw; **R** – 31, 185, 249, 389.

*Cletocamptus retrogressus* Schmankevitsch, 1875; **D** – 25, 26, 41, *69, 76, 77; **Z** – nam, (h); **E** – M-B, 60‰, eu, ps, s; **R** – 31, 185, 249, 389.

*Limnocletodes behningi* Borutsky, 1926; **D** – 25, 26, 35, 93; **Z** – pca, (po), Rc; **E** – M, bt (p), eh, ps-s, gw; **R** – 26, 31, 249, 252.

*Stylicletodes longicaudatus* (Brady et D. Robertson, 1880); **D** – 1, 7, 10, 26, 27, 28; **V** – 12-100; **Z** – SK, antamip; **E** – M, bt, hb, phs, s-ps, s; **R** – 31, 240, 249, 252.

*Miroslavia longicaudata* Apostolov, 1980; **D** – 25, 26, 95; **V** – -24; **Z** – ● p; **E** – M, bt, ep, ps, ps-s; **R** – 29, 31, 249, 252.

### Rhizothricidae

*Rhizothrix pubescens* Por, 1959; **D** – 1, 2, 3, 4, 5, 6, 7, 8, 9, 10, 11, 12, 13, 14, 15, 16, 17, 18, 19, 20, 21, 22, 25, 26; **V** – 20-30; **Z** – ● p; **E** – M, bt, ep-mb, pe, s-ps; **R** – 31, 240, 249, 252.

### Argestidae

*Eurycletodes latus* (T. Scott, 1892); **D** – 1, 2, 3, 4, 5, 6, 7, 8, 9, 10, 11, 12, 13, 14, 15, 16, 26, 27, 28; **V** – 20-150; **Z** – cm; **E** – M, bt, eb, phs, pe; **R** – 31, 240, 246, 249.

*Pontocletodes ponticus* Apostolov, 1980; **D** – 25, 95; **V** – 24; **Z** – adp; **E** – M, bt, ep, ps; **R** – 29, 31, 249, 252.

## Laophontidae

*Laophonte borceai* Jakubisiak, 1938; **D** – 4; **Z** – ● p; **E** – M, bt; **R** – 31, 157, 389.

*Laophonte elongata* Boeck, 1873 [*L. elongata triarticulata*]; **D** – 7, 11, 16, 19, 24, 25, 27; **V** – 3-40, 82; **Z** – neamep; **E** – M, bt, eb, ph, ps, s; **R** – 17, 31, 182, 249.

*Laophonte parvula* G. O. Sars, 1908; **D** – 7, 20, 25, 35; **V** – 10; **Z** – clm; **E** – M, bt, ep, ps, ph, gw; **R** – 14, 15, 31, 249, 252.

*Laophonte setosa* Boeck, 1865 [*L. similis*]; **D** – 7, 11, 16, 24, *69, 76, *79, *86, 96; **Z** – clm, ? ce; **E** – M-B, bt (p), eh, ph, et, co; **R** – 31, 182, 185, 249, 252, 389.

*Laophonte thoracica* Boeck, 1865; **D** – 20, 24, 25, 28, 35; **V** – 0-100; **Z** – clm; **E** – M, bt, eb, ph, ps, s, et; **R** – 14, 15, 31, 249, 252.

*Heterolaophonte curvata* (Dauwe, 1929) [*H. curvata micrarthros*]; **D** – 7, 16, 24; **Z** – hom; **E** – M, bt, ep, ph, slc; **R** – 31, 84, 240, 249.

*Heterolaophonte stroemii* (Baird, 1834) [*H. stroemii brevicaudata, H. stroemii paraminuta*]; **D** – 7, 10, 16, 20, 24, 27, 35; **V** – 0-50; **Z** – nam, (h); **E** – M, eh, mb, ph, s-ps, if, gw, eu, et, co; **R** – 14, 15, 16, 17, 31, 84, 182, 243, 249.

*Heterolaophonte uncinata* (Czerniavsky, 1868) [*Laophonte*]; **D** – 12, 13, 24, 49, *69; **Z** – lm; **E** – M, bt, ep, eh, l, ph; **R** – 31, 91, 249, 304.

*Paralaophonte brevirostris* (Claus, 1863) [*Laophonte*]; **D** – 5, 7, 9, 24; **Z** – namnei, ? SK; **E** – M, bt, ep, ph; **R** – 31, 157, 282, 249.

*Paralaophonte congenera* (G. O. Sars, 1908) [*P. c. mediterranea*]; **D** – 20, 24, 25; **V** – -30; **Z** – clmi; **E** – M, bt, ep, ph, slc, s-ps; **R** – 26, 31, 249, 252.

*Paralaophonte octavia* (Monard, 1935); **D** – 20, 21, 24, 35; **Z** – m, ? hom; **E** – M, bt, ep, l-sl, ph, gw; **R** – 17, 31, 249, 252.

*Asellopsis duboscqui* Monard, 1928; **D** – 7, 25, 26, 76; **V** – 0-20; **Z** – m, hom; **E** – M, bt, ep, ps, s-ps, s; **R** – 18, 31, 240, 252.

*Onychocamptus mohammed* (Blanchard et Richard, 1891) [*Laophonte*]; **D** – 42, *68, *69, *74, *79, 82, 83, 84, *86, 88, 90, 92, 94; **Z** – K, (k); **E** – M-B-L, bt, ep, eh; **R** – 31, 347, 389.

*Klieonychocamptus kliei* (Monard, 1935) [*K. kliei adriaticus, Onychocamptus*]; **D** – 6, 7, 10, 16, 25, 35; **Z** – lm; **E** – M, bt, ep, l, ps, gw, if; **R** – 14, 19, 31, 244, 349.

*Klieonychocamptus ponticus* (Serban et Plesa, 1957) [*Onychocamptus*]; **D** – 25, 35; **Z** – lmnei; **E** – M, bt, ep, l, ps, if; **R** – 31, 17, 244, 249.

## Normanellidae

*Normanella minuta* (Boeck, 1872); **D** – 7, 24; **V** – 10, 100; **Z** – ham; **E** – M, bt, eb, ph, pe, (ps); **R** – 12, 31, 249, 252.

*Normanella mucronata* G. O. Sars, 1909; **D** – 6, 7, 17, 20, 25, 27, 28; **V** – 10-100; **Z** – nam; **E** – M, bt, eb, ms, phs, ps; **R** – 12, 31, 246, 249.

*Normanella serrata* Por, 1959; **D** – 1, 2, 3, 4, 5, 6, 7, 8, 9, 10, 11, 12, 13, 14, 15, 16, 17, 18, 19, 20, 21, 22, 25, 27; **V** – 40-70; **Z** – m; **E** – M, bt, eb, ms, sg, s-ps; **R** – 31, 243, 246, 249.

## Latiremidae

*Delamarella karamani* Petkovski, 1957; **D** – 7, 24, 35; **Z** – adep; **E** – M, bt, l, gw, if, ph, r; **R** – 12, 14, 31, 249, 252.

COPEPODA: POECILOSTOMATOIDA

## Nereicolidae

*Vectoriella marinovi* Stock, 1968; **D** – 10, 13; **Z** – ● p; **E** – M, ec; **R** – 252, 336.

## Ergasilidae

*Ergasilus lizae* (Kroyer, 1863) [*E. nanus*]; **D** – *68; **Z** – namsp, (ha); **E** – M-B, ec; **R** – 389.

*Ergasilus sieboldi* von Nordmann, 1832; **D** – 7; **Z** – nam, (h); **E** – B-L, ec; **R** – 389, 412.

## Oncaeidae

*Triconia minuta* (Giesbrecht, 1893 [1892]) [*Oncaea*]; **D** – 7; **Z** – K, i; **E** – M, p, r; **R** – 160, 355, 360a.

COPEPODA: SIPHONOSTOMATOIDA

## Lernaeopodidae

*Clavellisa emarginata* (Kroyer, 1837) [*Anchorella*]; **D** – 3, 7; **Z** – cp, ? bap, (h); **E** – M-B-L, ec; **R** – 389, 412.

BRANCHIURA: ARGULOIDEA

## Argulidae

*Argulus foliaceus* (Linnaeus, 1758); **D** – 71, 89, 93; **Z** – (hno); **E** – B-L, ec, 3‰; **R** – 374, 389.

CIRRIPEDIA: THORACICA: VERRUCOMORPHA

## Verrucidae

*Verruca spengleri* Darwin, 1854; **Z** – lm; **E** – M, bt, r; **R** – 249.

CIRRIPEDIA: THORACICA: BALANOMORPHA

## Chthamalidae

*Chthamalus stellatus* (Poli, 1795); **D** – 3, 7, 16, 21, 22, 45, 50; **Z** – amip; **E** – M, bt, spr, lt, epi; **R** – 84, 249, 295, 360a, 389.

*Euraphia depressa* (Poli, 1795) [*Chthamalus*]; **D** – 45, 50; **Z** – m, em; **E** – M, bt, spr, r; **R** – 294.

## Chelonibiidae

*Chelonibia testudinaria* (Linnaeus, 1758); **D** – 17; **Z** – amip; **E** – M, bt, epi; **R** – 379, 389.

**Balanidae**

*Amphibalanus amphitrite* (Darwin, 1854) [*Balanus amphitrite communis*]; **D** – 13, 15; **Z** – amip, K, i; **E** – M, bt, sl, eu, is, epi; **R** – 84, 187a, 270e.

*Amphibalanus eburneus* (Gould, 1841) [*Balanus*]; **D** – 49, *68, *69, 77, *78, *79, 84, *86, 88, 96; **Z** – amip, i; **E** – M-B, bt-p, 7‰, is, sl; **R** – 103, 129, 155a, 249, 360a, 374, 389.

*Amphibalanus improvisus* (Darwin, 1854) [*Balanus*]; **D** – 1, 2, 3, 4, 5, 6, 7, 8, 9, 10, 11, 12, 13, 14, 15, 16, 17, 18, 19, 20, 21, 22, 49, 68, *69, *79, 96; **V** – 0-27; **Z** – amip, ? K, i; **E** – M, bt, sl, eu, is, epi; **R** – 84, 91, 103, 129, 249, 360a, 295, 317, 374.

CIRRIPEDIA: RHIZOCEPHALA: KENTROGONIDA

**Peltogastridae**

*Septosaccus rodriguezii* (Fraisse, 1877) [? *Peltogaster diogeni*]; **D** – 7, 25; **Z** – lm; **E** – M, pa; **R** – 71, 84, 389.

**Sacculinidae**

*Sacculina carcini* Thompson, 1836 [*S. benedeni*]; **D** – 1, 2, 3, 4, 5, 6, 7, 8, 9, 10, 11, 12, 13, 14, 15, 16, 17, 18, 19, 20, 21, 22, 23, 24, 50; **Z** – clmwi; **E** – M, pa; **R** – 71, 72, 389.

MALACOSTRACA: MYSIDA (MYSIDACEA)

**Mysidae**

*Siriella jaltensis* Czerniavsky, 1868 [*S. clausi*]; **D** – 2, 3, 5, 7, 17, 20, 24; **V** – 3, 0-4; **Z** – ham, amwi; **E** – M, bt, ep, lt, mc, ph; **R** – 35, 36, 370, 389.

*Gastrosaccus sanctus* (van Beneden, 1861) [? *H. normani*]; **D** – 6, 7, 16, 20, 25, 51, *69; **V** – 0-10; **Z** – ami; **E** – M, bt, ep, l-sl, ps; **R** – 35, 36, 249, 370.

*Leptomysis lingvura* (G. O. Sars, 1866) [*L. pontica, L. sardica*]; **D** – 3, 5, 7, 16, 20, 24, 29; **V** – 0-50; **Z** – ace; **E** – M, bt-p, me, lt, slc, r; **R** – 35, 36, 249, 389.

*Hemimysis anomala* G. O. Sars, 1907; **D** – 7, 24; **V** – -20; **Z** – pc, clm, (h); **E** – B-L, bt, ep, lt, ■; **R** – 36, 114, 249, 389.

*Hemimysis lamornae* (Couch, 1856) [*H. lamornae pontica, H. lamornae reducta*]; **D** – 7, 16, 24; **V** – 0.3-100; **Z** – clmm; **E** – M, bt, eb, slc, ro; **R** – 33, 36, 249, 389.

*Diamysis mecznikowi* (Czerniavsky, 1882) [*D. bahirensis, D. bahirensis mecznikowi*]; **D** – 7, 58, *69, *79, *86, 88, 92, 95; **Z** – ● p; **E** – M-B, bt, eh, 0.6‰; **R** – 35, 36, 249, 347.

*Limnomysis benedeni* Czerniavsky, 1882 [*Mysidella bulgarica*]; **D** – 58, 59, 65, *68; **V** – -9; **Z** – pc, (e), Rc; **E** – B-L, bt-p, 5‰, ep, ph; **R** – 35, 205, 249, 373.

*Haplostylus normani* (G. O. Sars, 1877) [*Gastrosaccus, ? G. sanctus*]; **D** – 5; **Z** – clmwi; **E** – M, bt; **R** – 35, 249, 389.

*Mesopodopsis slabberi* (van Beneden, 1861); **D** – 1, 2, 3, 4, 5, 6, 7, 13, 24, 29, *69, 76, *78, *79; **V** – 1-20, 50; **Z** – eami, (e); **E** – M-B, bt-p, ep-mb, eh, 1-40‰, l-sl; **R** – 35, 36, 198, 249, 317, 319, 370, 374.

*Paramysis agigensis* Bacescu, 1940; **D** – 3, 5, 7, 25, 26, 58, *69; **V** – 0-4; **Z** – ep; **E** – M, bt, ep, ps, ph, ps-s; **R** – 35, 36, 249, 389.

*Paramysis kroyeri* (Czerniavsky, 1882) [*Mesomysis*]; **D** – 7, 26, *68, *69; **V** – 1-22; **Z** – ● p; **E** – M-B, bt-p, ep, eh, pe; **R** – 36, 97, 249, 370.

*Paramysis pontica* (Bacescu, 1940); **D** – 7, 29; **V** – -100; **Z** – ● p; **E** – M, p-bt, eb, pe, ps; **R** – 36, 198, 249, 319.

MALACOSTRACA: AMPHIPODA

### Ampeliscidae

*Ampelisca diadema* (A. Costa, 1853); **D** – 2, 3, 5, 7, 10, 13, 17, *69; **V** – 1-27, 120; **Z** – eamwi, ? SK; **E** – M, bt, eb, ps, sg, pe; **R** – 81, 84, 249, 370.

### Ampithoidae

*Ampithoe gammaroides* (Bate, 1856) [*Pleonexes*]; **D** – 3, 24; **Z** – clm; **E** – M, bt, ep, ph, slc; **R** – 81, 249, 389.

124

*Ampithoe helleri* Karaman, 1975; **D** – 18; **V** – 2; **Z** – clmnei; **E** – M, bt, ep; **R** – 370.

*Ampithoe ramondi* Audouin, 1828 [*A. rubricata, A. vaillanti*]; **D** – 1, 2, 3, 4, 5, 6, 7, 8, 9, 10, 11, 12, 13, 14, 15, 16, 17, 18, 19, 20, 21, 22, 24; **V** – 1-20; **Z** – amip, ? SK; **E** – M, bt, ep, mc, slc; **R** – 81, 91, 368, 370.

*Cymadusa crassicornis* (Costa, 1857) [*Grubia*]; **D** – 1, 2, 3, 4, 5, 6, 7, 8, 9, 10, 11, 12, 13, 14, 15, 16, 17, 18, 19, 20, 21, 22, 24; **V** – 1-3; **Z** – ep; **E** – M, bt, seb, ph, mc; **R** – 91, 249, 370, 389.

## Aoridae

*Leptocheirus pilosus* Zaddach, 1844; **D** – 13, 25, 26, *69, *79, 84, *86, 88; **V** – 7; **Z** – clm; **E** – M, bt, ep, ps-pe, r; **R** – 249, 370, 374. 389.

*Microdeutopus anomalus* (Rathke, 1843); **D** – 3; **V** – 20-; **Z** – nam; **E** – M, bt, hb, pe; **R** – 81, 249, 389.

*Microdeutopus damnoniensis* (Bate, 1856) [nomen nudum]; **D** – 3, 7, 12, 16, 20, 25, 27, 69; **V** – 8-95; **Z** – bam; **E** – M, bt, eb, ps-pe, mc; **R** – 168, 249, 370, 392.

*Microdeutopus gryllotalpa* Costa, 1853; **D** – 7, 13, 16, 24, 25, *69; **V** – 5-20, 40; **Z** – nam; **E** – M, bt-p, mb, mc, ph, ps, sg, s, l-sl; **R** – 84, 91, 249, 368, 370, 374, 389.

*Microdeutopus stationis* Della Valle, 1893; **D** – 7, 25; **V** – -20; **Z** – lm; **E** – M, bt, ep, ps, sg, r; **R** – 171, 249, 389.

*Microdeutopus versiculatus* (Bate, 1856) [*Coremapus*]; **D** – 3, 13, 25, 26, 27, 68; **V** – 10-80; **Z** – clm; **E** – M, bt, eb, ps-s, pe, sg; **R** – 81, 249, 370, 389.

## Atylidae

*Nototropis guttatus* (A. Costa, 1853) [*Atylus*]; **D** – 7, 24, 25, *69; **V** – -5, 100; **Z** – clm, ? eam; **E** – M, bt (p), eb, ph, ps; **R** – 84, 249, 368, 370.

## Behningiellidae

*Cardiophilus baeri* G. O. Sars, 1896 [*C. marisnigrae*]; **D** – 7, 9, 10, *69; **V** – 10-20, 30; **Z** – pc, cpc, Rc; **E** – M-B-L, bt, ep, co; **R** – 84, 178, 249, 370.

**Biancolinidae**

*Biancolina algicola* Della Valle, 1893 [*B. cuniculus*]; **D** – 11, 12, 24; **V** – 1-3; **Z** – ep; **E** – M, bt, sep, ph, slc; **R** – 91, 249, 370, 389.

**Corophiidae**

*Monocorophium acherusicum* (Costa, 1857) [*Corophium*]; **D** – 7, 16, 25, 69; **V** – -10; **Z** – amip, ? SK; **E** – M, bt, ep, mc, ps; **R** – 367, 368, 370.

*Monocorophium insidiosum* (Crawford, 1937); **D** – 5, 16, 23, 24; **V** – 1-3; **Z** – amp, ? amip; **E** – M, bt, ep, ph, zc; **R** – 370.

*Crassicorophium bonellii* (Milne Edwards, 1830) [*Corophium*]; **D** – 3, 7, 16, 24, 68, *69; **V** – 0-12; **Z** – amp; **E** – M, bt-p, sep, ph, s; **R** – 81, 84, 249, 374.

*Crassicorophium crassicorne* (Bruzelius, 1859) [*Corophium*]; **D** – *68, *69, *78, *79, 84, *86; **Z** – abapbp; **E** – M, bt, eh-1‰, ep; **R** – 249, 368, 374, 389.

*Chelicorophium curvispinum* (G. O. Sars, 1895) [*Corophium*]; **D** – 59, 60, *68, *69, 76; **Z** – pc, (e), Rc; **E** – B-L, eh, bt, ep; **R** – 82, 84, 97, 249, 375.

*Chelicorophium sowinskyi* Martynov, 1924 [*Corophium curvispinum* var. *villosus*]; **D** – 58, 59, 60; **Z** – pc, Rc; **E** – B-L, 5-6‰, ps-pe, ph; **R** – 82. 83, 369.

*Corophium volutator* (Pallas, 1766); **D** – 7, 13, 26, 58, *68, 77, *79, 84; **V** – 7; **Z** – namswp; **E** – M, bt, 0.5‰, ep, pe; **R** – 249, 370, 374, 389.

*Medicorophium runcicorne* (Della Valle, 1893) [*Corophium*]; **D** – 3, 7, 26, 27, 28; **V** – 18-100; **Z** – ep, ? em; **E** – M, bt, hb, pe; **R** – 81, 171, 249, 367.

*Siphonoecetes dellavallei* Stebbing, 1899; **D** – 7, 25, 26, 27; **V** – -50; **Z** – lmsa; **E** – M, bt, mb, ps, pe; **R** – 171, 249, 392.

**Lysianassidae**

*Orchomene humilis* (Costa, 1853); **D** – 3, 28; **V** – 80-90; **Z** – clm; **E** – M, bt, shb, phs; **R** – 81, 249, 389.

## Cheluridae

*Chelura terebrans* Philippi, 1839; **D** – 7; **Z** – amip, ? SK; **E** – M, bt, ep; **R** – 252.

## Dexaminidae

*Dexamine spinosa* (Montagu, 1813); **D** – 11, 12, 13, 24, *69; **V** – 1-25, 85; **Z** – clmm; **E** – M, bt-p, eb, mc, ph; **R** – 84, 91, 249, 370.

## Gammarellidae

*Gammarellus angulosus* (Rathke, 1843) [*G. carinatus*]; **D** – 1, 2, 3; **V** – 0-40; **Z** – anam; **E** – M, bt, ep-mb, l-sl; **R** – 81, 249, 389.

## Eusiridae

*Apherusa bispinosa* (Bate, 1857); **D** – 3, 11, 12, 24, 27, 28, 69; **V** – -100; **Z** – anam, ? abam; **E** – M, bt, eb, ph, s; **R** – 81, 91, 249, 370.

## Gammaridae

*Gammarus aequicauda* (Martynov, 1931); **D** – 3, 7, 25, 51, 69, 77, 80, 84; **Z** – lm; **E** – M, eh, bt, ep, ps; **R** – 102, 252, 370.

*Gammarus crinicornis* Stock, 1966; **D** – 7; **V** – 0.5; **Z** – clm; **E** – M, bt, ep, r; **R** – 270.

*Gammarus insensibilis* Stock, 1966; **D** – 16, 23, 24, 76; **V** – 0-15, 30; **Z** – lm; **E** – M, bt, ep, ph, ps-s, ro; **R** – 252, 346, 370.

*Gammarus subtypicus* Stock, 1966 [*G. locusta*]; **D** – 1, 2, 3, 4, 5, 6, 7, 8, 9, 10, 11, 12, 13, 14, 15, 16, 17, 18, 19, 20, 21, 22, *66, *68, *69, *74, 76, 77, 78, 79, 80, 82, 84, *86, 88; **V** – 0-20; **Z** – m, ep, ? em; **E** – M, bt, eh-1-50‰, ep, mc, ro, l-sl; **R** – 84, 86, 91, 249, 295, 370, 374, 389.

*Echinogammarus foxi* (Schellenberg, 1928) [*Chaetogammarus*]; **D** – 16, 23; **V** – 2; **Z** – m, ep, ? nm; **E** – M, bt, ep, zc; **R** – 370.

*Echinogammarus marinus* (Leach, 1815) [*Gammarus*]; **D** – 7, 11, 13, 16, 25; **Z** – bam; **E** – M, bt, ep, ps, ro; **R** – 91, 249, 259, 389.

*Echinogammarus olivii* (Milne Edwards, 1830) [*Chaetogammarus, Gammarus*]; **D** – 7, 10, 16, 68, 69, 76; **V** – 0-10; **Z** – lm; **E** – M, bt, sep, mc, ro; **R** – 84, 249, 346, 370.

*Echinogammarus trichiatus* (Martynov, 1932) [*Chaetogammarus*]; **D** – 58, 59, 60; **Z** – pc, cpc, Rc; **E** – B-L, bt, ph; **R** – 130.

*Chaetogammarus ischnus* (Stebbing, 1899) [*Ch. ischnus major, Ch. tenellus behningi*]; **D** – 58, 59, 60, *69; **Z** – pc, cpc, Rc; **E** – M-B, eh, 0-12‰, sep, ro, l-sl, ■; **R** – 82, 114, 249, 374, 389.

*Dikerogammarus haemobaphes* (Eichwald, 1841) [*D. haemobaphes fluviatilis*]; **D** – 58, 59, 60, *69; **Z** – pc, cpc, Rc; **E** – B, bt, ep, ph, ro; **R** – 82, 249, 375, 389.

*Dikerogammarus villosus* (Sowinsky, 1894); **D** – 58, 59, 60, *68, 71, 92, 95; **Z** – pc, cpc, Rc, (ean); **E** – B-L, bt, 0.1-5‰, ep, sg, ro, ph, ■; **R** – 82, 114, 249, 374, 389.

*Iphigenella andrussowi* (G. O. Sars, 1896) [*Gammarus, Lanceogammarus*]; **D** – 59, 60; **Z** – pc, Rc; **E** – B-L, bt, ep, ps, ps-s, ■; **R** – 82, 114, 174, 249, 389.

*Shablogammarus shablensis* (Carausu, 1943) [*Gammarus, Iphigenella*]; **D** – 59, 60; **Z** – pc, Rc; **E** – B-L, bt, ep, ps, ps-s, ■; **R** – 82, 114, 249, 389.

## Pontogammaridae

*Pontogammarus borceae* Carausu, 1934 [*P. abreviatus borcea*]; **D** – *63, 65; **Z** – pc, Rc; **E** – B-L, bt, ep; **R** – 82, 249, 369, 389.

*Pontogammarus robustoides* (G. O. Sars, 1894) ; **D** – 58, 59, 60; **Z** – pc, cpc, Rc; **E** – B-L, ep, eu, ph; **R** – 82, 249, 369, 389.

*Euxinia maeoticus* (Sowinsky, 1894) [*Gammarus, Pontogammarus*]; **D** – 1, 2, 3, 4, 5, 6, 7, 8, 9, 10, 11, 12, 13, 14, 15, 16, 17, 18, 19, 20, 21, 22, 51, 58, 59, 60; **V** – 0.1-0.2; **Z** – pc, Rc; **E** – B-L, eh, l, ps; **R** – 91, 249, 369, 370.

*Stenogammarus carausui* Derzhavin & Pjatakova, 1962; **D** – 68; **Z** – pc, Rc; **E** – B-L, ep; **R** – 358.

## Hyalidae

*Micropythia carinata* (Bate, 1862) [*Hyale*]; **D** – 3, 7, 20; **Z** – ep; **E** – M, bt; **R** – 178, 392.

*Hyale crassipes* (Heller, 1866); **D** – 24; **Z** – lm; **E** – M, bt, ep-mb, mc; **R** – 370.

*Hyale perieri* (Lucas, 1849); **D** – 1, 2, 3, 4, 5, 6, 7, 8, 9, 10, 11, 12, 13, 14, 15, 16, 17, 18, 19, 20, 21, 22, 24, 26; **V** – -19; **Z** – klm; **E** – M, bt, ep, mc; **R** – 249, 368, 370.

*Hyale pontica* Rathke, 1837; **D** – 3, 7, 24; **V** – 0-50; **Z** – clm; **E** – M, bt, mb, ph, mc; **R** – 84, 167, 368, 370.

*Apohyale prevosti* (H. Milne Edwards, 1830) [*Hyale*]; **D** – 7, 13, 16, 21, 22, 24; **V** – 0-30; **Z** – anam; **E** – M, bt, ep, ph, lt; **R** – 84, 91, 249, 389.

## Talitridae

*Talitrus saltator* (Montagu, 1808); **D** – 1, 2, 3, 4, 5, 6, 7, 8, 9, 10, 11, 12, 13, 14, 15, 16, 17, 18, 19, 20, 21, 22, 49; **Z** – clm; **E** – M, bt, sep, l-sp; **R** – 171, 249, 392.

*Orchestia bottae* Milne Edwards, 1840; ? *O. cavimana* Heller, 1865; **D** – 7, 49, 52, 59, 77; **Z** – clm, ep; **E** – M-B-TL, bt, l-sp, eh; **R** – 84, 155a, 249, 389.

*Orchestia gammarellus* (Pallas, 1766); **D** – 3, 7, 8, 9, 10, 11, 12, 13, 14, 15, 16, 46, 49, 52; **Z** – ham; **E** – M-TL, bt, l-sp; **R** – 81, 84, 91, 249, 389.

*Orchestia mediterranea* (Pallas, 1766); Costa, 1853; **D** – 80; **Z** – clm; **E** – M-TL, bt, l-sp; **R** – 218.

*Orchestia montagui* Audouin, 1826; **D** – 7, 11, 12, 46, 51, 52; **Z** – ep, ? em; **E** – M-TL, bt, l-sp; **R** – 84, 91, 249.

*Pseudorchestoidea brito* (Stebbing, 1891) [*Talorchestia*]; **D** – 11, 21, 22; **Z** – clm; **E** – M-TL, bt, l-sp; **R** – 84, 249, 389.

*Deshayesorchestia deshayesii* (Audouin, 1826) [*Talorchestia*]; **D** – 7, 11, 16, 46, 49, 52, 76; **Z** – clm; **E** – M-TL, bt, l-sp; **R** – 84, 91, 249, 389.

## Isaeidae

*Microprotopus longimanus* Chevreux, 1887; **D** – 2, 20, 21, 23, 24, 27; **V** – 2-85; **Z** – clp; **E** – M, bt, eb, zc, pe, sg, r; **R** – 178, 249, 370, 392.

## Ischyroceridae

*Jassa ocia* (Bate, 1862); **D** – 1, 2, 3, 4, 5, 6, 7, 8, 9, 10, 11, 12, 13, 14, 15, 16, 17, 18, 19, 20, 21, 22, 24, 27; **V** – 0-20; **Z** – clm; **E** – M, bt, ep, ph, mc, ms; **R** – 84, 91, 367, 368.

*Ericthonius difformis* H. Milne Edwards, 1830; **D** – 1, 2, 3, 4, 5, 6, 7, 8, 9, 10, 11, 12, 13, 14, 15, 16, 17, 18, 19, 20, 21, 22, 24; **Z** – anamnp; **E** – M, bt, sep, mc, ph; **R** – 84, 91, 368, 370.

## Megaluropidae

? *Megaluropus agilis* Hoeck, 1889; **D** – 7, 69; **Z** – eami; **E** – - M-B, bt; R – 370.

*Megaluropus massiliensis* Ledoyer, 1976 [ ? *M. agilis*] ; **D** – 7, 25; **V** – 5-25; **Z** – m, adep; **E** – M-B, bt, ep, ps, r; **R** - 171, 249, 370, 392.

## Melitidae

*Melita palmata* (Montagu, 1804); **D** – 1, 2, 3, 4, 5, 6, 7, 8, 9, 10, 11, 12, 13, 14, 15, 16, 17, 18, 19, 20, 21, 22, 24, 27, *69; **V** – 0-20; **Z** – aam; **E** – M, bt, ep, mc-s, ph-ro; **R** – 84, 249, 370, 374.

*Cheirocratus sundevalli* (Rathke, 1843); **D** – 11, 16, 17, 20; **Z** – clm; **E** – M, bt; **R** – 173.

## Oedicerotidae

*Perioculodes longimanus* (Bate et Westwood, 1868); **D** – 1, 2, 3, 7, 13, 25, 69; **V** – 2-20, 100; **Z** – aami; **E** – M, bt, eb, ps, sls; **R** – 81, 249, 367, 370.

*Deflexilodes gibbosus* (Chevreux, 1888) [*Monoculodes*]; **D** – 13, 16, 23, 25; **V** – 16-100; **Z** – clm; **E** – M, bt, hb, zc, sg; **R** – 171, 249, 370.

*Synchelidium maculatum* Stebbing, 1906; **D** – 1, 2, 7, 13, 16, 27, 28; **V** – 10-105; **Z** – clm; **E** – M, bt, eb, pe, sg; **R** – 81, 249, 389.

## Bathyporeiidae (Pontoporeiidae)

*Bathyporeia guilliamsoniana* (Bate, 1857); **D** – 1, 2, 3, 4, 5, 6, 7, 8, 9, 10, 11, 12, 13, 14, 15, 16, 17, 18, 19, 20, 21, 22, 25; **V** – -6, 25; **Z** – clm; **E** – M, bt, ep, ps, sls; **R** – 84, 249, 367, 370.

## Stenothoidae

*Stenothoe monoculoides* (Montagu, 1815); **D** – 24, *69; **V** – 0-10; **Z** – clm, ? bam; **E** – M, bt, sep, ph, mc, lt; **R** – 249, 368, 370, 374.

## Photidae

*Megamphopus cornutus* Norman, 1869; **D** – 3; **V** – 10-50; **Z** – clmm; **E** – M, bt, mb, pe, sg; **R** – 81, 249, 389.

## Caprellidae

*Phtisica marina* Slabber, 1769 [*Ph. acaudata, Proto pedata*]; **D** – 2, 3, 7, 16, 26, 27, 28; **V** – 15-100; **Z** – amwi; **E** – M, bt-p, hb, ph, sg, pe; **R** – 91, 295, 351, 370.

*Pseudoprotella phasma* (Montagu, 1804); **D** – 3, 24, 27; **V** – -98; **Z** – clm, ? eam; **E** – M, bt, eb, ph, pe; **R** – 178, 249, 389.

*Caprella acanthifera* Leach, 1814 [*C. acanthifera ferox*]; **D** – 1, 2, 3, 5, 7, 11, 12, 16, 24, 27, 28; **V** – 0.2-100; **Z** – clm; **E** – M, bt, eb, ph, slc, pe, epi, eu; **R** – 41, 66, 81, 84, 87, 89, 249, 351, 370.

*Caprella danilevskii* Czerniavski, 1868; **D** – 3, 16, 23, 24; **V** – 0.5-8; **Z** – tam, ? cst; **E** – M, bt, ep, slc, ph, zc; **R** – 81, 249, 351, 370.

*Caprella mitis* Mayer, 1890; **D** – 7, 24; **V** – 1-12; **Z** – m, ep; **E** – M, bt, ep, slc, r; **R** – 84, 249, 389.

Malacostraca: Cumacea

## Pseudocumatidae (Pseudocumidae)

*Pseudocuma ciliatum* G. O. Sars, 1879; **D** – 5, 7, 25; **V** – 0.5-10; **Z** – hom; **E** – M, bt, ep, ps; **R** – 36, 249, 389.

*Pseudocuma longicorne* (Bate, 1858) [*P. longicorne pontica*]; **D** – 7, 21, 25, 69; **V** – 1-7; **Z** – clm, ? eamrs; **E** – M, bt, ep, ps; **R** – 36, 169, 249, 370.

*Pseudocuma tenuicauda* G. O. Sars, 1894 [*Stenocuma*]; **D** – 7; **Z** – pc, Rc; **E** – B-L, bt, ps, ps-pe, eh; **R** – 358.

## Bodotriidae

*Bodotria arenosa* Goodsir, 1843 [*B. arenosa mediterranea*]; **D** – 7, 20, 22, 23, 25; **Z** – clm; **E** – M, bt-p, ep, ps, zc; **R** – 36, 249, 389.

? *Bodotria scorpioides* (Montagu, 1804); **D** – 3, 5; **Z** – clmwi; **E** – M, bt-p, ep, ps; **R** - 274, 389.

*Iphinoe elisae* Bacescu, 1950; **D** – 1, 2, 3, 4, 5, 6, 7, 8, 9, 10, 11, 12, 13, 14, 15, 16, 17, 18, 19, 20, 21, 22, 25, 27, 28, 89; **V** – 9-95, 125; **Z** – lm; **E** – M, bt, eb, ps, pe, phs; **R** – 169, 249, 358, 370.

*Iphinoe maeotica* Sowinskyi, 1893 [*I. inermis*]; **D** – 7, 25, 26, 69; **V** – 6-25; **Z** – ep; **E** – M, bt, ep, ps-pe, s, eh; **R** – 36, 356, 370, 274.

*Iphinoe tenella* G. O. Sars, 1878; **D** – 7, 10, 13, 25, 28, 69; **V** – 7-78; **Z** – clm; **E** – M, bt, eb, ps, phs; **R** – 36, 249, 358, 370.

*Cumopsis goodsir* (van Beneden, 1861) [*C. longipes*]; **D** – 7, 25; **V** – 0-16; **Z** – clm; **E** – M, bt, ep, ps; **R** – 36, 249, 370, 389.

## Nannastacidae

*Nannastacus euxinicus* Bacescu, 1951; **D** – 7, 21, 24, 25; **Z** – m; **E** – M, bt, ep, ps, pe-lt; **R** – 37, 249, 389.

*Cumella limicola* G. O. Sars, 1879; **D** – 5, 7, 16, 20, 21, 23, 25; **V** – 0.5-1, 18; **Z** – lmwiwp; **E** – M, bt, ep, ps, zc; **R** – 249, 274, 370, 389.

*Cumella pygmaea* G. O. Sars, 1865 [*C. pygmaea euxinica*]; **D** – 1, 2, 3, 4, 5, 6, 7, 8, 9, 10, 11, 12, 13, 14, 15, 16, 17, 18, 19, 20, 21, 22, 27, 28, 69; **V** – 20-, 50-; **Z** – clm; **E** – M, bt, hb, pe, phs; **R** – 169, 249, 358, 392.

# Leuconidae

*Eudorella truncatula* (Bate, 1856); **D** – 1, 2, 3, 4, 5, 6, 7, 8, 9, 10, 11, 12, 13, 14, 15, 16, 17, 18, 19, 20, 21, 22, 27, 28; **V** – 40-150; **Z** – anamnep; **E** – M, bt, shb, pe, phs; **R** – 169, 249, 358, 392.

MALACOSTRACA: TANAIDACEA

## Apseudidae

*Apseudopsis acutifrons* (Sars, 1882) [? *Apseudes ostroumovi*]; **D** – 2, 3, 7, 13, 16, 27, 28; **V** – 2-120; **Z** – m, hom; **E** – M, bt, hb, ms, phs, zc; **R** – 41, 249, 370, 389.

*Apseudopsis ostroumovi* Bacescu & Carausu, 1947; **D** – 13; **V** – 27, -100; **Z** – ● p; **E** – M, bt, sg, ms, phs, ■; **R** – 114, 356.

## Tanaidae

*Tanais dulongii* (Audouin, 1826) [*T. cavolinii*]; **D** – 7, 14, 15, 23, 24, 69; **V** – 1; **Z** – aminwp; **E** – M, bt, ep, mc, zc, slc; **R** – 36, 249, 358, 370.

## Paratanaidae (Leptocheliidae)

*Heterotanais oerstedii* (Kroyer, 1842) [*H. gurneyi*]; **D** – 16, 23, *79, 84, 88; **V** – 1-25; **Z** – clm; **E** – M-B, ep, 6‰, zc; **R** – 249, 370, 374, 389.

*Leptochelia savignyi* (Kroyer, 1842) [*L. dubia*]; **D** – 3, 11, 12, 23, 24, 25; **V** – 1-2; **Z** – amip; **E** – M, bt, slc, zc, mc, ps; **R** – 91, 249, 370, 389.

MALACOSTRACA: ISOPODA

## Asellidae

*Asellus aquaticus* (Linnaeus, 1758); **D** – 58, 60, *68, *74, 77, 80, 89; **Z** – (h); **E** – L-B, bt, 5‰; **R** – 374, 389.

## Janiridae

*Jaera hopeana* Costa, 1853 [*J. charrieri*]; **D** – 7; **Z** – lm; **E** – M, bt, co; **R** – 249, 376, 389.

133

*Jaera nordmanni* (Rathke, 1837); **D** – 3, 7, 11, 16; **Z** – clm; **E** – M-B-L, bt, ep; **R** – 91, 93, 249, 377.

*Jaera sarsi* Valkanov, 1936; **D** – 58, 60, *68, *69, 84, 95; **Z** – cp, Rc; **E** – M-B-L, eh, bt, ro; **R** – 249, 374, 375, 377.

## Ligiidae

*Ligia italica* Fabricius, 1798 [*L. brandti*]; **D** – 20, 21, 22, 45; **Z** – lmm, (hom); **E** – M-T, spr, lt, ro; **R** – 34, 84, 91, 249, 389.

## Tylidae

*Tylos ponticus* Grebnitsky, 1874; **D** – 7, 46, 52, 69; **Z** – lm, (hom); **E** – M-T, sps; **R** – 249, 358.

*Tylos latreillei* Audouin, 1826; **D** – 7, 16, 46, 52; **Z** – klm, (hn); **E** – M-T, sps; **R** – 84, 389.

## Idoteidae

*Idotea balthica* (Pallas, 1772) [*I. balthica basteri, I. tricuspidata*]; **D** – 1, 2, 3, 4, 5, 6, 7, 8, 9, 10, 11, 12, 13, 14, 15, 16, 17, 18, 19, 20, 21, 22, 24, 25, 36, 37, 38, 39, 40, 49, 58, 59, 60, *69, 76, 77, *78, 84, 86, 88; **V** – 0-5, 30; **Z** – ham, ? SK; **E** – M, bt, ep, ph, l-sl; **R** – 84, 91, 249, 370, 374, 389.

*Idotea ostroumovi* Sowinsky, 1895 [*I. metalica, I. metalica elongata*]; **D** – 7, 10, 11, 29; **V** – 100-150; **Z** – em; **E** – M, p-bt, hb, r; **R** – 36, 84, 249, 370.

*Stenosoma capito* (Rathke, 1837) [*Idotea pontica, Synisoma*]; **D** – 1, 2, 3, 4, 5, 7, 11, 12, 13, 16, 20, 21, 23, 24, 25; **V** – 1-3, 92; **Z** – m, hom; **E** – M, bt, eb, mc, zc, slc, ps, sg, ms, phs; **R** – 36, 67, 84, 91, 93, 249, 370, 389.

## Limnoriidae

*Limnoria tuberculata* Sowinsky, 1884; **D** – 7; **Z** – amip, ? SK; **E** – M, bt, ep; **R** – 249, 252.

## Sphaeromatidae

*Dynamene bidentata* (Adams, 1800) [*Naesa*]; **D** – 1, 2, 3, 4, 5, 6, 7, 8, 9, 10, 11, 12, 13, 14, 15, 16, 17, 18, 19, 20, 21, 22, 24, 25, 49; **V** – -3; **Z** – clmi; **E** – M, bt, ep, mc, slc, ps; **R** – 36, 289, 370, 389.

*Lekanesphaera hookeri* (Leach, 1814) [*Sphaeroma pulchellum*]; **D** – 7, 16, 23, 24, *69; **V** – 0-40; **Z** – clm; **E** – M, bt, sp-sl, ph, ro, s; **R** – 84, 249, 389.

*Sphaeroma serratum* (Fabricius, 1787); **D** – 1, 2, 3, 4, 5, 7, 12, 13, 15, 16, 21, 24, 25, 46, 49, *69, 77, *78, *86; **V** – 0-10; **Z** – amiswp; **E** – M, bt, ep, l-sl, ps, slc, mc, ro, ph; **R** – 91, 93, 249, 370, 374, 389.

## Gnathiidae

*Gnathia maxillaris* (Montagu, 1804); **D** – 5; **Z** – ace ; **E** – M, bt-p, ec; **R** – 34, 389.

*Gnathia oxyuraea* (Lilljeborg, 1855); **D** – 7; **Z** – ace; **E** – M, bt-p, ec; **R** – 167, 249, 389.

## Cirolanidae

*Eurydice dollfusi* Monod, 1930 [*E. dollfusi maris-nigri*]; **D** – 7, 10, 20, 22, 25, 51; **V** – 0-3, 8; **Z** – m, hom; **E** – M, bt (p), sep, ps; **R** – 36, 249, 358, 370.

*Eurydice pontica* (Czerniavsky, 1868); **D** – 7, 51; **Z** – ep; **E** – M, bt (p), sep, l, ps; **R** – 36, 249, 389.

*Eurydice racovitzai* Bacescu, 1949; **D** – 24, 25; **V** – 2-5; **Z** – ● p; **E** – M, bt, ep, slc, lt; **R** – 249.

*Eurydice spinigera* Hansen, 1890; **D** – 7, 16, 20, 21, 22; **V** – -6, 30; **Z** – clm; **E** – M, bt-p, ep, pe; **R** – 36, 84, 249, 370.

*Eurydice valkanovi* Bacescu, 1949; **D** – 7, 24, 25; **V** – 2-14; **Z** – ● p; **E** – M, bt, ep, ps, ph; **R** – 36, 249, 389.

## Cymothoidae

*Livoneca sinuata* Kölbel, 1878 [? *L. mediterranea, Cymothoa carryensis*]; **D** – 7; **Z** – lm; **E** – M, bt, ec; **R** – 167, 389.

*Mothocya taurica* (Czerniavsky, 1868) [*Livoneca pontica, L. punctata, Cymothoa*]; **D** – 3, 7; **Z** – • p; **E** – M, bt, ec; **R** – 249, 389, 412.

## Bopyridae

*Anisarthrus pelseneeri* Giard, 1907; **D** – 16; **Z** – cp, ? clp; **E** – M, bt, pa; **R** – 94, 249, 389.

*Bopyrissa diogeni* (Popov, 1927) [*Pseudione*]; **D** – 7; **Z** – cp, ? clp; **E** – M, bt, pa; **R** – 71, 249, 389.

*Bopyrus squillarum* Ltreille, 1802; **D** – 7; **Z** – clmi; **E** – M, bt, pa; **R** – 91, 249, 389.

MALACOSTRACA: DECAPODA

## Alpheidae

*Alpheus dentipes* Guérin-Méneville, 1832; **D** – 16; **V** – 10-17, 40 ; **Z** – lmmg ; **E** – M, bt, mb, lt, r; **R** – 78, 249, 252, 360a, 371a.

*Athanas nitescens* (Leach,1814); **D** – 5, 7, 11, 12, 13, 16, 24; **V** – 3-25, 50; **Z** – eam; **E** – M, bt, mb, ph, mc, ro; **R** – 72, 249, 371, 371a, 389.

## Hippolytidae

*Hippolyte leptocerus* (Heller, 1863); **D** – 16, 24: **V** – 3; **Z** – eam; **E** – M, bt, ep, slc; **R** – 370, 371, 371a.

*Hippolyte sapphica* D'Udekem d'Acoz, 1993 [*H. gracilis, H. longirostris, Virbius*]; **D** – 7, 11, 13, 16, 20, 23, 24; **V** – 0-2, 25, 48; **Z** – lm, ? lmm; **E** – M, bt, mb, ph, lt, zc; **R** – 36, 72, 91, 249, 371, 371a.

## Crangonidae

*Crangon crangon* (Linnaeus, 1758) [*C. maculosus, C. vulgaris* var. *maculosus*]; **D** – 7, 11, 13, 25, 27, 28, 76, 77; **V** – 5-37, 100; **Z** – ce, ? namj; **E** – M, bt (p), eh, eb, ps, s, ph; **R** – 36, 72, 249, 371, 371a.

*Philocheras fasciatus* (Risso, 1816) [*Pontophilus*]; **D** – 13, 16, 24, 25; **V** – 1-2, 15; **Z** – clmm, ? eam; **E** – M, bt, ep, ps, ph, r; **R** – 36, 75, 249, 371a, 389.

*Philocheras trispinosus* (Hailstone, 1835) [*Pontophilus*]; **D** – 7, 16, 20, 22; **V** – -15; **Z** – clmm, ? eam; **E** – M, bt, ep, ps, r; **R** – 36, 249, 389.

## Palaemonidae

*Palaemon adspersus* Rathke, 1837 [*Leander rectirostris*]; **D** – 7, 12, 13, 23, 24, *69, *78, *79, 84, 88; **V** – -35; **Z** – eami; **E** – M, bt, ep, zc, ph, pe, s, r; **R** – 72, 249, 358, 370, 371a, 374, 389.

*Palaemon elegans* Rathke, 1837 [*Leander squilla* var. *elegans, L. elegans, Macrobrachium sintangense* (de Man, 1898)]; **D** – 1, 2, 3, 4, 5, 6, 7, 8, 9, 10, 11, 12, 13, 14, 15, 16, 17, 18, 19, 20, 21, 22, 24, *69, 77, 96; **V** – 0-30; **Z** – eam; **E** - M-B, bt, eh, 4-5‰, ep, ph, mc, s, eu; **R** – 72, 249, 370, 371, 371a, 389, 402a.

*Palaemon macrodactylus* Rathbun, 1902; **D** – 69; **Z** – j, SK, i; **E** – M, bt, eh, 1-35‰, et, is; **R** – 310a, 355, 360a, 371a.

*Palaemon serratus* (Pennant, 1777) [*Leander triellianus*]; **D** – 3, 7, 13, 20, 77; **V** – 7 ; **Z** – eamrs; **E** – M, bt, ep, lt, ro, r; **R** – 40, 91, 214, 249, 371a.

## Processidae

*Processa edulis* (Risso, 1816); **D** – 5, 7, 16; **Z** – eamrs; **E** – M, bt, mc, ph, r; **R** – 36, 249, 371a, 389.

## Porcellanidae

*Pisidia longimana* (Risso, 1816) [*Porcellanides, Porcellana longimana*]; **D** – 1, 2, 3, 4, 5, 6, 7, 8, 9, 10, 11, 12, 13, 14, 15, 16, 17, 18, 19, 20, 21, 22, 24, 50, 69; **V** – 0-22, 70; **Z** – eam; **E** – M, bt, mb, ph, mc, ro; **R** – 72, 73, 84, 249, 370, 371a.

## Diogenidae

*Diogenes pugilator* (Roux,1829) [*D. varians*]; **D** – 1, 2, 3, 4, 5, 6, 7, 8, 9, 10, 11, 12, 13, 14, 15, 16, 17, 18, 19, 20, 21, 22, 25, 36, 37, 38, 39, 40, 69; **V** – 0-15, 40; **Z** – eam, ? eami; **E** – M, bt, mb, ps, sg, ■; **R** – 72, 84, 91, 116a, 358, 371, 371a.

*Clibanarius erythropus* (Latreille, 1818) [*C. misanthropus*]; **D** – 1, 2, 3, 4, 5, 6, 7, 8, 9, 10, 11, 12, 13, 14, 15, 16, 17, 18, 19, 20, 21, 22, 24, 25; **Z** – lmmg, ? eami; **E** – M, bt, ep, lt, ro; **R** – 72, 73, 84, 249, 371a, 389.

## Astacidae

*Astacus leptodactylus* Eschscholtz, 1823; **D** – 58, 59, 60, *67, *68, 71, 79, 84, 92, 95; **Z** – (e), ? Rc; **E** – L-B, bt, ep, 0-9‰; **R** – 74, 91, 206, 249, 286, 389.

## Nephropidae

*Homarus gammarus* (Linnaeus, 1758) [*H. vulgaris*]; **D** – 3, 7, 12, 16, 17; **V** – 40-45; **Z** – clmm; **E** – M, bt, mb, r, ps; **R** – 72, 84, 103, 189, 249, 296, 371a, 389, 410.

## Callianassidae

*Pestarella candida* (Olivi, 1792) [*Callianassa pestae, C. pontica, C. tyrrhena*]; **D** – 4, 7, 11, 16, 20, 21, 25, 26; **V** – 0.5-20; **Z** – lm; **E** – M, bt, ep, ps-pe; **R** – 36, 77, 84, 249, 371a, 389.

*Necallianassa truncata* (Giard et Bonnier, 1890) [*Callianassa*]; **D** – 24, 25; **V** – 4.2; **Z** – lm; **E** – M, bt, r, pe-ps; **R** – 214, 371a.

## Grapsidae

*Pachygrapsus marmoratus* (Fabricius, 1787); **D** – 1, 2, 3, 4, 5, 6, 7, 8, 9, 10, 11, 12, 13, 14, 15, 16, 17, 18, 19, 20, 21, 22, 24, 45, 50; **V** – 0-7; **Z** – clmm; **E** – M, bt, sep, spr-slr, l, lt, ro, ■; **R** – 72, 84, 91, 114, 249, 370, 371, 371a, 389.

## Varunidae

*Eriocheir sinensis* (H. Milne Edwards, 1854); **D** – ? 96; **Z** – (esea, sk - i); **E** – L-B, eu, is; **R** – 214, 371a.

*Brachynotus gemmellaroi* (Rizza, 1839) [*Cleistostoma*]; **D** – 7; **V** – 15-35; **Z** – m, hom; **E** – M, bt, r; **R** – 370, 371, 371a.

*Brachynotus sexdentatus* (Risso, 1827) [*Heterograpsus (Brachynotus) lucasi*]; **D** – 7, 13, 25, 26, *69, 77; **V** – 1, 9-40; **Z** – hom, lm, clm; **E** – M, bt, mb, ps-pe, s, ph; **R** – 72, 73, 84, 155a, 249, 358, 370, 371, 371a, 389.

## Eriphiidae

*Eriphia verrucosa* (Forskål, 1775) [*E. spinifrons*]; **D** – 1, 2, 3, 4, 5, 6, 7, 8, 9, 10, 11, 12, 13, 14, 15, 16, 17, 18, 19, 20, 21, 22, 24, 45, 50; **V** – 0-30; **Z** – lmm, ? lmmg; **E** – M, bt, ep, sp-l-sl, lt, ro, ■, ▲; **R** – 72, 73, 91, 114, 249, 370, 371, 371a, 389.

## Potamidae

*Potamon ibericum* (Bieberstein, 1808) [*P. fluviatile, P. potamios, P. tauricum*]; **D** – 43, 74, 87, 92, 94; **Z** – (nmwca); **E** – L, eu, ro, ph, ■; **R** – 69, 114, 214.

## Inachidae

*Macropodia longirostris* (Fabricius, 1775) [*Stenorhynchus egyptius*]; **D** – 3, 5, 7, 11, 12, 16, 21, 23, 24; **V** – 0-20; **Z** – clmm; **E** – M, bt, ep, ph, zc, slc, r; **R** – 34, 72, 86, 249, 371a, 389.

## Pilumnidae

*Pilumnus hirtellus* (Linnaeus, 1761) [*P. hirtellus* var. *villosus, P. villosus*] ; **D** – 1, 2, 3, 4, 5, 6, 7, 8, 9, 10, 11, 12, 13, 14, 15, 16, 17, 18, 19, 20, 21, 22, 24, *69; **V** – 0-40; **Z** – clmm; **E** – M, bt, mb, ph, mc, lt, ro, slc, ■; **R** – 72, 73, 84, 91, 249, 358, 370, 371, 371a, 389.

## Carcinidae

*Carcinus aestuarii* Nardo, 1847 [*C. maenas, C. mediterraneus*]; **D** – 1, 2, 3, 4, 5, 6, 7, 8, 9, 10, 11, 12, 13, 14, 15, 16, 17, 18, 19, 20, 21, 22, 23, 24, 25, *69; **V** – 0-33, 70; **Z** – lm, ? amp; **E** – M, bt, ps, zc, ro, r, ■; **R** – 72, 73, 84, 370, 371, 371a. Common before 1990, this cancer is rare now. The reasons for its disappearance are unclear.

*Portumnus latipes* (Pennant, 1777); **D** – 7, 16, 21; **V** – 3; **Z** – clm; **E** – M, bt, ep, ps; **R** – 36, 72, 75, 249, 389.

## Polybiidae

*Liocarcinus depurator* (Linnaeus, 1758) [*Macropipus, Portunus*]; **D** – 7; **Z** – cem, ? eam; **E** – M, bt, r, ps; **R** – 84, 249, 370, 371, 371a.

*Liocarcinus navigator* (Herbst, 1794) [*L. arcuatus, Macropipus, Portunus*]; **D** – 7, 11, 16, 25, *69; **V** – 3-40, 80; **Z** – acem; **E** – M, bt, eb, ps, sg, s, ▪; **R** – 72, 73, 84, 249, 371a, 389.

*Liocarcinus vernalis* (Risso, 1827) [*Portunus holsatus, Macropipus*]; **D** – 7, 11, 12, 13, 16, 25, 36, 37, 38, 39, 40, *69, *78, *79, 84, *86, 88; **V** – 7-40, -80; **Z** – clmm, ? eam; **E** – M, bt, eb, eh, ps, sg; **R** – 72,73, 91, 249, 370, 371, 371a, 389.

## Portunidae

*Callinectes sapidus* Rathbun, 1896; **D** – 3, 5, 7, 13; **V** – 5-22; **Z** – amj, i; **E** – M, bt, eh, ep, ps, s, r; **R** – 76, 103, 249, 360a, 371, 371a.

## Panopeidae

*Rhithropanopeus harrisii* (Gould, 1841) [*R. harrisii tridentatus*]; **D** – 7, *68, *69, 87, 96; **V** – -10; **Z** – nam, i; **E** – M-B-L, bt, eh, ps, ps-s; **R** – 95, 164, 360a, 371, 371a, 386.

*Eurypanopeus depressus* (Smith, 1869); **D** – 4, 5, 7, 9, 11, 12, 13, 16, 69; **V** – 0.3-9 m; **Z** – nam, am, i; **E** – M, is, bt, ep, ps-s, sg, eh; **R** – 270c, 270d, 360a.

## Xanthidae

*Xantho poressa* (Olivi, 1792) [*X. hydrophilus, X. rivulosus*]; **D** – 1, 2, 3, 4, 5, 6, 7, 8, 9, 10, 11, 12, 13, 14, 15, 16, 17, 18, 19, 20, 21, 22, 24, 69; **V** – 1-15, 100; **Z** – lmm; **E** – M, bt, eb, ro, ▪; **R** – 72, 84, 91, 249, 371, 371a.

## Upogebiidae

*Upogebia pusilla* (Petagna, 1792) [*U. littoralis, Gebia*]; **D** – 7, 25, 26, 69, *79, 84, *86; **V** – -22; **Z** – clmm; **E** – M, bt, ep (p), ps-pe, ▪; **R** – 72, 84, 91, 249, 370, 371a.

## COLLEMBOLA

### Neanuridae

? *Anurida maritima* (Guérin-Méneville, 1836); **D** – 45, 46, 49, 52; **Z** – nam, (hnat); **E** – M-TL, l, sp; **R** – 47, 84, 249.

*Friesea acuminata* (Denis, 1925); **D** – 1, 2, 3, 4, 5, 6, 7, 8, 9, 10, 11, 12, 13, 14, 15, 16, 17, 18, 19, 20, 21, 22, 59, 60; **Z** – lm, (nm); **E** – M-TL, l; **R** – 247, 249, 252.

### Onychiuridae

*Protaphorura fimata* (Gisin, 1952) [*Onychiurus*]; **D** – 1, 2, 3, 4, 5, 6, 7, 8, 9, 10, 11, 12, 13, 14, 15, 16, 17, 18, 19, 20, 21, 22; **Z** – (e); **E** – TL, sp-l; **R** – 247, 249, 252.

### Isotomidae

*Archisotoma besselsi* (Packard, 1877); **D** – 1, 2, 3, 4, 5, 6, 7, 8, 9, 10, 11, 12, 13, 14, 15, 16, 17, 18, 19, 20, 21, 22, 51, 52; **Z** – ace, (h); **E** – M-TL, sp-l, ps; **R** – 247, 249, 252.

### Entomobryidae

*Seira ferrarii* Parona, 1888; **D** – 1, 2, 3, 4, 5, 6, 7, 8, 9, 10, 11, 12, 13, 14, 15, 16, 17, 18, 19, 20, 25, 51: **Z** – (m, hom); **E** – M-TL, l-sl, ps; **R** – 247, 249, 252.



## EPHEMEROPTERA

### Siphlonuridae

*Siphlonurus lacustris* (Eaton, 1870); **D** – 80; **Z** – (tp); **E** – L-TL; **R** – 218.

## Baetidae

*Cloeon dipterum* (Linnaeus, 1761); **D** – 58, 77, 80; **Z** – (h); **E** – L-TL, bt, ph; **R** – 218, 258.

*Centroptilum luteolum* (Müller, 1776); **D** – 58; **Z** – (tp, ? hes); **E** – L-TL, bt, ph; **R** – 218, 258.

## Caenidae

*Caenis horaria* (Linnaeus, 1758); **D** – 58, 77, 80; **Z** – (tp); **E** – L-TL, bt, ph; **R** – 218, 258.

*Caenis luctuosa* (Burmeister, 1839); **D** – 58, 59, 77, 80; **Z** – (wcp, ? hoes); **E** – L-TL, bt, ph; **R** – 218, 258.

*Caenis robusta* Eaton, 1884; **D** – 58; **Z** – (tp); **E** – L-TL, bt, ph; **R** – 218, 258.

ODONATA

## Calopterygidae

*Calopteryx splendens* (Harris, 1782); **D** – 65, 71, 73, 83, 84, *86, 91, 92; **Z** – (wcp); **E** – L-TL, ■, ♦; **R** – 114, 216, 217.

*Calopteryx virgo* (Linnaeus, 1758); **D** – 83, 91, 92, 94; **Z** – (tp); **E** – L-TL, ■, ♦; **R** – 114, 216.

## Euphaeidae

*Epallage fatime* (Charpentier, 1840); **D** – 92; **Z** – (om); **E** – L-TL, ■, ▲, ♦; **R** – 114, 216, 217.

## Lestidae

*Lestes barbarus* (Fabricius, 1798); **D** – 58, 60, 61, 63, 70, 72, 74, 76, 77, *78, 80, 81, 82, 84, 85, *86, 88, 92, 95; **Z** – (wpo); **E** – L-TL, 13‰; **R** – 216, 217, 218, 258.

*Lestes dryas* Kirby, 1890; **D** – 58, 80, *86, 88; **Z** – (h); **E** – L-TL; **R** – 215, 216, 217, 218.

*Lestes macrostigma* (Eversmann, 1836); **D** – 76, 80; **Z** – (mwca); **E** – L-TL, 14‰, ha; **R** – 59, 119, 216, 217.

*Lestes parvidens* (Artobolevski, 1929) [*Chalcolestes*]; **D** – 71, 77, 78, 81, 83, 84, 85, 91, 94; **Z** – (nemit); **E** – L-TL; **R** – 216, 217.

*Lestes sponsa* (Hansemann, 1823); **D** – 58, 63, *78; **Z** – (tp); **E** – L-TL; **R** – 59, 216, 217.

*Lestes virens* (Charpentier, 1825); **D** – 77, 78, 81; **Z** – (wp); **E** – L-TL; **R** – 216, 217, 258.

*Lestes viridis* (Vander Linden, 1825) [*Chalcolestes*]; **D** – 77, 86, 88, 90, 91; **Z** – (hom); **E** – L-TL, ▲, ♦; **R** – 217, 258.

*Sympecma fusca* (Vander Linden, 1820); **D** – 58, 60, *61, 69, 71, 80, 84; **Z** – (omca); **E** – L-TL; **R** – 59, 215, 216, 218.

## Platycnemididae

*Platycnemis pennipes* (Pallas, 1771); **D** – 60, 65, 68, 71, 72, 73, 74, 75, 76, 77, 78, 84, *86, 88, 90, 91, 92, 94, 95; **Z** – (wces); **E** – L-TL; **R** – 59, 119, 155a, 216, 217.

## Coenagrionidae

*Erythromma viridulum* (Charpentier, 1840); **D** – 58, 60, 63, 71, 73, *78, 80, 82, 83, 85, *86, 88, 92; **Z** – (mca, ? wp); **E** – L-TL; **R** – 59, 215, 216, 218.

*Pyrrhosoma nymphula* (Sulzer, 1776); **D** – 83, 89, 92; **Z** – (wp); **E** – L-TL; **R** – 216.

*Coenagrion ornatum* (Selys, 1850); **D** – 75, 76, 92, 95; **Z** – (nm); **E** – L-TL; **R** – 59, 215, 216.

*Coenagrion puella* (Linnaeus, 1758); **D** – 60, 63, 70, 71, *78, 80, 82, 83, *86, 88, 91, 92, 95; **Z** – (ena); **E** – L-TL; **R** – 59, 216, 217.

*Coenagrion pulchellum* (Vander Linden, 1820); **D** – 60, 63, 70, 71, 75, 77, *78, 80, 82, 83, 90; **Z** – (wes); **E** – L-TL; **R** – 59, 215, 216, 217, 218.

*Coenagrion scitulum* (Rambur, 1842); **D** – 71, 80, 88, 91; **Z** – (hom); **E** – L-TL; **R** – 59, 215, 216, 218.

*Cercion lindenii* (Sélys, 1840); **D** – 74, 78, 86, 92; **Z** – (hom); **E** – L-TL; **R** – 216.

*Enallagma cyathigerum* (Charpentier, 1840); **D** – 60, 61, 69, 74, 77, 78, 80; **Z** – (h); **E** – L-TL; **R** – 216, 217, 218.

*Ischnura elegans* (Vander Linden, 1820); **D** – 58, 60, 61, 63, 65, *68, 69, 70, 71, 72, 73, 74, 75, 76, 77, 78, 79, 80, 81, 82, 83, 84, 85, 86, 88, 89, 92, 95; **Z** – (po); **E** – L-TL; **R** – 119, 215, 216, 217, 218, 258, 259.

*Ischnura pumilio* (Charpentier, 1825); **D** – 58, 60, 63, 69, 70, 71, 77, 80, 92, 94; **Z** – (wcp, wcpo); **E** – L-TL; **R** – 215, 216, 217, 218, 258.

### Aeshnidae

*Anax imperator* Leach, 1815; **D** – 58, 60, 61, 63, *68, *69, 71, 73, 75, 77, 80, 83, 85, 86, 88, 92, 95; **Z** – (wppt); **E** – L-TL, ■, ♦; **R** – 114, 215, 216, 217, 218, 258.

*Anax parthenope* (Selys, 1839); **D** – 58, *68, *69, 73, 74, 75, 76, 77, 78, 79, 80, 82, 83, 84, 86; **Z** – (ptsp); **E** – L-TL; **R** – 216, 217, 218, 258.

*Hemianax ephippiger* (Burmeister, 1839); **D** – 58, 65, 77, 78, 80, 85; **Z** – (ptm); **E** – L-TL; **R** – 59, 216, 218, 258.

*Brachytron pratense* (Müller, 1764); **D** – 86; **Z** – (ean); **E** – L-TL; **R** – 216.

*Aeshna affinis* Vander Linden, 1820; **D** – 69, 74, 78, 80, 81, 83, 84, 86, 88; **Z** – (wcp); **E** – L-TL, 4‰; **R** – 216, 217, 218.

*Aeshna isoceles* (Müller, 1767) [*Anaciaeschna isosceles*]; **D** – 71, 76, 77, 78, 80, 83, 86, 89; **Z** – (hom, ? wp); **E** – L-TL; **R** – 119, 216, 258, 259.

*Aeshna mixta* Latreille, 1805; **D** – 72, 77, 80; **Z** – (tpo); **E** – L-TL; **R** – 215, 217, 218, 259.

## Corduliidae

*Somatochlora meridionalis* Nielsen, 1935; **D** – 69, 88, 89, 91; **Z** – (nm); **E** – L-TL; **R** – 59, 216.

## Libellulidae

*Libellula depressa* (Linnaeus, 1758); **D** – 70, 80, 83, 84, 86, 88, 89, 91, 92, 94, 95; **Z** – (esca); **E** – L-TL; **R** – 216, 217.

*Libellula fulva* Müller, 1764; **D** – 77, 78, 80, 88, 89; **Z** – (wesa); **E** – L-TL; **R** – 216, 217, 218, 258.

*Libellula quadrimaculata* Linnaeus, 1758; **D** – 84, 88; **Z** – (h); **E** – L-TL; **R** – 216, 217.

*Orthetrum albistylum* (Sélys, 1848); **D** – 60, 61, 68, 69, 72, 75, 77, 78, 80, 86; **Z** – (tp); **E** – L-TL; **R** – 215, 216, 217, 218.

*Orthetrum brunneum* (Fonscolombe, 1837); **D** – 63, 64, 68, 69, 70, 71, 72, 78, 79, 81, 83, 85, 88, 89, 91, 92; **Z** – (omca); **E** – L-TL; **R** – 216, 217.

*Orthetrum cancellatum* (Linnaeus, 1758); **D** – 58, 60, 61, 63, 73, 74, 75, 76, 77, 78, 80, 83, 84, 86, 88, 92, 94, 95; **Z** – (wp); **E** – L-TL, 13‰; **R** – 59, 215, 216, 217, 218.

*Orthetrum coerulescens* (Fabricius, 1798); **D** – 60, 61, 65, 68, 74, 76, 78, 83; **Z** – (e); **E** – L-TL; **R** – 119, 216, 217, 259.

*Crocothemis erythraea* (Brullé, 1832); **D** – 58, 60, 61, 63, 69, 71, 72, 74, 76, 77, 78, 79, 80, 81, 82, 83, 84, 85, 86, 88, 89, 92, 94, 95; **Z** – (ptmca); **E** – L-TL; **R** – 119, 217, 218, 259.

*Sympetrum depressiusculum* (Sélys, 1841); **D** – 81; **Z** – (tp); **E** – L-TL, ▲, ♦; **R** – 216, 217.

*Sympetrum fonscolombii* (Sélys, 1840); **D** – 58, 60, 71, 77, 80, 83, 84, 86; **Z** – (ptm, ptsp); **E** – L-TL; **R** – 216, 217, 218.

*Sympetrum meridionale* (Sélys, 1841); **D** – 58, 60, 61, 65, 68, 71, 72, 74, 75, 76, 77, 78, 80, 82, 83, 84, 85, 86, 88, 92; **Z** – (mca, omca); **E** – L-TL; **R** – 59, 119, 215, 216, 217, 218, 259.

*Sympetrum pedemontanum* (Allioni, 1766); **D** – 58, 60, 77; **Z** – (tp, ? dp); **E** – L-TL; **R** – 216, 217.

*Sympetrum sanguineum* (Müller, 1764); **D** – 60, 63, 71, 77, 80, 81, 82, 83, 84, 85, 86, 92, 95; **Z** – (wp); **E** – L-TL; **R** – 216, 217.

*Sympetrum striolatum* (Charpentier, 1840); **D** – *68, *69, 71, 74, 76, 77, 78, 79, 80, 81, 83, 85, 86, 92, 93, 95; **Z** – (tp); **E** – L-TL, eh; **R** – 119, 215, 216, 217, 218, 259.

*Sympetrum vulgatum* (Linnaeus, 1758); **D** – 60, 77, 80; **Z** – (ewca); **E** – L-TL, ▲, ♦; **R** – 216, 217, 218.

HETEROPTERA

## Corixidae

*Corixa affinis* Leach, 1817; **D** – 33, *69; **Z** – (po); **E** – L-TL, sp, ha; **R** – 152, 252.

*Corixa panzeri* Fieber 1848; **D** – 77; **Z** – (hom); **E** – L-TL, sp, eu; **R** – 155a.

*Hesperocorixa linnaei* (Fieber 1848); D – 77; Z – (dp); E – L-TL, sp, eu; R – 155a.

*Sigara lateralis* (Leach, 1817) [*S. hieroglyphica*]; **D** – 33, 41, 42, 59, 61; **Z** – (ppt); **E** – L-TL, sp, ha-6.8‰; **R** – 85, 158, 252, 389.

*Sigara mayri* (Fieber, 1860); **D** – 33, 77; **Z** – (em); **E** – L-TL, sp, ha-34‰; **R** – 155a, 158, 252.

*Sigara nigrolineata* (Fieber, 1848); **D** – 33; **Z** – (hop); **E** – L-TL, sp, ha-4.6‰; **R** – 158, 252.

*Sigara striata* (Linnaeus, 1758); **D** – 33, 77; **Z** – (tp, ? hop); **E** – L-TL, sp, hs-3.4‰; **R** – 155a, 158, 252.

## Notonectidae

*Notonecta glauca* Linnaeus, 1758; **D** – 58, 77; **Z** – (hes, ? hop); **E** – L-TL, sp, eu; **R** – 206, 258.

146

*Notonecta viridis* Delcourt, 1909; **D** – 33, 44; **Z** – (po); **E** – L-TL, sp, ha-4.5‰; **R** – 158, 252.

## Pleidae

*Plea minutissima* Leach, 1817; **D** – 80; **Z** – (wcp); **E** – L-TL, sp, eu; **R** – 218.

## Naucoridae

*Ilyocoris cimicoides* (Linnaeus, 1758); **D** – 80; **Z** – (tp); **E** – L-TL, sp, eu; **R** – 218.

## Nepidae

*Nepa cinerea* Linnaeus, 1758; **D** – 59; **Z** – (hop); **E** – L-TL, eu; **R** – 206.

*Ranatra linearis* (Linnaeus, 1758); **D** – 71; **Z** – (hop); **E** – L-TL, eu; **R** – 374, 389.

## Belostomatidae

*Lethocerus patruelis* (Stål, 1854) [*L. cordofanus*, *Belostoma niloticum*]; **D** – *78; **Z** – (om, ? osp); **E** – L-TL; **R** – 79, 389.

## Hebridae

*Hebrus pusillus* (Fallén, 1807); **D** – 80; **Z** – (wp); **E** – T, sp; **R** – 218.

## Miridae

*Deraeocoris serenus* (Douglas et Scott, 1868); **D** – 47, 76; **Z** – (hom); **E** – T, sp, ha; **R** – 158, 252.

*Teratocoris antennatus* (Boheman, 1852); **D** – 47; **Z** – (e, ? hop); **E** – T, sp, ha; **R** – 158, 252.

*Phytocoris insignis* Reuter, 1876; **D** – 77; **Z** – (e); **E** – T, sp, ha; **R** – 158, 252.

*Lygus italicus* Wagner, 1950 [*L. maritimus*]; **D** – 48, 77; **Z** – (atm); **E** – T, sp, ha; **R** – 158, 252.

*Polymerus cognatus* (Fieber,1858); **D** – 48; **Z** – (h); **E** – T, sp, ha; **R** – 158, 252.

*Orthotylus josifovi* Wagner, 1959; **D** – 48, *69, 76; **Z** – (Er); **E** – T, sp, ha; **R** – 158, 252.

*Orthotylus parvulus* Reuter, 1879; **D** – 47; **Z** – (nem); **E** – T, sp, ha; **R** – 158, 252.

*Orthotylus rubidus* Puton, 1874; **D** – 47, 48, 76; **Z** – (am); **E** – T, sp, ha; **R** – 158, 252.

*Orthotylus schoberiae* Reuter, 1876; **D** – 48, 76, *78; **Z** – (nm); **E** – T, sp, ha; **R** – 158, 252.

*Europiella albipennis* (Fallén, 1829) [*Plagiognathus lanuginosus*]; **D** – 47; **Z** – (et, ? mt); **E** – T, sp, ha; **R** – 158, 252.

*Compsidolon pumilum* (Jakovlev, 1876) [*C. atomosum, Psallus*]; **D** – 77; **Z** – (hom); **E** – T, sp, ha; **R** – 158, 252.

### Nabidae

*Nabis sareptanus* Dohrn, 1862 [*Halonabis*]; **D** – 48; **Z** – (omca, ? osp); **E** – T, sp, ha; **R** – 158, 252.

### Tingidae

*Catoplatus carthusianus* (Goeze, 1778); **D** – 48; **Z** – (csena); **E** – T, sp, ha; **R** – 158, 252.

*Agramma atricapillum* (Spinola, 1837) [*Serenthia*]; **D** – 48; **Z** – (mca); **E** – T, sp, ha; **R** – 158, 252.

*Agramma laetum* (Fallén, 1807) [*A. confusum, Serenthia*]; **D** – 48; **Z** – (wces); **E** – T, sp, ha; **R** – 158, 252.

### Saldidae

*Salda adriatica* Horvath, 1887 [*S. littoralis, S. littoralis adriatica*]; **D** – 47, 76; **Z** – (nem); **E** – T, sp, ha; **R** – 158, 252.

*Chartoscirta longicornis* (Jakovlev, 1882) [*Ch. elegantula longicornis*]; **D** – *69, 77; **Z** – (pm); **E** – T, sp, ha; **R** – 158, 252.

*Halosalda lateralis* (Fallén, 1807); **D** – 47; **Z** – (wp); **E** – T, sp, ha; **R** – 158, 252.

*Saldula arenicola* (Scholtz, 1847); **D** – 47, 76; **Z** – (hat); **E** – T, sp, ha; **R** – 158, 252.

*Saldula opacula* (Zetterstedt, 1838); **D** – 32, 33; **Z** – (ho); **E** – T, sp, ha; **R** – 158, 252.

*Saldula pallipes* (Fabricius, 1794); **D** – 47; **Z** – (hno); **E** – T, sp, ha; **R** – 158, 252.

*Saldula pilosella* (Thomson, 1871); **D** – 47, 76; **Z** – (tp); **E** – T, sp, ha; **R** – 158, 252.

## Piesmatidae

*Parapiesma salsolae* (Becker, 1867) [*Piesma*]; **D** – 48; **Z** – (wes); **E** – T, sp, ha; **R** – 158, 252.

## Lygaeidae

*Cymus claviculus* (Fallén, 1807); **D** – 48; **Z** – (h); **E** – T, sp, ha; **R** – 158, 252.

*Cymus grandicolor* Hahn, 1832; **D** – 48; **Z** – (tp); **E** – T, sp, ha; **R** – 158, 252.

*Cymus melanocephalus* Fieber, 1861; **D** – 48; **Z** – (hom); **E** – T, sp, ha; **R** – 158, 252.

*Henestaris halophilus* (Burmeister, 1835); **D** – 48; **Z** – (hom); **E** – T, sp, ha; **R** – 158, 252.

*Henestaris laticeps* (Curtis, 1836); **D** – 48; **Z** – (atm); **E** – T, sp, ha; **R** – 158, 252.

*Peritrechus meridionalis* Puton, 1877 [*P. ambiguus*]; **D** – 47, *69, 77; **Z** – (hom); **E** – T, sp, ha; **R** – 158.

## Pentatomidae

*Antheminia varicornis* (Jakovlev, 1876) [*Codophila*]; **D** – 47; **Z** – (tp); **E** – T, sp, ha; **R** – 158, 252.

*Eurydema spectabilis* Horváth, 1882; **D** – 48; **Z** – (nem); **E** – T, sps, ha; **R** – 158, 159, 252.

## Cydnidae

*Stibaropus henkei* (Jakovlev, 1874); **D** – 48; **Z** – (emca); **E** – T, sps; **R** – 47, 57.

COLEOPTERA

## Dytiscidae

*Hydroglyphus geminus* (Fabricius, 1792) [*Bidessus, Guignotus pusillus*] ; **D** – 42 ; **Z** – (po) ; **E** – L-TL, sp, eu, sw; **R** – 85, 389.

*Hygrotus confluens* (Fabricius, 1787) [*Coelambus*]; **D** – 42; **Z** – (wcp); **E** – L-TL, sp, sw, th; **R** – 85, 143, 389.

*Hygrotus lernaeus* (Schaum, 1857) [*Coelambus*]; **D** – 77; **Z** – (hom); **E** – L-TL, ha, sw; **R** – 143, 184, 389.

*Hygrotus pallidulus* (Aubé, 1850) [*Coelambus*]; **D** – 48; **Z** – (hom); **E** – L-TL, sp, ha; **R** – 143, 184, 389.

*Hygrotus parallellogrammus* (Ahrens, 1812) [*Coelambus*]; **D** – 48; **Z** – (wcp); **E** – L-TL, sp, ha; **R** – 143, 184, 389.

*Hydroporus jonicus* L. Miller, 1862; **D** – *69; **Z** – (hom); **E** – L-TL, sw; **R** – 143, 184, 389.

*Nebrioporus ceresyi* (Aubé, 1836) [*Deronectes, Potamonectes*]; **D** – 76; **Z** – (mca); **E** – L-TL, 7‰, ha, sw; **R** – 85, 86, 184, 389.

*Laccophilus minutus* (Linnaeus, 1758) [*L. obscurus*] ; **D** – 42 ; **Z** – (wpo); **E** – L-TL, sp, sw; **R** – 85, 184, 389.

150

*Rhantus suturalis* (MacLeay, 1825) [*R. pulverosus*]; **D** – 42; **Z** – (poa, ? sk, R); **E** – L-TL, sp, eu, sw; **R** – 85, 184, 389.

*Eretes sticticus* (Linnaeus, 1767); **D** – 42; **Z** – (k); **E** – L-TL, sp, sw; **R** – 85, 184, 389.

*Dytiscus marginalis* Linnaeus, 1758; **D** – 71, *78; **Z** – (h); **E** – L-TL, sw, eu; **R** – 184, 374, 389.

*Cybister lateralimarginalis* (DeGeer, 1774); **D** – 7, 59, 60; **Z** – (wcp); **E** – L-TL, sw, eu; **R** – 184.

## Gyrinidae

*Gyrinus distinctus* Aubé, 1836; **D** – 7; **Z** – (hop); **E** – L-TL, eu; **R** – 85, 184, 389.

*Gyrinus suffriani* Scriba, 1855; **D** – 7; **Z** – (wp); **E** – L-TL, eu, sw; **R** – 85, 184, 389.

## Hydrophilidae

*Helophorus aquaticus* (Linnaeus, 1758); **D** – 42; **Z** – (wp); **E** – L-TL, sp, sw; **R** – 85, 389.

*Helophorus brevipalpis* Bedel, 1881; **D** – 42; **Z** – (wcp, ? h); **E** – L-TL, sp, sw; **R** – 85, 389.

*Berosus bispina* Reiche et Saulcy, 1856; **D** – 42; **Z** – (mwca); **E** – L-TL, sp, sw, ha; **R** – 85, 389.

*Berosus spinosus* (Steven, 1808); **D** – *69; **Z** – (tp); **E** – L-TL, sw, eu; **R** – 184, 389.

*Cercyon arenarius* Rey, 1885; **D** – 46; **Z** – (se); **E** – L-TL, sp; **R** – 144.

*Paracymus aeneus* (Germar, 1824); **D** – 33, 41, *69, 76; **Z** – (tp); **E** – L-TL, sw, ha; **R** – 85, 86, 144, 184.

*Laccobius gracilis* Motschulsky, 1855; **D** – 42; **Z** – (hom); **E** – L-TL, sp, sw; **R** – 85, 389.

*Laccobius scutellaris* Motschulsky, 1855; **D** – 42; **Z** – (em); **E** – L-TL, sp, sw; **R** – 184, 389.

*Enochrus bicolor* (Fabricius, 1792) [*Philydrus*]; **D** – *69, 76, 77, *78; **Z** – (hop, ? h); **E** – L-TL, sw, eu; **R** – 85, 184.

*Hydrophilus piceus* (Linnaeus, 1758); **D** – 7, *78; **Z** – (tpo) ; **E** – L-TL, sw, eu; **R** – 85, 374, 389.

## Hydraenidae

*Ochthebius marinus* (Paykull, 1798); **D** – 33, 41, 42, 44, 76, 77, *78; **Z** – (h); **E** – L-TL, ha, sw, eu, sp; **R** – 85, 144, 184, 389.

*Ochthebius nanus* Stephens, 1829; **D** – 41; **Z** – (csena, hom); **E** – L-TL, 23‰, sp, sw; **R** – 184, 389.

*Ochthebius ponticus* Ienistea, 1956; **D** – 41, 42, 43; **Z** – (Er); **E** – L-TL, l, sp; **R** – V. Guéorguiev.

TRICHOPTERA

## Hydroptilidae

*Stactobia caspersi* Ulmer, 1950; **D** – 42; **Z** – (em); **E** – L-TL, sp, sw; **R** – 84, 211, 389.

DIPTERA

## Ceratopogonidae

*Dasyhelea halophila* Kieffer, 1911; **D** – 33; **Z** – (se); **E** – L-TL, 7-10‰, sw, ha; **R** – 386.

## Chironomidae

*Tanypus punctipennis* Meigen, 1818 [*Pelopia*]; **D** – 58, 59, 60, *68, *79; **Z** – (hno); **E** – L-TL, 6‰, sw; **R** – 96, 249, 389.

*Anatopynia plumipes* (Fries, 1823); **D** – *69; **Z** – (wces, ? wp) ; **E** – L-TL, 1‰, sw; **R** – 96, 389.

*Ablabesmyia monilis* (Linnaeus, 1758); **D** – 58; **Z** – (hnoa, ? sk); **E** – L-TL, sw; **R** – 96, 389.

*Thienemannimyia lentiginosa* (Fries, 1823) [*Ablabesmyia*]; **D** – *68; **Z** – (wp, ? wcp); **E** – L-TL, 1‰, sw; **R** – 96, 389.

*Thalassomyia frauenfeldi* Schiner, 1856; **D** – 2, 3, 7, 13, 14, 17, 32, 33; **Z** – (ase); **E** – M-TL, l-sp, lt; **R** – 85, 249, 380, 389.

*Clunio marinus* Haliday, 1855; **D** – 7, 16; **Z** – (ena); **E** – M-TL, sl-l, ph, lt; **R** – 84, 380.

*Clunio ponticus* Michailova, 1980; **D** – 5, 7, 11, 16, 20; **Z** – (Er); **E** – M-TL, sl-l, lt; **R** – 249, 252, 267.

*Cricotopus algarum* (Kieffer, 1911); **D** – 32, 33, 59, 60, *68, 78; **Z** – (eca, ? h); **E** – L-TL, 1‰, sw; **R** – 96, 249, 337, 353a, 389.

*Cricotopus annulator* Goetghebuer, 1927; D – 78; **Z** – (h); **E** – L-TL, po, sw, floral detritus, pe; **R** – 353a.

*Cricotopus bicinctus* (Meigen, 1718); **D** – 78; **Z** – (k); **E** – L-TL, sw; **R** – 353a.

*Cricotopus fuscus* (Kieffer, 1909); **D** – 78; **Z** – (h); **E** – L-TL, sw; **R** – 353a.

? *Cricotopus intersectus* (Staeger, 1839) [? *Cryptochironomus conjugens*] ; **D** – 58, 59, 60, *68; **Z** – (h); **E** – L-TL, 6‰, sw; **R** – 96, 389.

*Cricotopus ornatus* (Meigen, 1818); **D** – 32, 33, 59, 60; **Z** – (ho); **E** – L-TL, 0.8‰, sw; **R** – 249, 337.

*Cricotopus sylvestris* (Fabricius, 1794); **D** – 32, 33, 58, 59, 60, *68, *69, 78, 80; **Z** – (hno); **E** – L-TL, 8‰, l, sw; **R** – 96, 249, 337, 353a, 389.

*Halocladius mediterraneus* Hirvenoja, 1973; **D** – 76; **Z** – (em); **E** – M-TL, sw, ha; **R** – 119, 398.

*Halocladius varians* (Staeger, 1839) [*Cricotopus*]; **D** – 32, 33; **Z** – (ena, ? wp); **E** – M-TL, sp, ha; **R** – 249, 252, 266.

? *Halocladius vitripennis* (Meigen, 1818) [*Cricotopus, Trichocladius*], sp. dubia; **D** – 2, 7, 32, 33; **Z** – (e); **E** – M-TL, l-sp, ph, lt; **R** – 57, 85, 249, 266.

? *Limnophyes minimus* (Meigen, 1818) [*Camptocladius exiguus, Tanytarsus*]; **D** – *68; **Z** – (hptn, ? sk); **E** – L-TL, 1.5‰, sw; **R** – 96, 389.

*Limnophyes pentaplastus* (Kieffer, 1921) [*L. prolongatus*]; **D** – *68; **Z** – (h); **E** – L-TL, 1‰, sw; **R** – 96, 389.

*Orthocladius fuscimanus* (Kieffer, 1908) [*O. bipunctellus*]; **D** – 16; **Z** – (ena, ? wp); **E** – L-TL, l-sp; **R** – 252, 268.

*Orthocladius rubicundus* (Meigen, 1818) [*O. saxicola*]; **D** – 58, 59, 60; **Z** – (h); **E** – L-TL, 1‰, sw; **R** – 96, 389.

*Propsilocerus lacustris* Kieffer, 1923: **D** – *68; **Z** – (csee); **E** – L-TL, 1‰, sw; **R** – 353, 389.

*Psectrocladius obvius* (Walker, 1856) [*P. dilatatus*]; **D** – *68; **Z** – (h); **E** – L-TL, 1‰, sw; **R** – 96, 389.

*Psectrocladius psilopterus* (Kieffer, 1906); **D** – 58, 59, 60, *68; **Z** – (h); **E** – L-TL, 1‰, sw; **R** – 96, 389.

*Rheocricotopus atripes* (Kieffer, 1913) [*R. foveatus*]; **D** – 7, 11, 16, 32, 33; **Z** – (wcp); **E** – L-TL, sp; **R** – 249.

*Smittia duplicata* Strenzke, 1951: **D** – 16, 24; **Z** – (Er); **E** – M-TL, sl, ph; **R** – 249.

*Baeotendipes noctivagus* (Kieffer, 1911) [*Haliella caspersi*]; **D** – 76; **Z** – (hom); **E** – M-TL, 100‰, sw; **R** – 86, 119, 249, 389.

*Chironomus anchialicus* Michailova, 1974; **D** – 76; **Z** – (Ep); **E** – M-TL, 80‰, sw; **R** – 119, 252, 259, 265.

*Chironomus aprilinus* Meigen, 1818 [*Ch. halophilus*]; **D** – 33, 42, *69, 76, 80; **Z** – (wcp); **E** – M-B-TL, 12‰, l-sp; **R** – 84, 119, 249, 389.

*Chironomus plumosus* (Linnaeus, 1758); **D** – 16, 20, 33, 58, *68, 76, 78; **Z** – (hno); **E** – L-TL, 0-9‰, sw, sp; **R** – 84, 97, 119, 249, 353, 353a, 374, 389, 396.

*Chironomus riparius* Meigen, 1804 [*Ch. thummi*]; **D** – 42, 58-60, *68, *69, *74, 76, 77, 78, 80, 88; **Z** – (hn); E - L-TL, 0-16‰, sw, sp; **R** – 85, 96, 337, 353, 353a, 374, 389, 396.

*Chironomus salinarius* Kieffer, 1915; **D** – 33, 58, 59, 60, *69, *66, 76, 77, *78, *79, 80, *86; **Z** – (wcp); **E** – M-TL, 15-60‰, sw, sp; **R** – 84, 96, 119, 353, 259, 264, 374, 396.

*Chironomus valkanovi* Michailova, 1974; **D** – 76, 77; **Z** – (Ep); **E** – M-TL, 60‰, sw; **R** – 252, 264, 265, 398.

*Cryptochironomus defectus* (Kieffer, 1913); **D** – 58, 59, 60, *68, 76, 78; **Z** – (pa); **E** – L-TL, 2‰, sw; **R** – 96, 337, 353a, 389.

*Demicryptochironomus vulneratus* (Zetterstedt, 1838); **D** – 78; **Z** – (po); **E** – L-TL, sw, po, pe-ps; **R** – 353a.

*Dicrotendipes nervosus* (Staeger, 1839) [*Limnochironomus*]; **D** – 58, 59, 60, *68, 78, *79; **Z** – (h); **E** – L-TL, 2‰, sw; **R** – 96, 337, 353a, 389.

? *Endochironomus signaticornis* Kieffer, 1913, nom. dubia; **D** – 58, 60, 68; **Z** – (e); **E** – L-TL, 2‰, sw; **R** – 96, 389.

*Endochironomus tendens* (Fabricius, 1775); **D** – 59, 60, 78; **Z** – (tp, ? hop); **E** – L-TL, 0.8‰, sw; **R** – 337, 353a.

*Eukiefferiella clypeata* (Thienemann, 1919); **D** – 78; **Z** – (? tp); **E** – L-TL, po, sw, ps-pe; **R** – 353a.

*Eukiefferiella gracei* (Edwards, 1929); **D** – 78; **Z** – (tp, ? hop); **E** – L-TL, sw, po; **R** – 353a.

*Eukiefferiella similis* Goetghebuer 1939; **D** – 78; **Z** – (? tp); **E** – L-TL, po, sw, ps-pe; **R** – 353a.

*Glyptotendipes barbipes* (Staeger, 1839); **D** – 42, 46; **Z** – (ho); **E** – L-TL, sw, sp; **R** – 131, 252.

*Glyptotendipes caulicola* (Kieffer, 1913); **D** – 59, 60; **Z** – (cse); **E** – L-TL, 0.8‰, sw; **R** – 337.

*Glyptotendipes glaucus* (Meigen, 1818); **D** – 59, 60; **Z** – (tp); **E** – L-TL, 0.8‰, sw; **R** – 337.

*Glyptotendipes cauliginellus* (Kieffer, 1913) [*G. gripekoveni*] ; **D** – 58, 59, 60, *68, 76, *79, 80; **Z** – (po); **E** – L-TL, 2‰, sw; **R** – 96, 119, 337, 396.

*Glyptotendipes pallens* (Meigen, 1804) [*G. polytomus*]; **D** – 58, 59, 60; **Z** – (po); **E** – L-TL, 2‰, sw; **R** – 96.

? *Parachironomus pararostratus* Harnisch, 1923 [*Cryptochironomus*], sp. dubia; **D** – 58, 60; **Z** – (? tp); **E** – L-TL, 2‰, sw; **R** – 96, 389.

*Polypedilum nubeculosum* (Meigen, 1804); **D** – 2, 7, 58, 59, 60, *68; **Z** – (h); **E** – L-TL, 15‰, l, sw; **R** – 84, 96, 249, 337.

*Polypedilum nubifer* (Skuse, 1889) [*P. aberrans*, *P. pharao*]; **D** – 7, 24, 58, 59, 60, *68, 77; **Z** – (poa); **E** – L-TL, 1.5‰, l, ph; **R** – 97, 249, 389.

*Polypedilum pedestre* (Meigen, 1830); **D** – 78; **Z** – (tp, ? hop); **E** – L-TL, bt, po, sw, ph; **R** – 353a.

*Polypedilum scalaenum* (Schrank, 1803) [*P. breviantennatum*]; **D** – 58, 59, 60, *68; **Z** – (ho); **E** – L-TL, 2‰, sw; **R** – 97, 389.

*Cladotanytarsus mancus* (Walker, 1856) [*Tanytarsus*]; **D** – *68; **Z** – (h); **E** – L-TL, 1.5‰, sw; **R** – 97, 389.

*Stempellina bausei* (Kieffer, 1911); **D** – *68; **Z** – (wces); **E** – L-TL, 1‰, sw; **R** – 97, 389.

*Tanytarsus gregarius* Kieffer, 1909; **D** – *69, 78; **Z** – (h); **E** – L-TL, sw; **R** – 84, 353a, 389.

*Tanytarsus mendax* Kieffer, 1925 [*T. holochloris*]; **D** – 32; **Z** – (h); **E** – L-TL, sp; **R** – 249.

*Tvetenia calvrscens* (Edwards, 1929); D – 78; Z – (h); E – L-TL, po, ?sw; R – 353a.

## Chaoboridae

*Chaoborus crystallinus* (De Geer, 1776); **D** – 59, 60; **Z** – (h); **E** – L-TL, 0.8‰, ps, s, ro; **R** – 337.

## Stratiomyidae

*Stratiomys longicornis* (Scopoli, 1763); **D** – 59; **Z** – (tp); **E** – L-TL, 0.78‰, sw; **R** – 206.

## Hybotidae

*Chersodromia colliniana* Frey, 1936; **D** – 47; **Z** – (m); **E** – T, sps, ha; **R** – 46, 252.

*Chersodromia curtipennis* Collin, 1950; **D** – 47; **Z** – (pm, Ep); **E** – T, sps, ha; **R** – 46, 252.

## Dolichopodidae

*Aphrosylus fuscipennis* Strobl, 1909; **D** – 45; **Z** – (se); **E** – M-TL, sl-l-sp, lt; **R** – 50, 51, 249, 252.

*Aphrosylus piscator* Lichtwardt, 1902; **D** – 45; **Z** – (Eb); **E** – M-TL, sl-l-sp, lt, r; **R** – 50, 249, 252.

*Aphrosylus venator* Loew, 1857; **D** – 45, 46; **Z** – (hom); **E** – M-TL, sl-l-sp, lt; **R** – 46, 47, 50, 249, 252.

## Coelopidae

*Malacomyia sciomyzina* (Haliday, 1833); **D** – 46; **Z** – (e); **E** – T, sp, ha; **R** – 50, 54, 249, 252.

*Coelopa frigida* (Fabricius, 1805) [*C. eximia, Fucomyia*]; **D** – 46, 47; **Z** – (h); **E** – T, sp, ha; **R** – 46, 50, 249, 252.

## Helcomyzidae

*Helcomyza mediterranea* (Loew, 1854); **D** – 46; **Z** – (see); **E** – T, sp, ha, ps; **R** – 48, 53, 249.

## Tethinidae

*Tethina albosetulosa* (Strobl, 1900) [*T. griseola*]; **D** – 46, 47; **Z** – (wpat); **E** – T, l-sp, ha; **R** – 46, 52, 62, 249.

*Tethina czernyi* (Hendel, 1934); **D** – 46, 47; **Z** – (mca); **E** – T, l-sp, ha, r; **R** – 58, 60, 62.

*Tethina flavigenis* (Hendel, 1934); **D** – 47, 61, 62, 76; **Z** – (am); **E** – T, l-sp, ha, r; **R** – 46, 58, 62.

*Tethina grisea* (Fallén, 1823) [*T. cinerea*]; **D** – 46, 47; **Z** – (am); **E** – T, l-sps, ha; **R** – 46, 52, 62, 249, 252.

*Tethina pallipes* (Loew, 1865); **D** – 46, 48, 59, 60, 73, 74, 85; **Z** – (hpta, ? sk); **E** – T, l-sps, ha, r; **R** – 58, 62.

*Tethina strobliana* (Mercier, 1923); **D** – 58; **Z** – (wp); **E** – T, l-sp, ha, r; **R** – 58, 60, 62.

## Canacidae

*Canace salonitana* Strobl, 1900; **D** – 45, 50; **Z** – (nem); **E** – M-TL, ph, l-sp, lt; **R** – 48, 52, 62, 249.

## Sphaeroceridae

*Thoracochaeta brachystoma* (Stenhammar, 1855) [*Leptocera*]; **D** – 46, 47; **Z** – (k); **E** – T, l-sp, ha; **R** – 46, 50, 51, 249, 252.

## Ephydridae

*Hecamede albicans* (Meigen, 1830); **D** – 45, 46, 47; **Z** – (hat); **E** – T, l-spr, ha, lt; **R** – 46, 50, 51, 62, 249.

*Ephydra attica* Becker, 1896: **D** – 41, 44, 47, 76, 77; **Z** – (dp, ? mca); **E** – TL, 60‰, sl-l-sp; **R** – 46, 62, 85, 249, 283.

*Ephydra bivittata* Loew, 1860; **D** – 33, 41, 44, 59, 60, 64, 76, 77; **Z** – (hom); **E** – TL, l-sp, ha, r; **R** – 49, 62, 249.

*Ephydra flavipes* (Macquart, 1843); **D** – 33, 76, 77; **Z** – (ptm, ? atm); **E** – TL, l-sp, ha; **R** – 62.

*Ephydra murina* Wirth, 1975 [*E. macellaria*]; **D** – 32, 47, 62, 64, 76, 77, 80, 92; **Z** – (cseeit); **E** – TL, 60‰, sl-l-sp; **R** – 62, 85, 389.

*Ephydra riparia* Fallén, 1813; **D** – 47, 58, 61, 77, 91; **Z** – (h); **E** – TL, 60-80‰, l; **R** – 46, 62, 252.

*Schema acrosticale* (Becker, 1903); **D** – 47, 48, 61, 76, 77; **Z** – (wp); **E** – TL, ha; **R** – 62.

*Chlorichaeta albipennis* (Loew, 1848); **D** – 47, 76, 91, 92, 95; **Z** – (pata); **E** – TL, ha; **R** – 62, 63, 85.

*Glenanthe nigripes* Czerny, 1909; **D** – 47, 48, 61, 76, 77; **Z** – (nm); **E** – TL, ha, r; **R** – 45, 50, 53, 54, 55, 56, 61, 62, 63.

*Polytrichophora duplosetosa* (Becker, 1896); **D** – 47, 48, 58, 59, 60, 76, 91, 92, 93, 94, 95; **Z** – (wpat); **E** – TL, ha, r; **R** – 62, 63.

### Anthomyiidae

*Fucellia maritima* (Haliday, 1838); **D** – 46; **Z** – (ena); **E** – T, l-sp, ha; **R** – 46, 50, 51, 53.

*Fucellia tergina* (Zetterstedt, 1845); **D** – 46; **Z** – (hnata, ? sk); **E** – T, l-sp, ha; **R** – 249.

## MOLLUSCA

### POLYPLACOPHORA (LORICATA)

#### CHITONIDA (CHITONIFORMES)

### Lepidochitonidae (Tonicellidae, Ischnochitonidae)

*Lepidochitona caprearum* (Scacchi, 1836) [*L. corrugata, Chiton poli, Middendorffia*]; **D** – 13, 24; **V** – 17; **Z** – hom, ? lm; **E** – M, bt, ep, lt, r; **R** – 91, 249, 389.

*Lepidochitona cinerea* (Linnaeus, 1767) [*Chiton marginatus, Ch. variegatus*]; **D** – 7, 12. 16, 21, 24; **V** – 0-30; **Z** – clm; **E** – M, bt, ep, lt, ro; **R** – 84, 91, 249, 295.

### Acanthochitonidae

*Acanthochitona crinita* (Pennant, 1777) [*A. fascicularis*]; **D** – 21, 24; **V** – 0-20; **Z** – clmm; **E** – M, bt, ep, lt, ro, r; **R** – 270a.

GASTROPODA

PATELLOGASTROPODA (ARCHAEOGASTROPODA, DOCOGLOSSA)

### Patellidae

*Patella ulyssiponensis* Gmelin, 1791[*P. caerulea pontica, P. pontica, P. tarentina*]; **D** – 3, 11, 14, 16, 17, 50; **V** – 0-10; **Z** – clm; **E** – M, bt, sep, lt-ro, l-sl, r, ■; **R** – 31, 84, 91, 114, 249, 389.

VETIGASTROPODA (ARCHAEOGASTROPODA)

### Phasianellidae (Tricoliidae)

*Tricolia pullus* (Linnaeus, 1758) [*Phasianella pontica*]; **D** – 1, 2, 3, 4, 5, 6, 7, 8, 9, 10, 11, 12, 13, 14, 15, 16, 17, 18, 19, 20, 21, 22, 24; **V** – 0-10, 50; **Z** – eam; **E** – M, bt, ep, ph; **R** – 84, 91, 249, 389.

? *Tricolia tenuis* (Michaud, 1829) [*Phasianella*]; **D** – 11; **Z** – lmm; **E** – M, bt, ep, r; **R** – 91, 389.

### Trochidae

*Gibbula adriatica* (Philippi, 1844) [*G. euxinica, G. deversa, Trochus*]; **D** – 4, 24; **V** – -30, 50; **Z** – lm; **E** – M, bt, ep, ro, mc, ph; **R** – 67, 389.

*Gibbula albida* (Gmelin, 1791) [*G. albida* var. *pontica, ? Trochus fermonii*]; **D** – 24; **V** – 0-15, 40; **Z** – lm; **E** – M, bt, ep, ro, mc, ph; **R** – 249.

*Gibbula divaricata* (Linnaeus, 1758) [*Trochus*]; **D** – 1, 2, 3, 4, 5, 6, 7, 8, 9, 10, 11, 12, 13, 14, 15, 16, 17, 18, 19, 20, 21, 22, 24, 76; **V** – 0-10, 50; **Z** – lm; **E** – M, bt, sep, ro, lt, ph; **R** – 84, 91, 159a, 249, 389.

160

CYCLONERITIMORPHA (ARCHAEOGASTROPODA)

## Neritidae

*Theodoxus danubialis* (C. Pfeiffer, 1828) [*Neritina*]; **D** – *69; **Z** – (seepc); **E** – L, bt, 12‰, lt, pe; **R** – 84, 389.

*Theodoxus euxinus* (Clessin, 1887) [*Neritina*]; **D** – 96; **V** – -10; **Z** – Ep, Rc, Sf, +; **E** – M-B, bt, 5‰; **R** – 6, 126.

*Theodoxus fluviatilis* (Linnaeus, 1758) [*Neritina*]; **D** – 58, 59, 60, *68, *69, 71, 76, *78, *79, 84, *86, 88, 92, 94, 95, 96; **Z** – (e); **E** – L-B, bt, eh, 20‰, lt; **R** – 84, 159a, 305, 374, 389.

*Theodoxus pallasi* Lindcholm, 1924 [*Neritina*]; **D** – *69; **V** – 4-10; **Z** – pc, Rc, Sf, +; **E** – M-B, eh-14‰, ro, EX, ▲; **R** – 113, 164.

*Theodoxus pilidei* (Tournouer, 1879) [*Neritina*]; **D** – 28; **V** – 65-; **Z** – pc, Rc, Sf, +; **E** – M-B, bt, phs; **R** – 249.

ECTOBRANCHIA (HETEROSTROPHA, MESOGASTROPODA)

## Valvatidae

*Valvata cristata* O. F. Müller, 1774; **D** – *68, *69; **Z** – (wes); **E** – L, 0.5‰, bt, ph, s, r, ♦; **R** – 118, 305, 374, 389.

*Valvata piscinalis* (O. F. Müller, 1774) [*V. pulchella*]; **D** – *68, *69; **V** – 3-10, 80; **Z** – (wcp, i - h); **E** – L, 0.4‰, bt, sw, ph, pe, x, ♦; **R** – 84, 118, 305, 389.

SORBEOCONCHA (NEOTAENIOGLOSSA, MESOGASTROPODA)

## Cerithiidae

*Bittium reticulatum* (da Costa, 1778) [*Cerithiolum, Cerithium exile*]; **D** – 7, 8, 9, 10, 11, 12, 13, 14, 15, 16, 17, 18, 19, 20, 21, 22, 26, 27, 28, 69, 76, 77; **V** – 2-14-111; **Z** – clmm; **E** – M, bt, eb, ps, ps-s, pe; **R** – 66, 84, 155a, 159a, 249, 389.

*Bittium submamillatum* (de Reyneval et Ponzi, 1854) [*Cerithidium, C. pusillum*]; **D** – 3, 26, 27, 28; **V** – 20-140; **Z** – lm; **E** – M, bt, eb, ps-s, pe; **R** – 67, 389.

161

*Cerithium vulgatum* Bruguière, 1792 [*C. vulgatum ponticum*, *C. ponticum*]; **D** – 11, 13, 16; **V** – -15, 25; **Z** – lm; **E** – M, bt, ep; **R** – 91, 249, 389.

HYPSOGASTROPODA (NEOTAENIOGLOSSA, MESO-, NEOGASTROPODA)

### Caecidae

*Caecum armoricum* de Folin, 1869 [*C. tenue*, *Brochina*]; **D** – 7; **V** – 15-30; **Z** – lmm; **E** – M, bt, ep; **R** – 166, 249, 252.

*Caecum trachea* (Montagu, 1803) [*C. elegans*, *C. rugulosum*]; **D** – 7; **V** – 5-15, 50; **Z** – lm; **E** – M, bt, ep; **R** – 166, 167, 249, 352.

### Calyptraeidae

*Calyptraea chinensis* (Linnaeus, 1758); **D** – 7, 11, 13, 15, 16; **V** – 40, 70; **Z** – eam; **E** – M, bt, mb; **R** – 84, 91, 249, 389.

### Cerithiopsidae

*Cerithiopsis minima* (Brusina, 1865); **D** – 84; **V** – -30; **Z** – lmm; **E** – M, bt, ep, zc; **R** – 404, 405.

*Cerithiopsis tubercularis* (Montagu, 1803); **D** – 6, 7, 8, 26, 27, 28; **V** – 14-111; **Z** – eamswi; **E** – M, bt, eb, pe; **R** – 67, 249, 389.

### Cimidae (Aclididae)

*Graphis albida* (Kanmacher, 1798); **D** – 84; **V** – 0-50; **Z** – clmm; **E** – M, bt, ep; **R** – 404.

### Epitoniidae

*Epitonium turtonis* (Turton, 1819) [*Scalaria communis*, *S. tenuicostata*]; **D** – 7; **V** – 13-36, 60; **Z** – clm; **E** – M, bt, mb; **R** – 84, 249, 389.

? *Epitonium clathrus* (Linnaeus, 1758) [*Scalaria communis*]; **D** – 7; **Z** – clm, ce; **E** – M, bt, mb; **R** – 84, 389.

## Eulimidae

*Vitreolina incurva* (Bucquoy, Dautzenberg et Dollfus, 1883) [*Eulima*]; **D** – 3, 26, 27, 28; **V** – 15-100; **Z** – clm; **E** – M, bt, eb, pe; **R** – 67, 249, 389.

## Hydrobiidae

? *Hydrobia acuta* (Draparnaud, 1805) [*H. ventrosa*]; **D** – 7, 23, 24, 43, *66, *68, *69, *78, *79, *86; **V** – 0-22; **Z** – lm, ? hm; **E** – L-B, eh-60‰, bt, ep, sw, ph, ro, pe, ♦; **R** – 84, 118, 249, 305, 374, 389.

? *Hydrobia ulvae* (Pennant, 1777); **D** – 76; **Z** – clmm; **E** – M-B, bt, eh; **R** – 119, 396.

*Ecrobia maritima* (Milaschewitsch, 1916); **D** – 76; **Z** – ep; **E** – B, bt, eh; **R** – 159b, 159c, 159d, 291a.

*Ecrobia ventrosa* (Montagu, 1803) [*Hydrobia, H. acuta, Ventrosia*]; **D** – 13, 26, 27, 28, *66, *69, 76, 77, 78, *79, *86, 96; **V** – 0-111, 20; **Z** – clm; **E** – B-L, bt, eh, mb, ro, ph, zc, pe; **R** – 155b, 166, 249, 374, 389.

*Hauffenia lucidula* Angelov, 1967 [*Horatia*]; **D** – 53; **Z** – (El); **E** – L, 1‰, bt, cr, ▲, ♦; **R** – 6, 118.

*Lithoglyphus naticoides* (C. Pfeiffer, 1828); **D** – *69; **V** – 12; **Z** – (seep, Rc); **E** – L, 3‰, bt, lt, po, ♦; **R** – 84, 118, 249, 389.

*Lithoglyphus fuscus* (C. Pfeiffer, 1828) [? *L. pyramidatus* v. Mollendorf, 1873]; **D** – *69; **Z** – (Eb); **E** – L, bt, lt, rh, r; **R** – 84, 389.

*Potamopyrgus antipodarum* (J. E. Gray, 1843) [*P. jenkinsi*]; **D** – ? 96; **Z** – (nz, sk, i); **E** – B-L, 17‰, po-sw, eu, is; **R** – 118, 204.

## Bithyniidae

? *Bithynia leachii* (Sheppard, 1823) [? *B. transsilvanica* (Bielz, 1853)]; **D** – *68; **V** – 3; **Z** – (wp, ? e, wes); **E** – L, bt, 0.5‰, ph; **R** – 374, 389.

*Bithynia tentaculata* (Linnaeus, 1758); **D** – 60, *68, *69; **V** – 5; **Z** – (wp, ? h - i); **E** – L, bt, ph, ro, s, ps, ♦; **R** – 84, 118, 305, 389.

## Pyrgulidae (Micromelaniidae)

*Turricaspia lincta* (Milaschewitch, 1908) [*Micromelania, Pyrgula*]; **D** – 26, 28, *68, *79; **V** – 24-148; **Z** – pc, Rc, Sf, +; **E** – M-B, 8‰, bt, s, sw, ♦; **R** – 113, 166, 249.

## Littorinidae

*Melarhaphe neritoides* (Linnaeus, 1758) [*Littorina*]; **D** – 2, 3, 4, 5, 6, 7, 8, 9, 10, 11, 12, 13, 14, 15, 16, 17, 18, 19, 20, 21, 22, 45; **Z** – clmm; **E** – M, bt, lr-spr, lt; **R** – 34, 84, 249, 389.

## Rissoidae

*Alvania lactea* (Michaud, 1832) [*Massotia, Rissoa*]; **D** – 3, 7, 24; **Z** – clm; **E** – M, bt, ep, ph; **R** – 84, 282, 389.

*Rissoa membranacea* (J. Adams,1800) [*R. oblonga, R. pontica, R. venusta*]; **D** – 1, 2, 3, 4, 5, 6, 7, 8, 9, 10, 11, 12, 13, 23, 24, 76; **V** – -10; **Z** – clm; **E** – M, bt, ep, ph, zc; **R** – 84, 91, 249, 159a, 282.

*Rissoa parva* (da Costa, 1778) [*R. euxinica*]; **D** – 3, 13, 24; **V** – -20, 80; **Z** – clm; **E** – M, bt, ep-mb, ph; **R** – 249, 282, 389.

*Rissoa splendida* Eichwald, 1830; **D** – 7, 8, 9, 10, 11, 12, 13, 14, 15, 16, 17, 18, 19, 20, 21, 22, 23, 24, 77; **V** – -12; **Z** – m, hom; **E** – M, bt, sep, zc, ph, ro; **R** – 84, 91, 155a, 249, 389.

*Pusillina lineolata* (Michaud, 1832) [*Rissoa, ? R. membranacea*]; **D** – 7, 16, 17, 24, *69, 76; **V** – -47; **Z** – lm; **E** – M, bt, mb, ph, pe; **R** – 159a, 374, 389.

*Setia pulcherrima* (Jeffreys, 1848) [*S. turriculata*]; **D** – 17, 19, 24, 45; **Z** – lm; **E** – M, bt, sep, lr; **R** – 404, 405.

*Pontiturboella rufostrigata* (Hesse, 1916) [*Assiminea, Paludinella*]; **D** – *69; **Z** – ● p, Er; **E** – M, bt; **R** – 148.

## Tornidae

*Tornus subcarinatus* (Montagu, 1803) [*Adeorbis*]; **D** – 14; **Z** – clm; **E** – M, bt, ep; **R** – 166, 249, 392.

## Triphoridae

*Marshallora adversa* (Montagu, 1803) [*Trifora obesula, T. perversa* var. *parva*]; **D** – 16, 17, 19; **V** – -15, 80; **Z** – clm, ? eam; **E** – M, bt, mb, phc; **R** – 405.

## Muricidae

*Rapana venosa* (Valenciennes, 1846) [*R. bezoar, R. thomasiana*]; **D** – 1, 2, 3, 4, 5, 6, 7, 8, 9, 10, 11, 12, 13, 14, 15, 16, 17, 18, 19, 20, 21, 22, 24, 25, 69, 76; **V** – -70; **Z** – j, aminp, i; **E** – M, bt, mb-eb, eu, is; **R** – 159a, 165, 249, 360a, 392.

*Trophonopsis muricatus* (Montagu, 1803) [*T. breviatus, Trophon*] ; **D** – 3, 21, 22, 27, 28; **V** – 36-138; **Z** – clm; **E** – M, bt, shb, phs; **R** – 67, 249, 389, 410.

## Buccinidae

*Neptunea arthritica* (Valenciennes, 1858); **D** – 25, 26, 49; **Z** – j, pnep, i; **E** - M, bt , r; **R** – 355.

## Nassariidae

*Cyclope neritea* (Linnaeus, 1758) [*Cyclonassa kamischiensis, Nassa brusinai*]; **D** – 1, 2, 3, 4, 5, 6, 7, 8, 9, 10, 11, 12, 13, 14, 15, 16, 17, 18, 19, 20, 21, 22, 76; **V** – -54; **Z** – lm; **E** – M, bt, mb, ps-pe; **R** – 67, 84, 91, 159a, 249, 389.

? *Cyclope pellucida* Risso, 1826 [*C. donovani, Cyclonassa kamischiensis*]; D – 77; V – -30; **Z** – lm, ? m; **E** – M, bt, ep-mb, ps-pe; **R** – 117a.

*Nassarius reticulatus* (Linnaeus, 1758) [*N. nitidus, Nassa*]; **D** – 1, 2, 3, 4, 5, 6, 7, 8, 9, 10, 11, 12, 13, 14, 15, 16, 17, 18, 19, 20, 21, 22, 25, 36, 37, 38, 39, 40, *69, 76; **V** – -60; **Z** – clmm; **E** – M, bt, mb, ps, ps-pe; **R** – 84, 91, 159a, 249, 374.

## Conidae

*Bela nebula* (Montagu, 1803) [*Cythara fuscata, Mangelia, Raphitona*]; **D** – 7, *69; **V** – -20; **Z** – clm; **E** – M, bt, ep; **R** – 91, 249, 389.

*Mangelia costata* (Pennant, 1777) [*M. pontica, M. taeniata, Cythara*]; **D** – 7, *69; **V** – -50; **Z** – clm; **E** – M, bt, mb; **R** – 84, 249, 389.

## Omalogyridae

*Omalogyra atomus* (Philippi, 1841) [*Homalogyra, Truncatella*]; **D** – 17, 84; **Z** – aam; **E** – M, bt, ep; **R** – 405.

*Ammonicera fischeriana* (Monterosato, 1869) [*Homalogyra*]; **D** – 17, 84; **Z** – lm, ? lmm; **E** – M, bt, mb; **R** – 405.

## Pyramidellidae (Ebalidae, Turbonillidae, Murchisonellidae)

*Chrysallida emaciata* (Brusina, 1866) [*Parthenina, Turbonilla*]; **D** – 17, 19, 84; **Z** – lm; **E** – M, bt, ep; **R** – 405.

*Chrysallida incerta* (Milaschewitsh, 1916) [*Ch. brusinai, Odostomia, Parthenia*]; **D** – 17, 19, 69, 84; **V** – -18; **Z** – m, ? lmmg; **E** – M, bt, ep; **R** – 405.

*Chrysallida interstincta* (J. Adams, 1797) [*Ch. obtusa, Parthenina tenuistriata*]; **D** – 12, 17, 19, 69, 84; **V** – 6-80; **Z** – clmm; **E** – M, bt, mb; **R** – 164, 405.

*Chrysallida terebellum* (Philippi, 1844) [*Parthenina costulata, Odostomia*]; **D** – 7, 69; **V** – -10, 42; **Z** – lm; **E** – M, bt, ep, pe; **R** – 166, 167, 249, 392.

*Eulimella acicula* (Philippi, 1836) [*E. laevis, Belonidium, Odostomia*]; **D** – 17, 84; **V** – -18, - 47; **Z** – clm; **E** – M, bt, ep, r; **R** – 405.

*Ebala pointeli* (de Folin, 1868) [*Anisocycla, Eulimella, Turbonilla*]; **D** – 7, 69; **V** – -20; **Z** – lm, ? lmm; **E** – M, bt, ep, ps, s; **R** – 166, 167.

*Odostomia erjaveciana* Brusina, 1869 [*O. nitens*]; **D** – 11, 17, 19, 84; **Z** – lm; **E** – M, bt, ep; **R** – 405.

*Odostomia eulimoides* Hanley, 1844 [*O. novegradensis, O. pallida*]; **D** – 7, 16, 17, 19, 84; **V** – -70; **Z** – clm; **E** – M, bt, mb; **R** – 405.

*Odostomia plicata* (Montagu, 1803); **D** – 16, 17, 19, 84; **V** – -20; **Z** – clm; **E** – M, bt, ep; **R** – 405.

*Odostomia scalaris* MacGillivray, 1843 [*O. albella, O. rissoiformis*]; **D** – 1, 2, 3, 4, 5, 6, 7, 8, 9, 10, 11, 12, 13, 14, 15, 16, 17, 18, 19, 20, 21, 22; **V** – -15, 50; **Z** – clm; **E** – M, bt, mb; **R** – 166, 178, 249.

*Noemiamea dolioliformis* (Jeffreys, 1848) [*Odostomia*]; **D** – 17, 84; **Z** – lm; **E** – M, bt, ep, r; **R** – 405.

*Turbonilla delicata* Monterosato, 1874 [*Odostomia*]; **D** – 7; **V** – -35; **Z** – clm; **E** – M, bt, ep-mb; **R** – 166, 249, 392.

*Turbonilla pusilla* (Philippi, 1844) [*T. elegantissima, T. pupaeformis*]; **D** – 16, 17, 84; **Z** – kclm; **E** – M, bt, ep, r; **R** – 405.

C ephalaspidea

## Retusidae

*Cylichnina umbilicata* (Montagu, 1803) [*C. strigella, C. variabilis, Retusa*]; **D** – 7, 11, 13, 16, 69; **V** – -50; **Z** – clm; **E** – M, bt, mb, s, sg; **R** – 91, 249, 389.

? *Cylichnina robagliana* (Fischer, 1869) [*C. ovoides, Bulla, Retusa*]; **D** – 7; **V** – -40; **Z** – lm; **E** – M, bt, mb; **R** – 84, 249, 389.

*Retusa truncatula* (Bruguière, 1792) [*R. truncatula opima, Bulla, Cylichna*]; **D** – 7, 13, 17, 24, 26, 27, 28, *69; **V** – 0-140; **Z** – eam; **E** – M, bt, eb, pe, ph, mc; **R** – 84, 249, 389.

S acoglossa

## Stiligeridae (Hermaeidae)

*Calliopaea bellula* (Orbigny, 1837) [*Stiliger*]; **D** – 7, 24, 34, *69; **Z** – clm; **E** – M, bt, ep, slc; **R** – 249, 374, 389, 392.

## Limapontiidae (Stiligeridae)

*Limapontia capitata* (O. F. Müller, 1774); **D** – 7, 24; **Z** – clm, ce; **E** – M, bt, ep, slc; **R** – 249, 392.

## Microhedylidae (Parhedylidae)

*Parahedyle tyrtowii* (Kowalewsky, 1901) [*Microhedyle*, *Hedyle*]; **D** – 7, 11, 16, 25; **Z** – m; **E** – M, bt, ep, ps; **R** – 249, 392.

NUDIBRANCHIA

## Pseudovermidae

*Pseudovermis paradoxus* Perejaslawtzeva, 1890; **D** – 7, 11, 25; **Z** – lm; **E** – M, bt, ep, ps; **R** – 249, 392.

## Tergipedidae

*Embletonia pulchra* (Alder et Hancock, 1844) [*Aeolis*, *Eolis*, *Pterochilus*]; **D** – 23, 24, *69, *79; **V** – -40; **Z** – clm; **E** – M, bt, mb, slc, zc; **R** – 249, 374, 386, 389.

## Corambidae

*Corambe obscura* (Verrill, 1870) [*Doridella*]; **D** – 7, 13; **V** – -5; **Z** – vck, i; **E** – M, bt, ep; **R** – 328, 360a.

## Dorididae

*Doris ocelligera* (Bergh, 1881); **D** – 7; **Z** – lm; **E** – M, bt; **R** – 358.

EUPULMONATA

## Ellobiidae

*Myosotella myosotis* (Draparnaud, 1801) [*Alexia*, *Auricula*, *Ovtella*]; **D** – 3, 4, 7, 11, 16, 19, 20, 21, 22, 39, 45, 48, 64, 65, 76, 87, 91, 95; **Z** – amp; **E** – M-TL, spr, r; **R** – 159a, 166, 249, 270b.

BASOMMATOPHORA

## Acroloxidae

*Acroloxus lacustris* (Linnaeus, 1758); **D** – 58, 59, 60, 77, 80; **Z** – (wes, ? hoes); **E** – L, bt, ph, sw, ♦; **R** – 155a, 206, 218, 259.

## Lymnaeidae

*Lymnaea stagnalis* (Linnaeus, 1758); **D** – *68; **V** – 0-4; **Z** – (h); **E** – L, 7‰, bt, ph, pe, ♦; **R** – 118, 305, 389.

*Stagnicola corvus* (Gmelin, 1791) [*S. palustris* var. *corvus*, *Lymnaea*, *Galba*]; **D** – 58, *68, *69; **V** – 0-50; **Z** – (hop, ? e); **R** – 118, 206, 305, 389.

*Radix auricularia* (Linnaeus, 1758) [*Lymnaea*]; **D** – 58, *68, *69, 71; **V** – 0.2-25; **Z** – (h, ? hop); **E** – L, 6‰, bt, ph, rh, s, ♦; **R** – 84, 118, 374, 389.

*Radix balthica* (Linnaeus, 1758) [*R. ovata*, *Lymnaea*]; **D** – 59, 60, 80, 96; **Z** – (hop); **E** – L, 3-10‰, bt, ph, eu, ♦; **R** – 206, 218.

*Galba truncatula* (O. F. Müller, 1774) [*Lymnaea*]; **D** – *68, *69; **V** – -6; **Z** – (h); **E** – L, bt, eu, pe, ph, ar; **R** – 305, 389.

## Planorbidae

*Ferrissia fragilis* (Tryon, 1862) [*F. wautieri*, *F. clessiniana*]; **D** – 68, 83; **V** – 0-8; **Z** – (h, sk, i); **E** – L, bt, eu, th, ph, r, is, ♦; **R** – 6, 118.

*Planorbis carinatus* O. F. Müller, 1774; **D** – 77; **V** – -10, -18; **Z** – (wes, ? h); **E** – L, bt, sw, ph, pe, r, ♦; **R** – 257, 258.

*Planorbis planorbis* (Linnaeus, 1758) [*Tropidiscus umbilicatus*]; **D** – 7, 58, 59, *68, *69, 70, 71, 76, 77; **Z** – (h); **E** – L, bt, 2‰, sw, ph, pe, ♦; **R** – 6, 84, 159a, 258, 389.

*Anisus septemgyratus* (Rossmaessler, 1835) [*Paraspira*]; **D** – *68, *69, 71 ; **Z** – (wes, ? e) ; **E** – L, 8‰, sw, ph, α-β, r, ♦; **R** – 6, 305, 389.

*Anisus vortex* (Linnaeus, 1758) [*Planorbis*, *Spiralina*]; **D** – *68; **Z** – (wces); **E** – L, bt, 8‰, ph, α-β, r, ♦; **R** – 6, 374, 389.

*Anisus vorticulus* (Troschel, 1834); **D** – 60, 83; **Z** – (wces, ? wes); **E** – L, bt, pe, ph, rh, sw, r, ♦, HD; **R** – 6, 118.

*Gyraulus crista* (Linnaeus, 1758) [*Armiger*, *Planorbis*]; **D** – *68; **Z** – (h); **E** – L, 1.5‰, eu, ph, α-β, ♦; **R** – 6, 118, 374, 389.

*Hippeutis complanatus* (Linnaeus, 1758) [*Segmentina*]; **D** – 59, 60; **Z** – (wces, ? wcp); **E** – L, bt, ph, s-ar, sw, α, r, ♦; **R** – 6, 118.

*Segmentina nitida* (O. F. Müller, 1774); **D** – *68, *69, 83; **Z** – (wcp); **E** – L, bt, sw, ph, α-β, ♦; **R** – 6, 118, 305, 389.

*Planorbarius corneus* (Linnaeus, 1758) [*Coretus corneus* var. *ammonoceras*]; **D** – *68, *69, 71, 77; **V** – -9; **Z** – (wces); **E** – L, 5‰, sw, po, α-β, ♦; **R** – 6, 305, 389.

## Physidae

*Physella acuta* (Draparnaud, 1805); **D** – 58, 75, 76, 77, 78, 87, 90; **Z** – (na, sk, i); **E** – L, bt, eu, pe, ph, po, sw, tx, α-β, is, ♦; **R** – 6, 118, 159a, 257, 259, 353a.

*Physa fontinalis* (Linnaeus, 1758); **D** – *68, *69; **Z** – (tp, ? h); **E** – L, bt, sw, po, cr, ph, ps-pe, β, r, ♦; **R** – 6, 84, 389.

## BIVALVIA

### ARCOIDA

## Noetiidae

*Striarca lactea* (Linnaeus, 1758) [*Arca, Arcopsis, Galactella*]; **D** – 7, 16, 24; **V** – -20; **Z** – eami, +; **E** – M, bt, ep, lt, r; **R** – 84, 155c, 389.

## Arcidae

*Anadara kagoshimensis* (Tokunaga, 1906) [*A. inaequivalvis, Cunearca cornea, Scapharca inaequivalvis, Anadara* sp.]; **D** – 3, 6, 7, 12, 13, 25, 26, 69, 76, 77; **V** – -25, -40; **Z** – miwp, i; **E** – M, bt, ep, ps, s, sg, is; **R** – 155a, 155c, 159a, 183, 249, 255, 288, 360a, 408a.

### MYTILOIDA

## Mytilidae

*Mytilaster lineatus* (Gmelin, 1791) [*M. lineatus pontica, M. minimus, M. monterosatoi*]; **D** – 1, 2, 3, 4, 5, 6, 7, 8, 9, 10, 11, 12, 13, 14, 15, 16, 17, 18, 19, 20, 21, 22, 24, 68, 69, 76, 77; **V** – 0-30, 50; **Z** – lm; **E** – M, bt, 5‰, sep, lt-slc; **R** – 61, 84, 91, 155a, 155c, 159a, 249, 295.

*Mytilus galloprovincialis* Lamarck, 1819 [*M. edulis* var. *galloprovincialis*] ; **D** – 1, 2, 3, 4, 5, 6, 7, 8, 9, 10, 11, 12, 13, 14, 15, 16, 17, 18, 19, 20, 21, 22, 24, 27, 68, 69, 76, 77, 155c; **V** – 0-80; **Z** – eamp; **E** – M, bt, eb, lt, s, s-ps; **R** – 84, 91, 155a, 159a, 389.

*Gibbomodiola adriatica* (Lamarck, 1819) [*Modiola, Modiolus*]; **D** – 7, 11, 13, 16, 25, 26; **V** – 20-36, 75; **Z** – lm, ? clm; **E** – M, bt, mb, pe, s-ps; **R** – 91, 155c, 249, 389.

*Modiolula phaseolina* (Philippi, 1844) [*Modiola, Modiolus*]; **D** – 1, 2, 3, 4, 5, 6, 7, 8, 9, 10, 11, 12, 13, 14, 15, 16, 17, 18, 19, 20, 21, 22, 28; **V** – 30-140; **Z** – clmm; **E** – M, bt, shb, pe, phs; **R** – 249, 389, 410.

*Arcuatula senhousia* (W. H. Benson, 1842); D – 12, 13, 16; V – 11-12, 0-7; Z – iwp, sk i; E – M, bt, ep, sg; R – 88a (Todorova in Chartosia et al. 2018), 155c, 270f, 360a.

Ostreoida

## Pectinidae

*Flexopecten glaber* (Linnaeus, 1758) [*Chlamys, Pecten glaber* var. *pontica*]; **D** – 1, 2, 3, 4, 5, 6, 7, 8, 9, 10, 11, 12, 13, 14, 15, 16, 17, 18, 19, 20, 21, 22, 25, 76, 77; **V** – 3-40, 70; **Z** – lm; **E** – M, bt, sep, ps, ps-s; **R** – 84, 91, 155a, 155c, 159a, 249, 355f, 389.

? *Pecten jacobaeus* (Linnaeus, 1758); **D** – 7; **Z** – lmmg; **E** – M, bt; **R** – 84, 389.

? *Pecten maximus* (Linnaeus, 1758); **D** – 11; **Z** – eam, ? clm; **E** – M, bt; **R** – 84, 389.

Shells of *P. jacobaeus* and *P. maximus* have been established by Caspers (1951). These species are not found by other authors in the Black Sea (Valkanov 1957a; Kaneva-Abadjieva 1960; Skarlato & Starobogatov 1972). Arbelieved are have been thrown from the ships.

## Anomiidae

*Anomia ephippium* Linnaeus, 1758; **D** – 7; **V** – -30; **Z** – ami; **E** – M, bt, ep, ro; **R** – 84, 155c, 389.

171

## Ostreidae

*Ostrea edulis* Linnaeus, 1758 [*O. lamellosa, O. sublamellosa, O. taurica*]; **D** – 3, 5, 7, 12, 16, 17, 24, 76, 77; **V** – 7-23, 65; **Z** – anamnep, i; **E** – M, bt, sep, ro, ◼; **R** – 67, 84, 91, 114, 155a, 155c, 159a, 249.

*Magallana gigas* (Thunberg, 1793) [*Crassostrea gigas* (Thunberg, 1793); *Ostrea gigas* Thunberg, 1793]; **D** – 4, 7, 13, 15, 17, 19; **V** – 0-7; **Z** – nwp, sk, i; **E** – M, B, bt, ep, ro; **R** – 43a, 155c, 270g, 360a.

UNIONOIDA

## Unionidae

*Unio pictorum* (Linnaeus, 1758) [*U. pictorum gaudioni, U. gemtilis*]; **D** – *68, *69, 92; **V** – -11; **Z** – (e, ? ena); **E** – L, bt, 2‰, pe-ar-ps, β; **R** – 6, 84, 305, 374, 389.

*Unio tumidus* Retzius, 1788; **D** – *69; **V** – -9; **Z** – (e, ? wes); **E** – L, bt, pe-ar, β; **R** – 6, 84, 389.

*Anodonta cygnea* (Linnaeus, 1758) [*A. cygnea piscinalis, A. piscinalis*]; **D** – *68, 78, 92; **V** – -17; **Z** – (e); **E** – L, bt, 2‰, pe-ps, β; **R** – 6, 84, 374, 389.

*Pseudanodonta complanata* (Rossmaessler, 1835) [*Anodonta*]; **D** – 92; **V** – -9; **Z** – (e); **E** - L, bt, ps-pe, β; **R** – 6.

VENEROIDA

## Lucinidae

*Lucinella divaricata* (Linnaeus, 1758) [*Divaricella, Lucina commutata, Tellina*]; **D** – 1, 2, 3, 4, 5, 6, 7, 8, 9, 10, 11, 12, 13, 14, 15, 16, 17, 18, 19, 20, 21, 22, 25, 26, 69; **V** – -30; **Z** – clm, ? lm; **E** – M, bt, ep, ps, ps-s; **R** – 84, 155c, 249, 389.

*Loripes lacteus* (Linnaeus, 1758; Poli, 1791) [*L. lucinalis, Lucina leucoma, Tellina*]; **D** – 3, 7, 11, 16, 17, 20, 25, 76; **V** – -25; **Z** – lmm, ? clmm; **E** – M, bt, sep, ps, zc; **R** – 67, 84, 155c, 159a, 249, 389.

## Leptonidae

*Hemilepton nitidum* (Turton, 1822) [*Kellia compressa, Erycina, Lepton*];
**D** – 25, *69, 84; **V** – 0-25, 34; **Z** – clm; **E** – M, bt, ps, ep-mb; **R** – 155c, 164, 403.

? *Kellia suborbicularis* (Montagu, 1803) [? *Hemilepton nitidum*]; **D** – 69;
**V** – 2; **Z** – namnep; **E** – M, bt, eb, ps; **R** – 155c, 356.

## Montacutidae

*Kurtiella bidentata* (Montagu, 1803) [*Mysella*]; **D** – 16, 17; **V** – -50; **Z** –
eam; **E** – M, bt, sg, ro; **R** – 155c, 403.

## Cardiidae

*Acanthocardia paucicostata* (G. B. Sowerby, 1834) [*Cardium paucicostatum* var. *impedita*]; **D** – 3, 7, 13, 16, 17, 25, 26, 27, 69; **V** – 15-100; **Z** – lm; **E** – M, bt, mb-eb, pe, sg; **R** – 67, 84, 91, 155c, 249, 389.

*Cerastoderma glaucum* (Poiret, 1789) [*Cardium edule, C. clodiense, C. lamarcki*]; **D** – 1, 2, 3, 4, 5, 6, 7, 8, 9, 10, 11, 12, 13, 14, 15, 16, 17, 18, 19, 20, 21, 22, 25, 26, 27, 68, *69, 76, 77, *78; **V** – -80; **Z** – clm; **E** – M, bt, eh-3.9‰, eb, ps, ps-s, pe; **R** – 84, 91, 155c, 159a, 249, 295, 374, 389.

*Parvicardium exiguum* (Gmelin, 1791) [*Cardium*]; **D** – 3, 7, 13, 16, 25, 27, 69; **V** – 10-120; **Z** – clm; **E** – M, bt, eb, ps, ps-s, pe; **R** – 67, 84, 91, 155c, 249, 389.

*Papillicardium papillosum* (Poli, 1791) [*Cardium fasciatum, C. simile*]; **D** – 3, 7, 17, 25, 27, 28; **V** – -30, 100; **Z** – lm; **E** – M, bt, eb, pe, pe-ps; **R** – 67, 84, 249, 389.

*Monodacna colorata* (Eichwald, 1829) [*Didacna, Hypanis*]; **D** – 3, 7, 25, 26, 27, 28; **V** – 13, -148; **Z** – pc, Rc, Sf, +; **E** – M-B, bt, ps-pe, s, ep; **R** – 66, 68, 155c, 249, 389.

*Hypanis plicatum* (Eichwald, 1829) [*Adacna relicta, H. plicatum relicta*];
**D** – 3, 27, 28; **V** – 35-90; **Z** – pc, Rc, Sf, +; **E** – M-B, bt, pe, ar, s-ps; **R** – 68, 155c, 172, 249, 389.

# Mactridae

*Spisula subtruncata* (da Costa, 1778) [*Mactra subtruncata triangula, M. triangula*]; **D** – 1, 2, 3, 4, 5, 6, 7, 8, 9, 10, 11, 12, 13, 14, 15, 16, 17, 18, 19, 20, 21, 22, 25, 26, 27, 28, 69, 76; **V** – 5-140; **Z** – clmm; **E** – M, bt, eb, pe, ps-s, sg; **R** – 67, 84, 155c, 159a, 249, 259a, 389.

# Mesodesmatidae

*Donacilla cornea* (Poli, 1795) [*Mesodesma*]; **D** – 1, 2, 3, 4, 5, 6, 7, 8, 9, 10, 11, 12, 13, 14, 15, 16, 17, 18, 19, 20, 21, 22, 51, 76; **V** – 0.2, -2; **Z** – lm; **E** – M, bt, sep, ps, ■; **R** – 114, 155c, 159a, 166, 249, 389.

# Solenidae

*Solen marginatus* Pulteney, 1799 [*S. vagina*]; **D** – 1, 2, 3, 4, 5, 6, 7, 8, 9, 10, 11, 12, 13, 14, 15, 16, 17, 18, 19, 20, 21, 22, 25, 76; **V** – 0-10; **Z** – clmm; **E** – M, bt, sep, ps, ■; **R** – 84, 91, 114, 155c, 159a, 249.

# Tellinidae

*Gastrana fragilis* (Linnaeus, 1758) [*Psammobia, Tellina*]; **D** – 1, 2, 3, 4, 5, 6, 7, 8, 9, 10, 11, 12, 13, 14, 15, 16, 17, 18, 19, 20, 21, 22, 25; **V** – -36; **Z** – clm; **E** – M, bt, ep, ar, ar-ps, sg; **R** – 84, 155c, 249, 389.

*Tellina donacina* Linnaeus, 1758 [*Angulus, Moerella*]; **D** – 1, 2, 3, 4, 5, 6, 7, 8, 9, 10, 11, 12, 13, 14, 15, 16, 17, 18, 19, 20, 21, 22, 25, 26; **V** – -20, 30; **Z** – clmm; **E** – M, bt, sep, ps-pe; **R** – 84, 91, 155c, 249, 389.

*Tellina fabula* Gmelin, 1791 [*Angulus, Fabulina, Moerella*]; **D** – 7, 25; **V** – -10, 40; **Z** – clm; **E** – M, bt, ep, ps, ps-s; **R** – 84, 155c, 249, 389.

*Tellina tenuis* da Costa, 1778 [*T. exigua, Angulus exiguus, Moerella*]; **D** – 1, 2, 3, 4, 5, 6, 7, 8, 9, 10, 11, 12, 13, 14, 15, 16, 17, 18, 19, 20, 21, 22, 25, 26, 77, *79; **V** – 0.2-24; **Z** – clm; **E** – M, bt, sep, ps, ps-s; **R** – 84, 91, 155c, 249, 389.

# Donacidae

*Donax trunculus* Linnaeus, 1758 [*D. trunculus julianae, D. julianae*]; **D** – 1, 2, 3, 4, 5, 6, 7, 8, 9, 10, 11, 12, 13, 14, 15, 16, 17, 18, 19, 20, 21, 22, 25, 76; **V** – 2-15; **Z** – clm; **E** – M, bt, sep, ps; **R** – 84, 91, 155c, 159a, 249, 295.

*Donax semistriatus* Poli, 1795 [*D. fabagella, D. venustus*]; **D** – 16, 25; **V** – 10-15, 20; **Z** – lm; **E** – M, bt, ep, ps; **R** – 84, 155c, 249, 389.

## Semelidae

*Abra alba* (W. Wood, 1802) [*A. alba pontica, Syndesmya*]; **D** – 1, 2, 3, 4, 5, 6, 7, 8, 9, 10, 11, 12, 13, 14, 15, 16, 17, 18, 19, 20, 21, 22, 26, 27, 28, 69; **V** – - 135, 170; **Z** – clmm; **E** – M, bt, eb, pe, phs; **R** – 84, 155c, 249, 389.

*Abra prismatica* (Montagu, 1808) [*A. fragilis, A. milaschevichi, A. nitida*]; **D** – 1, 2, 3, 4, 5, 6, 7, 8, 9, 10, 11, 12, 13, 14, 15, 16, 17, 18, 19, 20, 21, 22, 26, 27, 69; **V** – 5-90; **Z** – clm, ? clmm; **E** – M, bt, eb, pe, s-ps; **R** – 67, 84, 155c, 249, 389.

*Abra segmentum* (Récluz, 1843) [*A. ovata, Syndesmya*]; **D** – 7, 13, 23, 26, *66, 68, *69, 76, 77, *79, 85, *86; **V** – -15, 40; **Z** – ml; **E** – M, bt, eh-6-60‰, ep, et, pe, ps-s; **R** – 84, 91, 155c, 159a, 249, 374, 389.

## Dreissenidae

*Dreissena polymorpha* (Pallas, 1771) [*D. (Dreissena) polymorpha polymorpha*]; **D** – 58, 59, 60, *68, *69, 71, 76, 92, 96; **V** – -20; **Z** – (h - i), Rc; **E** – B-L, bt, eh, is; **R** – 84, 91, 152, 249, 305, 374, 375, 389.

*Dreissena grimmi* Andrusov, 1890 [*Dreissena rostriformis, D. (Pontodreissena) rostriformis distincta*]; **D** – 1, 2, 3, 4, 5, 6, 7, 8, 9, 10, 11, 12, 13, 14, 15, 16, 17, 18, 19, 20, 21, 22, 27, 28; **V** – 36-148, 50-200; **Z** – pc, Rc, Sf, +; **E** – B, bt, mb-eb; **R** – 155c, 166, 172, 178, 249.

## Pisidiidae

*Sphaerium corneum* (Linnaeus, 1758); **D** – *67; **Z** – (h); **E** – L, bt, 5‰, sw, ph, s; **R** – 6, 305, 389.

*Musculium lacustre* (O. F. Müller, 1774); **D** – 42, 43, *67, 92; **Z** – (tp); **E** – L, bt, 3‰, sw, ph, s; **R** – 4, 6, 305, 389.

## Veneridae

*Gouldia minima* (Montagu, 1803) [*Circe, Venus*]; **D** – 5, 16, 25, 26, 27; **V** – -25, 70; **Z** – clm; **E** – M, bt, eb, ps-pe, mc; **R** – 67, 91, 155c, 249, 389.

*Pitar rudis* (Poli, 1795) [*Meretrix rudis* var. *ochropicta, Cytherea, Venus ochropicta*]; **D** – 3, 7, 13, 16, 17, 25, 26, 27; **V** – -40, 200; **Z** – lmmg; **E** – M, bt, mb, ps-pe, sg; **R** – 67, 84, 91, 155c, 249, 389.

*Irus irus* (Linnaeus, 1758) [*Venerupis*]; **D** – 3, 7, 24; **V** – 0-10; **Z** – eamiswp; **E** – M, bt, sep, lt, slr; **R** – 67, 84, 155c, 249, 389.

*Chamelea gallina* (Linnaeus, 1758) [*Chione gallina corrugata, Venus*]; **D** – 1, 2, 3, 4, 5, 6, 7, 8, 9, 10, 11, 12, 13, 14, 15, 16, 17, 18, 19, 20, 21, 22, 25, 69, 76; **V** – -30, 90; **Z** – clm, ce; **E** – M, bt, mb, ps, s-ps, sg; **R** – 84, 91, 155c, 159a, 249, 389.

*Polititapes aureus* (Gmelin, 1791) [*Paphia aurea, Polititapes petalina, Venus petalina , Tapes aureus var. rugata* Bucquoy, *T. rugatus, T. discrepans, T. lineatus, T. proclivis*]; **D** – 1, 2, 3, 4, 5, 6, 7, 8, 9, 10, 11, 12, 13, 14, 15, 16, 17, 18, 19, 20, 21, 22, 25, 26, 27, 76; **V** – 2-65; **Z** – clm; **E** – M, bt, mb, ps-pe, s; **R** – 67, 84, 91, 155c, 159a, 249, 389.

## Petricolidae

*Petricola lithophaga* (Philippson, 1788); **D** – 7, 11, 12, 13, 24; **V** – -10, 26; **Z** – lmi; **E** – M, bt, sep, lt; **R** – 84, 91, 155c, 249, 389.

MYOIDA

## Myidae

*Mya arenaria* Linnaeus, 1758 [*M. truncata*]; **D** – 1, 2, 3, 4, 5, 6, 7, 8, 9, 10, 11, 12, 13, 14, 15, 16, 17, 18, 19, 20, 21, 22, 25, 26, 68, 69, 76, 77; **V** – 0.30-12; **Z** – cbm, i; **E** – M, bt, ep, ps, ps-s, is; **R** – 155a, 155c, 159a, 175, 249, 252, 253, 360a.

## Corbulidae

*Lentidium mediterraneum* (O. G. Costa, 1829) [*Corbulomya maeotica, Corbula*]; **D** – 7, 25, 69; **V** – 7, 25, 69; **Z** – m, ? lm; **E** – M, bt, sep, ps, ps-s; **R** – 84, 155c, 249, 389.

## Pholadidae

*Pholas dactylus* Linnaeus, 1758; **D** – 5, 7, 10, 11, 12, 13, 24, 76; **V** – -15; **Z** – eamrs, amrs; **E** – M, bt, sep, lt, BA, BC; **R** – 84, 91, 155c, 159a, 249, 389.

*Barnea candida* (Linnaeus, 1758) [*Pholas*]; **D** – 7, 24; **V** – -15; **Z** – namnei, ami; **E** – M, bt, ep, lt, eh; **R** – 84, 155c, 249, 389.

## Teredinidae

*Teredo navalis* Linnaeus, 1758; **D** – 1, 2, 3, 4, 5, 6, 7, 8, 9, 10, 11, 12, 13, 14, 15, 16, 17, 18, 19, 20, 21, 22, 69; **V** – -70; **Z** – amip, K, i; **E** – M, bt, ep-me, c, is; **R** – 84, 91, 155c, 249, 360a, 374.

*Nototeredo norvagica* (Spengler, 1792) [*Teredo utriculus*]; **D** – 10, 17; **V** – -70; **Z** – bam; **E** – M, bt, ep-me; **R** – 75, 155c, 252.

ANOMALODESMATA (PHOLADOMYOIDA)

## Thraciidae

*Thracia papyracea* (Poli, 1791); **D** – 7, 25; **V** – 0.2-22; **Z** – clm, ? nam; **E** – M, bt, ep, ps, ps-pe; **R** – 84, 155c, 249, 389.

# BRYOZOA

## GYMNOLAEMATA

### CHEILOSTOMATIDA

## Membraniporidae

*Membranipora tenuis* Desor, 1848; **D** – 7, *69; **Z** – nap; **E** – M, bt; **R** – 84, 249, 389.

*Conopeum reticulum* (Linnaeus, 1767) [*Membranipora denticulata*]; **D** – 3; **V** – -40; **Z** – ham; **E** – M, bt, eh, mb; **R** – 67, 249, 389.

*Conopeum seurati* (Canu, 1928); **D** – 84; **Z** – neamswp; **E** – M-B, bt, eh, 8-10‰; **R** – 141, 249, 252.

**Electridae (Membraniporidae)**

*Einhornia crustulenta* (Pallas, 1766) [*Electra, Membranipora*]; **D** – *69, 96; **Z** – abapbp; **E** – M-B, bt, eh, 3‰; **R** – 84, 249, 389.

*Electra monostachys* (Busk, 1854); **D** – 13; **V** – 0-60; **Z** – ham; **E** – M, bt, ro, sg; **R** – 252, 330.

*Electra pilosa* (Linnaeus, 1767) [*Membranipora*]; **D** – 7; **V** – 10-50; **Z** – namswp; **E** – M, bt, mb; **R** – 84, 249, 389.

*Electra pontica* Gruncharova, 1980; **D** – 2, 84; **V** – 0.2-0.7; **Z** – ● p, Er; **E** – M, bt, ep; **R** – 142, 252.

**Tendridae (Membraniporidae)**

*Tendra zostericola* Nordmann, 1839 [*Electra, Membranipora*]; **D** – 7, 11, 12, 13; **Z** – adp; **E** – M, bt; **R** – 67, 91, 249, 389.

**Cryptosulidae (Smittinidae)**

*Cryptosula pallasiana* (Moll, 1803) [*Lepralia*]; **D** – 3, 7, 11, 16, *69; **Z** – amp; **E** – M, bt, ep; **R** – 66, 67, 91, 249, 389.

**Bitectiporidae (Schizoporellidae)**

*Schizomavella linearis* (Hassall, 1841) [*Schizoporella*]; **D** – 7; **V** – -10; **Z** – clm, bamnep; **E** – M, bt, mb-hb; **R** – 84, 249, 389.

Ctenostomatida

**Vesiculariidae**

*? Bowerbankia caudata* (Hincks, 1877); **D** – *69, *79, 84, *86; **Z** – aamip; **E** – M, bt, ep, 6‰; **R** – 141, 249, 374, 384.

*Bowerbankia gracilis* Leidy, 1855 [? *B. caudata*]; **D** – *69, *79, 84, *86; **V** – -10; **Z** – amp; **E** – M, bt, ep, 6‰; **R** – 141, 249, 374, 384.

*Bowerbankia imbricata* (Adams, 1798); **D** – 84; **Z** – amp; **E** – M, bt, eh, 10‰; **R** – 141, 249, 252.

## Victorellidae

*Victorella pavida* Saville Kent, 1870; **D** – *68, 71, *78, *79, 84, *86, 88, 90, 93, 94; **V** – 5; **Z** – pc, amp, Rc; **E** – M, bt, ep, eh, 0.4-20‰; **R** – 141, 249, 374, 389.

Phylactolaemata

Plumatellida

### Fredericellidae

*Fredericella sultana* (Blumenbach, 1779); **D** – 84, 96; **Z** – (k); **E** – L, bt, 0-10‰; **R** – 141, 249, 252.

### Plumatellidae

*Plumatella casmiana* Oka, 1907; **D** – 84, 96; **Z** – (hptn, ? sk); **E** – L, bt, 3-6‰; **R** – 141, 249, 252.

*Plumatella emarginata* Allman, 1844; **D** – 84, 96; **Z** – (k); **E** – L, bt, 3-6‰; **R** – 141, 249, 252.

*Plumatella fungosa* (Pallas, 1768); **D** – 60, *68, *79, 84, *86, 88; **Z** – (h); **E** – L, bt, 1‰; **R** – 141, 249, 374, 389.

*Plumatella repens* (Linnaeus, 1758); **D** – 60, *68, 84, 92, 96; **Z** – (k); **E** – L, bt, 3-6‰; **R** – 141, 249, 374, 389.

## PHORONIDA

### Phoronidae

*Phoronis psammophila* Cori, 1889 [*Ph. euxinicola*]; **D** – 1, 2, 3, 4, 5, 6, 7, 8, 9, 10, 11, 12, 13, 14, 15, 16, 17, 18, 19, 20, 21, 22, 25, 26, 27, 28, 35, 51, 69; **V** – 0-100; **Z** – amip, ? K; **E** – M-B, bt, eb, ps-pe, s; **R** – 247, 249, 252.

# ENTOPROCTA (CAMPTOZOA)



COLONIALES (URNATELLIDA)

## Barentsiidae

*Barentsia benedeni* (Foettinger, 1886) [*Arthropodoria kowalewskii*]; **D** – *69, *79, 84, *86, 88; **Z** – bapp; **E** – M-B, bt, 8-16‰; **R** – 249, 374, 389.

# CHAETOGNATHA

SAGITTOIDEA

APHRAGMOPHORA

## Sagittidae

*Parasagitta setosa* (Müller, 1847) [? *Sagitta euxinica*]; **D** – 7, 11, 12, 17, 29, *69; **Z** – bam; **E** – M, p, 12‰; **R** – 91, 198, 319, 374.

# ECHINODERMATA

HOLOTHUROIDEA

DENDROCHIROTIDA

## Cucumariidae

*Stereoderma kirchsbergii* (Heller, 1868) [*Cucumaria*, ? *C. orientalis* – n. nudum]; **D** – ?; **V** – 30-70; **Z** – m, hom, ? lm; **E** – M, bt, hb, pe; **R** – 198, 249.

APODIDA

## Synaptidae

*Leptosynapta inhaerens* (O. F. Müller, 1776) [? *Synapta hispida* – n. nudum]; **D** – 1, 2, 3, 4, 5, 6, 7, 8, 9, 10, 11, 12, 13, 14, 15, 16, 17, 18, 19, 20, 21, 22, 27, 28; **V** – 60-100; **Z** – anamnp; **E** – M, bt, hb, pe; **R** – 223, 249, 392.

*Labidoplax digitata* (Montagu, 1815) [*Ostergrenia adratica*]; **D** – ?; **V** – -70; **Z** – neamal; **E** – M, bt, eb, ps, pe; **R** – 198, 249.

? *Labidoplax thompsoni* (Herapath, 1865) [*Ostergrenia*]; **D** – ?; **V** – -70; **Z** – lm; **E** – M, bt, eb, ps; **R** – 198, 249.

OPHIUROIDEA

OPHIURIDA

## Amphiuridae

? *Amphipholis squamata* (Delle Chiaje, 1828) [*Amphiura*]; **D** – 11, 13, 21, 22; **Z** – K; **E** – M, bt, eb; **R** – 91, 389, 410.

*Amphiura stepanovi* Djakonov, 1954; **D** – 1, 2, 3, 4, 5, 6, 7, 8, 9, 10, 11, 12, 13, 14, 15, 16, 17, 18, 19, 20, 21, 22, 27, 28; **V** – 20-175; **Z** – ● p; **E** – M, bt, eb-hb, phs; **R** – 198, 249.

## CHORDATA

ASCIDIACEA

STOLIDOBRANCHIA

## Molgulidae

*Eugyra adriatica* Drasche, 1884; **D** – 26, 27, 28; **V** – 25-100; **Z** – m, hom; **E** – M, bt, hb, pe; **R** – 198, 249.

*Molgula appendiculata* Heller, 1877 [*Ctenicella*]; **D** – 7, 25, 26, 27, 28; **V** – 20-120; **Z** – lm; **E** – M, bt, hb, pe; **R** – 198, 249, 358.

*Molgula euprocta* Drasche, 1884 [*M. impura, M. ompura* ?]; **D** – 5, 7, 69, *79; **Z** – lm; **E** – M, bt, eh; **R** – 67, 84, 249, 374.

## Styelidae (Botryllidae)

*Botryllus schlosseri* (Pallas, 1766); **D** – 1, 2, 3, 4, 5, 6, 7, 8, 9, 10, 11, 12, 13, 14, 15, 16, 17, 18, 19, 20, 21, 22, 24; **V** – 0-60; **Z** - namp, ? SK; **E** - M, bt, ep-mb, slc, eu; **R** - 84, 91, 249, 295, 298, 374.

## Ascidiidae

*Ascidiella aspersa* (O. F. Müller, 1776); **D** – 3, 7; **V** – 20-50; **Z** – amiswp; **E** – M, bt, mb, sg, phc; **R** – 67, 249, 295, 298.

## Cionidae

*Ciona intestinalis* (Linnaeus, 1758); **D** – 2, 3; **V** – 15-100; **Z** – antamp; **E** – M, bt, mb-eb, phc; **R** – 67, 249, 389.

### LARVACEA (APPENDICULARIA)

#### COPELATA

## Oikopleuridae

*Oikopleura dioica* Fol, 1872 [*O. cophocerca*]; **D** – 7, 11, 13, 29, 69; **Z** – amip, ? SK; **E** – M, p, et, eh; **R** – 91, 198, 319, 389.

### LEPTOCARDII

#### AMPHIOXIFORMES

## Branchiostomidae

*Branchiostoma lanceolatum* (Pallas, 1774) [*Amphioxus*]; **D** – 6, 7, 8, 11, 13, 16, 17, 18, 19, 20, 21, 25; **V** – 17, 21; **Z** – amip; **E** – M, bt, ep, ps, ■; **R** – 84, 91, 249, 389.

Most marine invertebrates have been established throughout the Bulgarian Black Sea coast. Species, which are distributed either in the northern or in the southern part of the coast (with 1 to 3 localities), number 20-25%. Most of them are benthic forms that belong to poorly explored or rare taxa. A part of recently reported Black Sea endemics could be added to this group as well. In most cases, the species distribution is related to the level of study of the corresponding coastal region. It can be seen under juxtaposition of the found species in localities (Table 1). Three areas of good research are outlined (over 200 species established). Firstly are the vicinities of Varna (601 species) where the investigations of the Black Sea began a centure ago. The popular resort centers – Nesebar, Pomorie, Burgas and Sozopol (from 220 to 274 species) form an area of good research. Owing to Bulgarian and Romanian specialists, the region of Kaliakra Cape (230 species) is also well studied. Of coastal basins, the most studied (about 100 species recorded) are the lakes Durankulak, Ezerets-Shabla, Beloslav, Varna, Pomorie, Atanasovsko, Burgas, Mandra and the firth of Ropotamo River. The rich species composition of the lakes Varna and Beloslav is changed after the connection with sea (1909 and 1923) and their transformation into ports (after 1976). Fauna of the lakes Burgas (anthropogenic impact after 1960) and Mandra (dam since 1963) became considerably poor.

The Black Sea below the depths of 180-200 m is enriched with released $H_2S$, which makes the real deep-sea life impossible. The groups of **steno-** (epi- and hypo-), **meso-**, and **eurybathic** species are presented. The vertical distribution is analyzed for 800 species (75.9%) – marine and marine-brackish forms, according to available data (Table 3). Most species are found from 0 to 25 m on sand (396 species) and rocky (257 species) bottom.

The most numerous are the **stenoepibathic** species (465 species – 58.1%). The inhabitants of the supralittoral zone, shallow coastal zone (to 5-10 m), as well as species, which reach the depth of 15-30 m and approach to mesobathic forms, belong to this group. An intermediate niche, closer to mesobathic forms, is occupied by some representatives of the group, which reach higher depths. They are presented in 16 types, of which the most numerous are the representatives of Arthropoda, Annelida and Mollusca.

The group of **Eurybathic** (168 species – 21.0%) species includes Black Sea species, which are found in both little and great depths. Most eurybathic forms reach the depth of 130-150 m. They are presented in 12 types, of which the representatives of Arthropoda, Nematoda, Annelida and Mollusca predominate.

The **Mesobathic** (115 species – 14.4%) species reach the depth over 40 m in the Black Sea. Some species could be considered stenomesobathic forms. They are presented in 10 types, of which the representatives of Arthropoda and Mollusca prevail.

The **Stenohypobathic** species are the smallest group (52 species – 6.5%). They rarely could be found at smaller depth up to 25 m and usually reach highest density below 60-120 m. They are presented in 10 types, of which the representatives of Arthropoda and Nematoda predominate. Recently some species of Nematoda, Polychaeta and Harpacticoida have been estsblished in hypoxic habitats at depth below 200-250 m (Sergeeva & Zaika, 2013).

FORMATION OF THE BLACK SEA FAUNA is connected with the origin of the Black Sea basin itself. The Upper Miocene Sarmatian Sea (18-30‰, a descendant of Tethys) gave rise to the Pontian Sea-Lake, from which two separate basins were formed later, the Black Sea and the Caspian Sea. Initially, the Black Sea basin had been inhabited by fauna similar to the Caspian one [Chaudian Sea (12-14‰) and Paleoeuxinian Sea (6-8‰)]. Then, it had been connected with the Mediterranean Sea and became saline, so the Mediterranean fauna penetrated into it, whereas the Caspian fauna retreated to the brackish coastal parts [time of Uzunlar Sea (16‰) and Karangat Sea (22-30‰)]. Later, the connection with the Mediterranean Sea had been severed, and the brackish basin [the New Euxinian Sea (7‰)] originated, where the Mediterranean fauna disappeared. Recently, 7000-8000 years ago, this basin had been again connected with the Mediterranean Sea and its level increased. The marine fauna invaded it and the current Black Sea had been formed (Mischev & Popov, 1978; Shopov, 1993; Dimitrov et al., 1998; Evlogiev, 2009; Studencka & Jasionowski, 2011).

Then, before the last glaciation, a connection with the Caspian basin arose (via Manych channel), and Caspian interglacial immigrants invaded the Black Sea (Mordukhay-Boltovskoy, 1960; Nevesskaya, 1965; Starobogatov, 1970; Shopov, 1996). Most authors accept these species as Caspian relicts (known also as Sarmatian, Pontian, Pontian-Caspian, or autochthonous faunal elements). They are concentrated mainly in the coastal lakes-firths and the mouths of the Black Sea rivers and inhabit the freshwater and brackish basins. Part of them is subfossils for the sea itself. The Caspian relicts usually have Pontian or Pontian-Caspian ranges. Some of them have entered the river systems of Central and Western Europe (spread to other continents) where they are considered invasive species. According to Mordukhay-Boltovskoy (1960) the evolution of the Caspian fauna gave rise to the origin of eurybiontic oligohaline and freshwater forms, which began to acquire new habitats with their pervasion in Black Sea. The „relicts" *Dreissena polymorpha* Pallas and *D. bugensis* – one of the most invasive recent mollusk, are a typical example. Recent data for the distribution of many relict taxa (mainly in the latest electronic editions) contradict their relict nature. It has been established that these taxa are widespread outside the Pontian-Caspian region. These may be invasive relict forms (a small number of species) or species with uncleared distribution, accepted as relicts. The main portion of the Caspian relicts (41 species, or 3.9%) is benthic brackish forms (38 species, or 92.7%).

THE MARINE AND MARINE-BRACKISH FORMS are divided into 162 zoogeographical (areographical) categories (Table 4), combined into 4 main groups and 16 subgroups (Hubenov 2014, 2015).

The main portion of the Black Sea fauna (740 species, or 70.2 %) has an Atlantic–Mediterranean origin and represents the impoverished Atlantic-Mediterranean fauna. As this fauna was becoming impoverished, the stenobiotic Lusitanian-Mediterranean species were eliminated, so this category is defined by the eurybiontic forms, often distributed along the European coast up to Scotland, North Sea and Scandinavia. Thus an impression is created of the atlantization of this fauna, manifested differently in the various taxonomic groups, benthic and planktonic forms. The atlantization is poorly presented in the planktonic forms (50 species, or 42.4%) and most presented in the benthic forms (712 species, or 72.8%). There is no a considerable difference (in percentages) in the atlantization of the marine (716 species, or 69.9%) and the brackish (101 species, or 65.6%) species.

A portion of the Arctic- and Antarctic-Atlantic (high-latitude boreal and antiboreal) species (43 species, or 4.1%) is not presented in the Mediterranean Sea. Its percentage varies slightly as their number is insignificant in the brackish and planktonic forms. The Arctic-North-Atlantic-, Arctic-Boreal-Atlantic-Mediterranean and circumeuropean species predominate (total of 26 species, or 2.5%). Not all of the Holatlantic and North Atlantic species (99 species, or 9.4%) are presented in Mediterranean Sea. The marine and benthic forms prevail, of which the North-Atlantic-, Boreal-Atlantic- and Holatlantic-Mediterranean and Boreal Atlantic-Pontian species (total of 76 species, or 7.2%) predominate. Between the tropical- and subtropical Atlantic species (109 species, or 10.3%), the Lusitanian-Mediterranean forms (79 species, or 7.5%) predominate. The East and Northeast Atlantic species are most numerous (251 species, or 23.8%). The main portion of them is the Celtic-Lusitanian-Mediterranean (148 species, or 14.1%) and Celtic-Pontian (44 species, or 4.2%) forms. The Mediterranean species (90 species, or 8.5%) are poorly presented in the planktonic and brackish communities. Almost all Pontian-Caspian species (28 species, or 2.7%) are benthos and brackish (27 species, or 17.5%) forms.

The **PONTIAN SPECIES** (Black Sea endemics) are 118 (11.2%). The benthic (115 species, or 97.5%) and marine (114 species, or 96.6%) forms predominate. The brackish species (11 species, or 9.3%) most often are Caspian relicts. Some of the Pontian species is likely to be found in other seas under better research. This refers mainly to the groups of Nematoda, Ostracoda and Copepoda which are well-studied in the Black Sea. The most Black Sea endemic species are concentrated in several groups – Porifera, Nemertini, Nematoda, Rotifera, Ostracoda and Copepoda. Many endemic forms, known from previous data, are brought to synonyms or downgraded to subspecies today. Thus there are no data on the Black Sea endemic species in the recent malacological literature, of which there were more in the old literature [Kaneva-Abadjieva (1960a) recorded 24 species endemic mollusks from the Bulgarian Black Sea coast]. Many species have changed after their penetration into the Black Sea; they are described as new taxa. An interesting example is the hermit-crab *Clibanarius erythropus*. Because of the lack of large mollusks in the Black Sea, it was forced to use the shells of snails of the genus *Gibbula* and greatly reduces its size (became known as *C. misanthropus*). With the appearance of *Rapana venossa* in Black Sea, the crab begins to use large shells and again reaches the normal size for crabs in Mediterranean Sea.

The number of forms with Cosmopolitan (121 species – 11.5%), Atlantic-Pacific (120 species – 11.4%) and Atlantic-Indian (72 species – 6.8%) type of distribution is considerably smaller than the forms of Atlantic type. They predominate as species composition in the marine and benthic species but their percentage is highest in the plankton forms (from 8.5 to 28.0%). The differences in the percentages are not big for the benthic, marine and brackish forms (from 5 to 11%) with the exception of the brackish Atlantic-Pacific species (18.2%). Most Cosmopolitan forms (2/3) have Atlantic-Indian-Pacific distribution. The Holatlantic- and East Atlantic-Indian-Pacific species (58 species, or 5.5%) predominate, of which the Atlantic-Mediterranean-Indo-Pacific forms are the most numerous (37 species). The number of North Atlantic-Indian-Pacific (15 species) and the tropical and subtropical Atlantic-Indian-Pacific (6 species) species is smaller. About 1/3 of the Cosmopolitan forms have Arctic-Antarctic-

Atlantic-Indian-Pacific distribution as the typical Cosmopolitans and Subcosmopolitans (a total of 25 species, or 2.4%) predominate. About 1/4 of the species with Atlantic-Pacific type of distribution belong to Arctic-Antarctic-Atlantic-Pacific forms (23 species – 2.2%). The Atlantic-Pacific species (97 species, or 9.2%) predominate, the main parts of which are with North and South Atlantic-, Hol- and North Atlantic- and North Atlantic-Pacific distribution (a total of 73 species, or 6.9%). The Atlantic-Mediterranean-Pacific species (14 species) are the most numerous. A small number of  Hol- and South Atlantic-, East and West Atlantic- and Tropical and Subtropical Atlantic-Pacific species (a total of 24 species, or 2.3%) occur as well. Most forms with Atlantic-Indian type of distribution include Hol- and North Atlantic-Indian and East and Northeast Atlantic-Indian species (a total of 61 species, or 5.8%). The North Atlantic-Mediterranean-Indian and Celtic-Lusitanian-Mediterranean-Indian species (7 species each) are the most numerous. Seven tropical and subtropical Atlantic-Indian and 4 Arctic-Antarctic-Atlantic-Indian forms occur as well.

THE FRESHWATER-BRACKISH, FRESHWATER AND TERRESTRIAL FORMS, connected with water, recorded from the Bulgarian Black Sea coast, are divided into 80 zoogeographical categories (Table 5), combined into 2 main groups and 5 subgroups (Hubenov 2014, 2015).

**Species distributed in Palaearctic and beyond it**. This group (296 species, or 58.3%) includes 36 zoogeographical categories, of which 29 combine species of **Northern type** (widely distributed in Holarctic or Palaearctic) and 7 – species of **Southern type** (distributed only in southern Palaearctic. This group is important for the zoogeography of the coastal fauna because of its great species diversity. It is connected with the typical for the sea coasts natural habitats, optimum for the development of its representatives and is poorly presented in the interior. The difference among the brackish, freshwater and terrestrial forms is from 8.7% to 37.0% (from 74 to 271 species). The species of northern type have vast areas and ecological flexibility. The Cosmopolitan, Subcosmopolitan and Holarctic species (a total of 184 species, or 36.2%) are the most numerous. These species, except the Holarctic forms, are almost not presented in the terrestrial forms. In the brackish communities, the Holarctic species are poorly represented (17 species). The species of southern type are best represented in the terrestrial forms (11 species, or 5.5%). The presence of this group in different taxa depends on whether they include highly mobile and widely distributed forms or combine less mobile and more closely connected with certain conditions species. In the latter case, more important are the specific natural habitats to which species are adapted.

**Species distributed only in Palaearctic but in more than one subregion (Palaearctic type)**. A total of 79 species (15.5%) that belong to this group, combined into 11 zoogeographical categories, has been established along the coast. The group includes from 8.1% to 25.9% (from 14 to 65 species) of the brackish, freshwater and terrestrial forms. Its character is determined by Transpalaearctic, West Palaearctic, West and Central Palaearctic and Holopalaearctic species (a total of 63 species, or 12.4%) that are the most numerous. This correlation remains almost the same and varies from 1.9% to 8.5% (from 4 to 18 species) in the freshwater and terrestrial forms. In brackish forms the group is poorly presented – from 0.6% to 2.3% (from 1 to 4 species). The number of the European-North African species (from 3 to 5 species in the group) varies insignificantly. Two species have a longitudinal disjunction of their areas that includes parts of Siberia and Central Asia.

**Species distributed within one subregion of Palaearctic**. This group (126 species, or 24.8%) includes species with **Eurosiberian** and **Mediterranean type** of distribution. The Mediterranean-Central Asian species are also included here according to many authors who combine Mediterranean and Central Asian subregions. The species with Mediterranean type of distribution are accepted in a general way and include elements (Submediterranean, Subiranian, and Pontian), that could be considered separately as well (Gruev & Kusmanov, 1994, 1999; Gruev 1995, 2000). The **Eurosiberian species** are 55 (10.8%), of which the European species (31 species, or 6.1%) are the most numerous. They are combined into 9 zoogeographical categories and include from 6.9% to 11.5% (from 12 to 48 species) of the brackish, freshwater and terrestrial forms. The Eurosiberian species are best represented in freshwater forms and poorly represented in brackish forms. Thirty-six species have European distribution only, of which 31 are wide-spread in Europe and 5 – in its separate regions (Central and South Europe). The **Mediterranean species** are 71 (13.9%), of which the Holomediterranean species (19 species – 3.7%) are most numerous. This group combines 24 zoogeographical categories with different origin, distribution and ecological peculiarities. It includes from 7.7% to 28.4% (19-57 species) of the brackish, freshwater and terrestrial forms. The group is best represented in terrestrial forms and poorly represented in freshwater forms. The endemics (11 species – 2.2%) are poorly represented – their number varies from 4 to 7 species. More are the regional endemics that have been found in the terrestrial forms. The specific conditions along the coast do not favor the formation of endemic taxa, which mostly are newly described forms or rare species with unclear distribution. The only local endemic (*Hauffenia lucidula*) is a crenobiontic species of the family Hydrobiidae – 95.5% of the species of this group are freshwater endemic forms.

**ZOOPLANKTON**. It includes representatives of Protozoa, Coelenterata, Ctenophora, Rotifera, Annelida (larvae), Arthropoda, Mollusca (larvae) and Chordata (Valkanov, Dimov & Naidenov, 1978; Konsulov, 1991, 1998; Konsulov & Konsulova, 1993, 1998). In regard to species diversity, the Black Sea zooplankton is characterized as poor (in comparison with the Mediterranean Sea). According to thermal conditions in the water mass the zooplankton is divided to thermophilic, eurythermic and cryophilic. The thermophilic representatives are found mainly in surface water layers to 25 m. The cryophilic and eurythermic forms dominate below the zone of thermowedge. The vertical distribution depends on the oxygen penetrating in the depth (100-175 m). The zooplankters have a specific distribution according to seasons and depth, which is determined by their temperature and trophic requirements. The eurythermic forms occur throughout the year as they are not influenced by temperature. The cryophilic forms occur in all depths in winter and fall below 50-60 m in summer. The thermophilic forms are found mainly in summer and disappear in winter. The bottom larvaton (meroplankton) is of great importance for species diversity of the benthic fauna. It is represented by Mollusca – veliger, Cyripedia – nauplius and cypris, Polychaeta – larvae, Pisces ova and larvae. The average annual biomass of the zooplankton in front of the Bulgarian coast is 74.25 mg/m$^3$ and varies by seasons: in winter – 27 mg/m$^3$, in spring – 77 mg/m$^3$, in summer – 135 mg/m$^3$ and in autumn – 48 mg/m$^3$ (Valkanov, Dimov & Naidenov, 1978). After 10 years, these values are lower. The average annual spring biomass of zooplankton in the coastal zone is highest in front of Cape Galata (72.58 mg/m$^3$), lower in front of Cape Kaliakra (58.79 mg/m$^3$) and the lowest in front of Cape Emine (35.39 mg/m$^3$). In the autumn zooplankton, almost all summer zooplankters are established but with lower average value of the biomass (36.13 mg/m$^3$) in comparison with the summer value (102.10 mg/m$^3$). The highest average value of the autumn biomass (78.17 mg/m$^3$) is recorded in coastal waters east of Cape Galata and the lowest one (16.76 mg/m$^3$) – east of Cape Maslen Nos (Konsulov & Konsulova, 1993, 1998). The layer, richest in plankton, is at a depth to 23 m (31.1% of the total biomass). The layer up to 50 m depth contains plankton equivalent to 51.8% of the total biomass. The increase of biomass at a depth of 100-125 m is due to the accumulation of cryophilic plankters. With highest amounts in the period 1970-1988 are the species *Oithona minuta* (to 2813 ind/m$^3$), *Acartia clausi* (to 2388 ind/m$^3$), *Penilia avirostris* (2044 ind/m$^3$) and *Pleopis polyphaemoides* (to 767 ind/m$^3$) in summer and *Paracalanus parvus* (364 ind/m$^3$) in autumn (Konsulov & Konsulova, 1993, 1998). Significant impact on the planktonic cenoses in the 80s and 90s has *Mnemiopsis leidyi*, with maximum numbers of 450 ind/m$^3$ recorded (Konsulov, 1989, 1990; Konsulov & Konsulova, 1993, 1998; Kamburska, 2004). Maximum values of 12 kg/m$^3$ are established in the shelf water areas in April of 1990 (Bogdanova & Konsulov, 1993). As a determining factor for zooplankton development, the species has become an indicator of the pelagic ecosystem and threat to the species diversity of planktonic cenoses. After 2000 the number of *M. leidyi* decreased as a result of the predatory pressure of *Beroe ovata* and the structure of the dominant groups began to recover. Most appreciable are changes in the coastal zone. The Crustacea (predominantly Copepoda and Cladocera) usually comprise 70-80% of the zooplankton biomass. Of these crustaceans, about 11280 ind/m$^3$ were established in the summer of 2005 in front of Cape Galata (Shigalova et al., 2008).

**ZOOBENTHOS.** The number of zoobenthos species is about 1000 (1370 species with Protozoa and parasitic forms), belonging to 19 types. It is studied better in Bulgaria than in other Black Sea countries. Arthropoda, Annelida, Mollusca and Nematoda have the greatest species diversity. Since some taxonomic groups have not yet been sufficiently investigated, it could be accepted that the species composition is higher. Three main zones are established in the Bulgarian Black Sea coast – **supralittoral, mediolittoral** (littoral, pseudolittoral) and **sublittoral** (infralittoral, circalittoral). In these zones 12 biocenoses and a great number of series are differentiated, which are characterized by definite species composition (Kaneva-Abadjieva 1960, 1962; Valkanov & Marinov 1978; Marinov 1990; Konsulov & Konsulova 1993, 1998; Todorova 2005, 2011, 2015, 2017; Revkov et al. 2008; Todorova et al. 2008a, 2008b, 2022; Konsulova et al. 2010; Todorova & Panayotova 2011, 2015). Thirty-two biotopes have been localized as national habitats (Todorova & Moncheva 2013; Todorova & Milkova 2017). Recently, biogenic reefs built by *Ostrea edulis* "ostrak" have been established (Todorova et al. 2008a, 2009).

**Rocky supralittoral.** A characteristic species are *Chthamalus stellatus*, *Melarhaphe neritoides* and *Ligia italica*, found up to 2-3 m above the water on the rocks. The highest settled specimens of *Ch. stellatus* and *M. neritoides* are active only during rough sea. The density of *Ch. stellatus* reaches thousands ind/m$^2$. *Myosotella myosotis* is found sometimes under the stones in that biocenosis and *Pachygrapsus marmoratus* temporarily goes out (Valkanov & Marinov, 1978; Marinov, 1990).

**Sandy supralittoral and washed out algae**. The species of family Talitridae are typical forms, of which *Orchestia bottae* (to 3500 ind/m$^2$ and 90 g/m$^2$) is a typical mass species (Stoykov, 1975). Many insects, inhabiting the coast, of the orders Collembola, Heteroptera, Coleoptera and Diptera are presented. A part of their larvae grow up in washed out algae. Marine forms as *Orchestia gammarellus* (to 2831.7 mg/100 g algae), connected with algae and sea grass and other washed ashore species, are presented. The highest biomass (8661.6 mg/m$^2$) is established in autumn (Beschovski, 1964a, 1975a, 1978; Marinov, 1990).

**Rocky mediolittoral** (Enteromorpha zone). The most abundant species are *Spirorbis pussilla*, *Chthamalus stellatus*, *Balanus improvisus*, *Idotea balthica*, *Mytilus galloprovincialis* and the larvae forms of *Thalassomyia frauenfeldi*. The species *Patella ulyssiponensis* is a typical inhabitant which is rare now but formerly it occurred in great quantity along the southern coastal zone, according to older literature (Kaneva-Abadjieva, 1960a, 1962). *Eriphia verrucosa* and *Pachygrapsus marmoratus* are observed during the warm months (Valkanov &. Marinov 1978; Marinov, 1990; Konsulov, 1998; Todorova 2011, 2015).

**Sandy mediolittoral**. A very typical and abundant species is *Donacilla cornea*, the maximum density of which reaches 9800 ind/m$^2$ at Alepu. A comparatively high density is established at Stomoplo (6000 ind/m$^2$) and Ahtopol (2000 ind/m$^2$) (Valkanov &. Marinov 1978; Marinov 1990; Konsulov 1998; Konsulov & Konsulova 1993, 1998). *Ophelia bicornis* is also a mass species with density of 2000 ind/m$^2$; it is found on the beach at Lozenetz Village. The Caspian relict *Euxinia maeoticus* occurs rarely. Over 60 species have been found in the subterranean beach waters, of which Harpacticoida (29 species), Polychaeta (10 species), Turbellaria and Halacaridae (7 species each) are best represented. The average density of the meiobenthos varies from 14552 to 74000 ind/m$^2$, as Oligochaeta and Harpacticoida predominate. The maximum density of Oligochaeta reaches 76760 ind/m$^2$ (Marinov, 1990; Konsulov & Konsulova, 1993, 1998).

**Cystoseira overgrowths**. An exceptionally rich biocenosis, connected to the algae *Cystoseira barbata* and *C. crinita* that develop on the rocky bottom from 0.5 to 20 m of depth. Over 130 species have been established, of which Crustacea

(68 species) and Polychaeta (31 species) are best represented. Characteristic species are *Caprella acanthifera*, *Palaemon adspersus*, *Clibanarius erythropus*, *Macropodia longirostris*, *Pilumnus hirtellus*, *Nereis zonata*, *Spirorbis pussila*, etc.. Eight mollusk species develop on the thallus of the algae, the majority of which are *Tricolia pullus*, *Bittium reticulatum*, *Rissoa splendida* and *Mytilaster lineatus* (which densely overgrows the base of the thallus). Some Cnidaria (*Sarsia tubulosa*, *Aglaophenia pluma*, *Lucernariopsis campanulata*, etc.) and Bryozoa (*Electra pilosa*) are attached to the algae. The average density of the macro- and meiobenthos is 709424 ind/m$^2$, the average biomass – 20.9 g/kg algae. The maximum density reaches 1040000 ind/m$^2$, and the biomass – 400 g/m$^2$. Harpacticoida has a high density (113396 екз./kg algae). The density and biomass are highest in spring and summer (212668 ind/m$^2$ and 58.5 g/kg algae). The biomass is formed mainly from the mollusks (Kaneva-Abadjieva & Marinov, 1977; Marinov, 1990).

**Rocky sublittoral**. This biocenosis spreads in depth from 0.5 to 30 m and comprises over 130 species with average density of 105105 ind/m$^2$. Representatives of Crustacea (40 species), Mollusca (36 species) and Polychaeta (31 species) predominate. Many Porifera (*Petrosia ficiformis*, *Dysidea fragilis*, genus *Haliclona*, etc..), cnidarians (*Aglaophenia pluma*, *Orthopyxis integra*, *Actinia aquina*, etc..), sedentary Polychaeta (*Sabellaria taurica*, *Spirobranchus triqueter*, *Vermiliopsis infundibulum* and *Spirorbis pusilla*), Bryozoa (*Membranipora* and *Cryptosula*) and Ascidiacea (*Botryllus schlosseri*) are presented. The decapods (*Palaemon adspersus*, *Palaemon elegans*, *Hippolyte sapphica*, *Clibanarius erythropus*, *Eriphia verrucosa*, *Pachygrapsus marmoratus*) are most represented. However, the mass species *Carcinus aestuarii* has declined sharply in numbers lately. The species *Lepidochitona cinerea* and *Rapana venosa* are permanent inhabitants of the rocky bottom. The most abundant snails are *Tricolia pullus*, *Gibbula divaricata* and *Rissoa splendida*. The mussels *Mytilus galloprovincialis* and *Mytilaster lineatus* occur in large quantities. The karst and mergel rocks are pierced by the holes of *Petricola lithophaga*, *Pholas dactylus* and *Barnea candida* (Marinov, 1990). Inhabitants of the rocky bottom are *Ostrea edulis* (living forms are not established – Todorova et al. 2008a, 2009) and *Magallana gigas* (Mitov et al. 2020).

**Sandy sublittoral**. Extends from 0 to 17-20 m depth and is the richest in terms of species biocenosis. Above 300 species have been established, some of which penetrated from neighboring biocenoses. Polychaeta (above 60 species), crustaceans (about 50 species), mollusks (above 30 species) dominate as well as great number of other groups, in which psammophilous and psammobiontic species are presented. The zoocenosis has been divided into 5 subcenoses according to the qualitative composition and ground characteristics (Marinov, 1990). Some Polychaeta (*Scolelepis squamata, Glycera convoluta* and *Prionospio cirrifera*), crustaceans (*Bathyporeia guilliamsoniana, Ampelisca diadema, Diogenes pugilator*) and common mollusks (*Chamelea gallina, Lucinella divaricata* and *Bittium reticulatum*) are dominated forms. Characteristic psammophilous Polychaeta are *Arenicola marina, Pisione remota, Prionospio malmgreni* and *Polygordius lacteus*, some Cumacea from crustaceans (*Pseudocuma, Bodotria, Cumopsis* and *Cumella*) and many mollusks (*Loripes orbiculatus, Solen marginatus, Moerella donacina, Macomangulus tenuis, Fabulina fabula, Donax trunculus* and *Gouldia minima*). *Actinothoe clavata* and *Branchiostoma lanceolatum* (related to a certain structure of the sand) are typical representatives. The density and biomass are determined by psammobiontic forms and as exception by some eurybionts (*Balanus improvisus* and *Ampelisca diadema*). The highest is the density of Polychaeta; Mollusca have the highest biomass. The average density varies from 1484 ind/m$^2$ to 2576 ind/m$^2$; as Polychaeta comprise 55%, followed by crustaceans – 16%, mollusks – 27% and other groups – about 1.5-2%. The average biomass is 136.4 g/m$^2$ as the mollusks comprise 92.8% (126.5 g/m$^2$) of it (Valkanov et al., 1978; Marinov, 1990). The average density of mollusks is 398 ind/m$^2$; it is the highest for *Lentidium mediterraneum* (168 ind/m$^2$) which can reaches a maximum density up to 21000 ind/m$^2$. For Polychaeta, *Spio filicornis* reaches a maximum density of 8320 ind/m$^2$ (Valkanov et al., 1978; Marinov, 1990). The maximum biomass for mollusks, from 1542 to 1787 g/m$^2$ (density about 2700 ind/m$^2$), is established for *Ch. gallina* (Kaneva-Abadjieva & Marinov, 1966; Marinov, 1990). The invasive immigrants *Anadara kagoshimensis* (to 4282 g/m$^2$) and *Mya arenaria* (to 4596 g/m$^2$, 4862 ind/m$^2$) are also represented here (Marinov 1990; Todorova & Panayotova 2011, 2015).

**Coastal silt**. This zone begins from where the sandy bottom ends. It extends from 15-20 to 30-40 m depth and is better developed in the northern coastal zone. Relatively poor biocenosis, in which about 50 species have been established. It can be divided into 2 subcenoses depending on the abundance of *Melinna palmata* (Polychaeta) – high occurrence of *M. palmata* and low occurrence of *M. palmata*. The former comprises 2/3 of the entire cenosis, has a richer species diversity, density and biomass. The maximum density of *M. palmata* reaches to 2010 ind/m$^2$ and the biomass – to 44.9 g/m$^2$ (Nguen Suan Li, 1984; Marinov, 1990). The average density of the biocenosis is 564 ind/m$^2$ and the maximum one – 2543 ind/m$^2$. The average biomass is 76.3 g/m$^2$ and the maximum one – 280.7 g/m$^2$. Polychaeta comprises 82.4% of the density and Mollusca forms about 77.9% of the biomass of this zoocenosis. *Spisula subtruncata* and *Chamelea gallina* have the highest density. *Polititapes aureus* and *Ch. gallina* have the highest biomass, to 84.7 g/m$^2$ and 56.1 g/m$^2$ respectively. In some places the biomass is determined by the presence of *Mytilus galloprovincialis*. Of Crustacea, *Ampelisca diadema* and *Upogebia pusilla* have the highest density as the latter forms the biomass as well. Characteristic species of the sandy-oozy ground are *Actinothoe clavata*, *Lagis koreni*, *U. pusilla* and many mollusks of the orders *Caecum, Calyptraea, Cyclope, Nassarius, Lucinella, Spisula, Abra* and *Gouldia* (Valkanov & Marinov, 1978; Nguen Suan Li, 1984; Marinov, 1990).

**Mytilus silt**. This zone begins to the north of Cape Kaliakra from the depth of 45 m and reaches the depth of about 70 m; at the Cape Emine – 80 m. The width of this zone varies from 2-3 miles to 10-15 miles. Ahead of Burgas Bay the zone is the widest and begins from the depth of 13-20 m. In the southern part of the Bulgarian coast (from Sozopol to Rezovo) this biocenosis begins very close to the shore. Typical for the Black Sea is a zone at a depth of 40 to 140 m, composed of *Mytilus galloprovincialis* (to about 70 m) and *Modiolula phaseolina* (from 70 to 140 m). Nowhere *M. galloprovincialis* reaches such quantity as in the Black Sea. In the southern half of the Bulgarian shelf the higher biomass is concentrated to 50 m isobath, while the northern half of it reaches 80 m isobath. Whenever the biomass is over 500 g/m$^2$ it is caused by *Mytilus galloprovincialis* which is the main dominant. In front of Pomorie a biomass to 2694 g/m$^2$ is established, in front of Tsarevo – 3354 g/m$^2$, in front of Burgas – 4865 g/m$^2$ and in front of Krapets – 5900 g/m$^2$ (Kaneva-Abadjieva, 1962; Marinov, 1990; Marinov & Stoykov, 1995; Stoykov & Uzunova, 1999). The biomass of *M. galloprovincialis* shows 3 maximums, of which the maximum at 45 m depth is the biggest. The species composition (over 100 species) is represented by eurybathic as well as typical for this depth species. This biocenosis is considered one of the richest in species and also ranks highest in biomass. Polychaeta is best presented, followed by Crustacea and Mollusca. *Aglaophenia pluma, Sertularella polyzonias, Cerebratulus ventrosulcatus, Aricidea claudiae, Ciona intestinalis* and *Ascidiella aspersa* are characteristic inhabitants of this biocenosis. Of mollusks, *Spisula subtruncata, Abra alba* and *Polititapes aureus* have a very high density. Biomass of the zoocenosis is formed mainly by the mollusks (94.3%). *M. galloprovincialis* alone gives 63.6% from the biomass. However, as a result of its patchy distribution, the coefficient of permanency of this species is only 25.6%, followed by *S. subtruncata* with the coefficient of permanency 79.2% but with a twice less biomass. The average density is 666 ind/m$^2$; the maximum one – 4185 ind/m$^2$. The average biomass is 134.4 g/m$^2$; the maximum one – 3817 g/m$^2$. About 57.2% of the density is formed by Polychaeta as *A. claudiae* predominates (Valkanov & Marinov, 1978; Marinov, 1990; Konsulov & Konsulova, 1993, 1998; Konsulov, 1998).

**Phaseolina silt**. This zone extends from 60-65 m to 184 m of depth. In the southern region, due to the larger slope of the bottom, the zone is narrower. *Modiolula phaseolina* is a dominant species, with the coefficient of permanency from 22.2% (south of Kamchiya River) to 96.5% (in the north). The higher biomass (200-900 g/m$^2$) below 60 m isobath is due to this species. In the south *M. phaseolina* is poorly represented, therefore the average biomass is often below 50 g/m$^2$. Its density and biomass show many peaks, of which the density maximum (4963 ind/m$^2$) is at 110-120 m of depth and the biomass maximum (389 g/m$^2$) – at 65 m (Kaneva-Abadjieva & Marinov, 1960b). Recently, this species was recorded at the depth of 55 m in front of Cape Kaliakra, with a density up to 13040 ind/m$^2$ and biomass up to 995.6 g/m$^2$ (Marinov & Stoykov, 1995). The greater species diversity of this biocenosis (over 60 species) is up to 100 m depth. Below 150 m of depth occur only *Sycon ciliatum, Actinothoe clavata, Pachycerianthus solitarius, Nephtys hobergii* and *Leptosynapta inchaerens* (Marinov, 1990). The average density is 853 ind/m$^2$ and the maximum one is 5125 ind/m$^2$. Mollusks comprise 81.7% of the density, and *M. phaseolina* alone, forms 79.9% of the total density. The average biomass is 44 g/m$^2$ and the maximum one is 394 g/m$^2$. The mollusks account for 86.5% of the whole biomass, and *M. phaseolina* alone, for 79.2% (Marinov, 1990). Of Gastropoda, *Trophon muricatus* is the only representative. This conspicuous predator feeds on the mussels with thinner shells, such as *Modiolula phaseolina, Parvicardium exiguum, Papillicardium papillosum, Spisula subtruncata, Abra alba* and *A. prismatica*. Characteristic of this biocenosis are *Suberites carnosus, P. solitarius, L. inchaerens, Amphiura stepanovi*, etc. The mass species *Terebellides stroemii* and *A. stepanovi* are widely distributed. (Valkanov & Marinov, 1978; Marinov, 1990; Konsulov, 1998).

**Meiobenthos in the sublittoral zone**. Nematoda, Polychaeta and Harpacticoida dominate on sand and silt bottoms from 5 to 150 m of depth. Cephalorhyncha, Chalacaridae, Cyclopoida are often found; Oligochaeta and Ostracoda are rarely found. Over 100 species have been established as the most representatives have a low frequency of occurrence. The total number is 164152 ind/m$^2$ and the maximum number reaches up to 1181000 ind/m$^2$. The density is formed mainly by Nematoda (77.5%); its maximum number reaches up to

1034000 ind/m$^2$. The maximum number of Oligochaeta and Harpacticoida is 65000 and 102000 ind/m$^2$ respectively. Ostracoda is rarely found; its number amounts to 1000-2000 ind/m$^2$. Due to the spotty distribution of meiobenthos depth is not always determinative for the number that decreases below 35 m (Marinov, 1990; Konsulov, 1998).

COASTAL BASINS. There are about 40 lakes, marshes and areas, flooded by rivers along the Bulgarian coast (Varbanov, 2002; Hristova, 2012; Table 1). Most common are firth lakes (blocked estuaries) and lagoon lakes (areas separated from the open sea). The presence of brackish elements is a special feature of their fauna. A "saline wedge" is formed at the lower parts of the rivers, which is situated below the fresh waters. In this "wedge" the bottom inhabitants are marine or brackish, whereas those in upper water layers are freshwater species. A specific fauna inhabits the lakes, firths and marshes along the coast. The marine brackish species endure water down to 1 ‰ salinity and the freshwater forms withstand water salinization from 1.5 ‰ to 8 ‰. Many euryhaline sea species also take part in the formation of the coastal basins's fauna, which could vary from marine to freshwater, depending on the water salinity. Nineteen rivers enter the Bulgarian Black Sea (12 rivers enter the Black Sea and 8 rivers discharge into the coastal lakes). Their mouth areas have firth nature where oligo- or euryhaline forms are presented. It is therefore difficult the fauna of these rivers to be scrutinized separately from the coastal stagnant basins (Valkanov, 1935, 1936; Drensky, 1947; Petrbok, 1947; Kaneva-Abadjieva, 1957, 1976; Zaschev & Angelov, 1959; Mihailova-Neikova, 1961; Kaneva-Abadjieva & Marinov, 1967; Stoykov, 1979; Marinov, 1990; Trayanova, 2003, 2008).

The mass development of the mussel *Dreissena polymorpha* that reaches to 3-4 m depth and forms a ring around the shore is characteristic for the lakes **Durankulak**, **Ezerets** and **Shabla**. The Caspian relict *Hypania invalida* has a high density (thousands ind/m$^2$) and in the zone of *D. polymorpha* Ostracoda reaches 8595 ind/m$^2$ (Cvetkov, 1958; Kaneva-Abadjieva & Marinov, 1967; Valkanov et al., 1978; Kovachev & Uzunov, 1981; Marinov, 1990; Naidenov, 1998; Stoichev, 1998; Kovachev et al., 1999, 2002; Petrova & Stoykov, 2002). *Cordylophora caspia*, many species of the families Corophiidae and Gammaridae, *Astacus leptodactylus* and *Theodoxus fluviatilis* are common species. In Shabla Lake the average biomass is about 19.4 g/m$^2$ and the maximum biomass reaches 842 g/m$^2$. In Durankulak Lake the maximum biomass is 60 g/m$^2$ (Valkanov et al., 1978). The lakes have been investigated recently as protected areas and have a rich fauna (from 101 to 137 species).

Fauna of the **Beloslav** Lake, before its transforming into port, was formed mainly by freshwater Crustacea and Chironomidae larvae (Valkanov, 1935, 1936, 1937; Cvetkov, 1955a, 1955b, 1957). In the Chironomidae complex haloxenes are presented but there is a lack of typical halophils and halobionts. The freshwater species are best represented in most crustaceans as there are brackish and marine forms. Ostracoda comprises a significant percentage (15 species, or 35.7%). Copepoda includes more haloxenes and halobionts. All Amphypoda, Mysida, *Iera sarsi* and *Rhitropanopeus harrissi* are brackish forms. *Cordilophora caspia*, *Leander adspersus*, *Astacus leptodactylus*, *Potamon ibericum*, *Theodoxus fluviatilis*, *Unio pictorum* and *Dreissena polymorpha* occur. In quantitative terms Oligochaeta, Ostracoda and Chironomidae predominate (Cvetkov, 1957). After 1964-1966 the state of the lake got worse and it became to a chironomid type. Typical are periodic extinctions due to oxygen deficiency. In 1990-1991 the species composition is poor, there are lack of mollusks, dead zones and single species communities. After 2000 the species diversity increases, 5 mollusk species appear and there are no permanent dead zones (Trayanova, 2003, 2008).

After the connection of **Varna** Lake with the sea, the freshwater species disappeared there and the fauna was formed by euryhaline marine and brackish forms (Kaneva-Abadjieva, 1957). The following species dominate: *Hydrobia ventrosa* (up to 24520 ind/m$^2$), *Mytilaster lineatus* (up to 37400 ind/m$^2$ and 620 g/m$^2$), *Mytilus galloprovincialis* (to 1837 ind/m$^2$ and 6023 g/m$^2$), *Cerastoderma glaucum* (to 968 ind/m$^2$) and *Abra segmentum* (to 1177 ind/m$^2$). The reserves of *M. galloprovincialis* were estimated as ca. 5000 t and were of great economical importance. Mass overgrowths, causing problems of TPP Varna in the 70s, are *Ficopomaticus enigmaticus*, *Amphibalanus improvisus* and *M. galloprovincialis*. The maximum overgrowth reaches 39260 g/m$^2$ for 3 months (Dimov et al., 1970). In the last decades, the lake underwent profound transformations and its fauna changed (Marinov, 1990; Konsulov & Konsulova, 1993, 1998). In the 50s crustaceans and mollusks prevail. In the 90s the number of mollusks decreased, whereas Polychaeta increased and dominated in numbers (70.6%). In recent years, the number of species is increased (especially in crustaceans). An increase in numbers (from 5787 ind/m$^2$ to 18841 ind/m$^2$) and a shift of Polychaeta (38%) by Crustacea (50.9%) was established. The biomass is formed mainly by mollusks (58.7%) and crustaceans (39.0%) (Trayanova, 2003, 2008; Tododrova et al, 2008b).

The halobionts *Artemia parthenogenetica* and *A. salina* are typical for the hyper-saline lagoon lakes **Pomorie** and **Atanasovsko** [salinity from 30-60‰ to

100-250‰ (Ivanov et al., 1964)]. In spring with the increasing of the water's temperature (20-22°C) and salinity (24-26‰), the quantity of *Artemia* increases (to 300-350 ind/l adults and 800-1000 ind/l juveniles) and reaches to 1800 ind/l adults and to 3000 ind/l juvenile forms in summer. It reaches to a high density – 3400 ind/l water (17 g/l) (Andreev, 1997, 2003) which is caused by the abundant phytoplankton blooming. Under the salinity of 250-260‰ *Artemia* maintain life processes (over 170-180‰ they are not propagated) and die under salinity of 340‰ (Caspers, 1952). Of marine forms, *Ecrobia ventrosa* [from 6924-10000 ind/m$^2$ (Gecheva et al., 2013) to 19800 ind/m$^2$ (Cvetkov, 1958)], *Nassarius reticulatus*, *C. glaucum* and *A. segmentum* are presented. The density and biomass of *Cyprideis torosa* reach 312532 ind/m$^2$ and 37.1 g/m$^2$; of *C. glaucum* – 3234 ind/m$^2$ (maximum to 134376 ind/m$^2$) and 338.7 g/m$^2$ (Cvetkov, 1958). The number of *Acartia clausi* in the Pomorie Lake reaches 130000 ind/m$^3$ (Vassilev, 1994). The lakes are protected areas, recently explored in connection with the plans of their management (Michev, 1997, 2003; Georgiev & Nikolov, 2010; Pechlivanov, 2010; Varadinova et al., 2010; Hubenov et al., 2015; Kalcheva et al. 2015; Kalcheva 2017, 2018, 2019). In these lakes specific rich fauna is stored (106-157 species).

In **Burgas** Lake, *A. segmentum* in 1954 reached average density 5544 ind/m$^2$ and biomass 790 g/m$^2$; its total biomass was estimated as 16957 t (Zaschev & Angelov, 1959). The average density of *E. ventrosa* during the same period was 3256 ind/m$^2$, biomass – 10.4 g/m$^2$ and the total biomass – 216 t. For *C. glaucum* these data are, respectively: 1840 ind/m$^2$, 201.8 g/m$^2$ and 3081 t. Later, because the salinity decreased from 11‰ to 0.7-1.9‰ these mollusks completely disappeared and currently no living specimens have been found (Pandourski, 2001). The density of the following crustaceans was high: *Crassicorophium crassicorne* (10806 ind/m$^2$ – 31.1 g/m$^2$) and *Gamarus subtypicus* (748 ind/m$^2$ – 3.7 g/m$^2$), with total biomass 549 t and 41.1 t respectively (Kaneva-Abadjieva & Marinov, 1967; Kaneva-Abadjieva, 1976; Stoykov, 1979). According to Kaneva-Abadjieva & Marinov (1967), 60% of the biomass of the zoobenthos is formed by Chironomidae and 40% – by Olygochaeta. Nowadays 93% of the total biomass of the zoobenthos (16.6 g/m2) is formed by Chironomidae larvae (Valkanov et al., 1978). According to Todorova (2014) the composition of the zooplankton and zoobenthos [Rotifera, Olygochaeta, Crustacea (Cyclopoida and Cladocera) and Diptera (Chironomidae)] differs significantly from the previous studies (only 4 common species were found). The highest values of the abundance and biomass are in the central parts of the lake. In Chironomidae, the maximum biomass reaches 485.6 g/m$^2$ (average – 9.53 g/m$^2$), and in Olygochaeta it varies from 11.0

g/m² to 152.1 g/m². In most parts of the lake, the ecological condition is bad (Todorova 2014).

In **Mandra** Lake, before the dam construction, many marine forms (*Hediste diversicolor, Corophium volutator, Palaemon adspersus, Upogebia pusilla, Liocarcinus vernalis, H. ventrosa, C. glaucum* and *A. segmentum*) have been recorded, which can be found now in the lakes **Uzungeren** and **Poda**, remaining outside of the dam (Valkanov, 1936; Mihailova-Neikova, 1961; Kaneva-Abadjieva & Marinov, 1967; Kaneva-Abadjieva, 1976). Over 100 species benthos forms have been found, which average biomass is 14 g/m² (57.5% Chironomidae). After the dam construction, the zoobenthos (11.8 g/m²) is formed mainly by Oligochaeta and Chironomidae (60%). Changes in biomass (2.4-4.2 g/m² to 10.7-15.7 g/m²) and a dominance of Chironomidae larvae are observed (Kaneva-Abadjieva, 1976; Stoykov, 1979).

The firth of **Ropotamo River** is well-studied and over 100 species have been established there (Valkanov, 1934, 1935, 1936; Cvetkov & Gruncharova, 1976, 1979; Gruncharova, 1977). It is abundant in the invasive forms *Ficopomaticus enigmaticus* (up to 18400 ind/dm² – 360 g/dm²), *Amphibalanus eburneus* and *Rhithropanopeus harrisii*. The following species are often found: *Blackfordia virginica, Cordylophora caspia*, some Chironomidae larvae, *Palaemon adspersus, Hydrobia acuta, Mytilus galloprovincialis, Mytilaster lineatus, Cerastoderma glaucum, Abra segmentum, Barentsia benedeni* (to 8000 zooids/dm²) and many Bryozoa.

In the firth of **Veleka River**, *Dreissena polymorpha* covers the shells of *Anodonta cygnaea* and *Unio pictorum*. A significant number have the populations of *Astacus leptodactilus, Potamon ibercum* and *T. fluviatilis* (often and *P. planorbis*); of the marine forms *H. ventrosa, C. glaucum* and *A. segmentum* are presented.

Some coastal basins were explored comparatively long ago and the investigations do not reflect the recent condition of their fauna (Shabla Tuzla, Nanevska Tuzla, Balchishka Tuzla, Batovsko Swamp, Nesebarsko Swamp, the swamps Alepu, Arkutino, Stomoplo and Dyavolsko, the firths of the rivers Batova, Kamchiya, Dvoynitsa, Dyavolska, Karaagach, Veleka, Silistar and Rezovska). Furthermore, the coast is exposed to strong anthropogenic presence and changes that reqiure a continuous update of the information.

**ALIEN IMMIGRANTS**. The penetration of alien species through ballast waters and/or as fouling organisms on ship hulls is one of the greatest threats for the world's oceans. The assessment whether a certain species is invasive and harmful to the Black Sea ecosystem is complex. During a certain point of time the species may be harmful and subsequently to get a positive effect – food base for other species, an element for increasing the diversity or resource. Lists of species, introduced in the Black Sea, have been published by several authors (Cvetkov & Marinov, 1986; Gomoiu & Scolka, 1996; Konsulov, 1998; Shadrin, 2000; Zaitsev & Öztürk, 2001; Gomoiu et al., 2002; Moncheva & Kamburska, 2002; Kamburska & Moncheva, 2003; Zaitsev et al., 2004; Konsulova & Stefanova, 2007). Thirty-one invertebrate species are known from the Bulgarian Black Sea coast, occurring at different times (Table 6). The most significant changes in the Black Sea communities are caused by 5 species, introduced in the last 60 years – *Mnemiopsis leidyi*, *Beroe ovata*, *Rapana venosa*, *Anadara kagoshimensis* and *Mya arenaria* (Cvetkov & Marinov, 1986; Marinov, 1990; Konsulova & Stefanova, 2007; Todorova & Moncheva, 2013).

The introduction of the ctenophore **Mnemiopsis leidyi** in the Black and Azov Seas in the 1980s caused the decline of anchovies' and sprat' stocks, estimated at 200 millions USD per year (Zaitzev & Mamaev, 1997). This species becomes a determinative factor for the development of zooplankton and indicator of the pelagic community. *M. leidyi* is a self-fertile hermaphrodite with broad food spectrum (from microzooplankton to crustaceans, eggs and fish larvae) and a high tolerance to temperature and salinity. It is reproductively mature 12 days after hatching and produces over 2000-3000 eggs. In the 80s *M. leidyi* caused a drop in pelagic fish reserves, by competing for the same food resources. A maximum number of *M. leidyi* (450 ind/m$^3$) was recorded in Varna Lake in 1986 (Konsulov, 1986; Konsulov & Konsulova, 1993, 1998; Kamburska, 2004). After 2000 the number of *M. leidyi* decreases, affected by the predatory pressure from *B. ovata*.

**Beroe ovata** is known to feed on planktivorous ctenophores and, in particular, on *M. leidyi*. There is a trophic type predator-prey relationship between *B. ovata* and *M. leidyi*. The development of *B. ovata* depends on the seasons; the species occurs mostly in summer and autumn. Its reproductive potential is closed to the potential of *M. leidyi*. The dynamics of *B. ovata*, its seasonal and annual variations, development and survival depend on *M. leidyi* (Konsulov & Kamburska, 1998). Like *M. leidyi*, this species was transported in ships ballast waters from the Northwest Atlantic. The arrival of *B. ovata* appears to have stabilized the Black Sea ecosystem, leading to a reduction in *M. leidyi* populations and subsequent recovery of plankton and fish populations.

The first specimen of **Rapana venosa** in the Bulgarian aquatory was found in 1956 in Varna Bay, near Cape Galata (Kaneva-Abadjieva, 1958). Development of this snail in the rocky sublittoral has a substantial impact on *Mytilus* and *Ostrea*, and in the sand sublittoral – on *Chamelea gallina*. The great eurybiontness, high fecundity and lack of competitors allowed this predator to reach mass development in the Black Sea and aroused discussion for eventual measures for a struggle with it. In a single trawling, up to 1500 specimens have been caught, and in some regions between Balchik and Kavarna the entire bottom was covered with Rapa whelks. Very high numbers was observed in Byala, in the region of Cape Cherni Nos (Klissurov, 2008). During the last 20 years, the snail was gathering for food with all possible means. After conquest of the Black Sea the species penetrated the Aegean, Adriatic and Mediterranean Seas, Atlantic coast of France, North Sea, East coast of USA, the mouth of the Rio de la Plata River between Uruguay and Argentina and around New Zealand. The way how the species was transported in the Black Sea is unclear. *R. venosa* is an eurythermal and euryhaline species that develops in the coastal zone on solid substrate and sandy and silty bottom at a depth to 30-40 m. The snail withstands temperature changes (from 0 to 30°C), water pollution and reduced oxygen content. There is a huge fertility (a snail delayed approximately 220000 eggs) which compensates its exploitation by man. It lives about 10 years. There are no precise data on the population of Rapa on the Bulgarian coast (Konsulov & Konsulova, 1993, 1998; Konsulov, 1998).

The first specimens of **Anadara kagoshimensis** for the Bulgarian coast were found in 1982 in Varna Bay (Marinov et al. 1983; Kaneva-Abadjieva & Marinov 1984; Nguen Suan Li 1984). A little later, a high density of the species has been found in Burgas Bay (up to 400 ind/m$^2$ and biomass 4280 g/m$^2$) (Marinov 1990). This mussel is a eurythermal and euryhaline species that endures very low oxygen concentrations in the water due to the presence of hemoglobin in the haemolimph. It has a long life cycle and low coefficient of mortality. In a short time, *A. kagoshimensis* became a significant element of psammo- and pelophilous zoocenoses and started to displace some local species. Thus the „*Chamelea gallina*" group in front of Balchik, Varna and Burgas transforms into „*A. kagoshimensis*" group. The distribution of this species in the Bulgarian part of the Black Sea is restricted from Balchik to the south part of Burgas Bay (Cvetkov & Marinov 1986; Marinov 1990; Konsulov 1998).

*Mya arenaria* has been first reported for the Bulgarian coast in the Bay of Burgas in 1973 (Kaneva-Abadjieva 1974). The mussel inhabits the sandy sublittoral and reaches the surf zone. It has a high ecological plasticity and easy endures variations of the salinity and temperature, and oxygen deficiency. It reaches a high density (over 300-400 to 4862 ind/m$^2$) in the bays in front of the river mouths (Marinov 1990; Todorova & Panayotova 2011, 2015). *M. arenaria* is found along the beaches all over the Bulgarian coast but the greatest number of it occurs in front of Durankulak and Albena, in the Varna Bay, Varna Lake, at the influx of the Kamchiya River and Burgas Bay (Stoykov 1983; Cvetkov & Marinov 1986; Marinov 1990). Spawning by eggs thrown straight into the water during the summer months. From fertilized eggs planktonic larvae develop which 5-6 days after egg hatching convert to mussels. In many areas of the Black Sea shelf *M. arenaria* was a dominant species in new zoocenosis, called her name. In the 1970s, this mussel is a dominant species in the Romanian coastal zone as 4-5 years after its appearance reaches biomass 16 kg/m$^2$ and numbers more than 8000 ind/m$^2$ (Gomoiu & Porumb, 1969).

RESOURCE SPECIES AND OVERGROWERS. Industrial catch of *Rapana venosa* in Bulgaria began in 1991 when to 500 t of meat were exported. In the following years the catch increased and during 1992-1994 period was about 27000 t (weight with shells). The maximum was reached in 2004 (1195 t meat of Rapa), then a decrement of catches was observed (Konsulova & Stefanova, 2007). According to Ministry of Agriculture and Food, in 2012 the harvest reached 3100 t, and the export of Rapa meat is 1000-1500 t. In 1978 a decline in the population of *Mytilus galloprovincialis* was found although the harvesting was gradually halted (Marinov 1978, 1990; Konsulov & Konsulova 1993, 1998). The main reasons for mass mortalities appear to have been the eutrophication and destruction of natural mussel fields by trawling and predatory pressure from *Rapana venosa*. A decrease in predatory pressure in deeper waters and stopping of trawling has resulted in a progressive restoration of mussel resources. In 1990s, the commercial farming of *M. galloprovincialis* on artificial substrata was mostly near the town of Sozopol, where the harvest had reached 150 tons per year (Konsulov & Konsulova 1993, 1998; Konsulov 1998). In 2009, the harvest reached 812 t. Today the number of mussel farms is 17; the farm near Kavarna alone has the capacity of 2000 t per year. There is an interest in the populations of *Mya arenaria* and *Chamelea gallina* but it is not clear how much their exploitation is realized. The foreign interest in these mussels is great (Konsulova & Stefanova, 2007).

Essential roles as overgrowers along the Bulgarian Black Sea coast have *Ficopomaticus enigmaticus*, *Amphibalanus improvisus* and *Mytilus galloprovincialis* (Marinov, 1990). These species cause problems to port facilities, craft and TPP Varna (Dimov et al., 1970). Surveys conducted in the old channel connecting Varna Lake with the sea show that for six months *A. eburneus* and *M. galoprovincialis* give 8205 $g/m^2$ (Kaneva-Abadjieva & Marinov, 1965, 1977). In spring the biomass reaches 3357.5 $g/m^2$ and is composed mostly by *A. improvisus* and *M. galoprovincialis*. In summer the total biomass reaches 17504 $g/m^2$ and is composed mainly by *F. enigmaticus* and *A. improvisus*. For a year, the total overgrowth (and other overgrowers as well) can reach 84 $kg/m^2$ (Marinov, 1990). The total biomass of the overgrowers in the region of TPP Varna in June reaches 6480 $g/m^2$; in August the overgrowth is mainly of *F. enigmaticus*. The number of attached specimens of this species reaches 3000000 $ind/m^2$ for 15 days (Dimov et al., 1970). In 2 months the overgrowth by *F. enigmaticus* exceeds a thickness of 15 cm. In the 90s, the species mass developed in the Varna Bay. The first reported problems with *Dreissena polymorpha* in the country is related to the glass factory, located by the canal between Beloslavsko Lake and Varna Lake (Russev, 1965). Since 90 years the mussel invades freshwater basins inside the country and now is presented almost in all river systems (Trichkova et al., 2009). There are problems in TPP Maritsa Iztok 2 where losses of several million leva are registered. Temperature conditions in Ovcharitsa dam created by this plant are optimum for the mussel development which reaches from 300000 to 500000 $ind/m^2$.

CONSERVATION SIGNIFICANSE of the fauna. A total of 35 species from the Bulgarian Black Sea invertebrates are included in the Black Sea Red Data Book (1 of Porifera, 1 of Cnidaria, 2 of Annelida, 26 of Arthropoda, 4 of Mollusca and 1 of Chordata); of which 7 are included in the European Red List as well (Table 7, 8). *Pholas dactylus* is included in the international conventions for European and Mediterranean fauna. There are differences in the levels of threat of the species in the separate Black Sea countries (Dumont et al. 1999). Most species (34 species) belong to the categories endangered (EN) and vulnerable (VU). Five species have a Pontian-Caspian distribution, the Caspian relicts also are 5 species and *Apseudopsis ostroumovi* is a Black Sea (Pontian) endemic. Most often the species are widely distributed as *Halichondria panicea* is a Cosmopolitan species. Some taxa, included in the Black Sea Red Data Book, have stable populations along the Bulgarian coast and are not threatened at this stage. Red Data Book of the Republic of Bulgaria includes 8 species invertebrate animals from the Bulgarian Black Sea (*Hirudo verbana, Eriphia verrucosa, Epallage fatime, Lestes viridis, Sympetrum depressiusculum, S. vulgatum, Theodoxus pallasi* and *Hauffenia lucidula*). The species *T. pallasi* – Pontian-Caspian brackish relict species, is accepted as extincted. The species *Carcinus aestuarii* and *Pholas dactylus* which are exceptionally rare recently are not included. Twenty species of freshwater Gastropoda with Eurosiberian, Palaearctic or Holarctic distribution, included in the European Red List and IUCN Red List are known from the coastal brackish basins (Table 7). These include the North American invasive species *Physella acuta* and *Ferrissia fragilis*, which until recently were considered European forms. The snail *Turricaspia lincta*, a Caspian subfossil relict, established only by shells, has a Pontian-Caspian area. A total of 118 Black Sea endemic forms have been found along the Bulgarian coast, of which 12 species are accepted as regional endemics. The local endemics are an exception in the marine forms and usually are newly described taxa with unclear distribution. A total of 98 rare species and 41 Caspian relict forms have been established. The number of rare species depends on the level of study of the respective groups. Some relitcts are eurybiontic invasive forms with secondary anthropogenic areas.

In the Red Data Book of the Republic of Bulgaria 11 marine natural habitats with different conservation status are included – critically endangered (CR), endangered (EN), vulnerable (VU) and nearly threatened (NT) (Gusev, 2015; Petrov & Pandourski, 2015; Todorova, 2015; Todorova & Panayotova, 2015; Tzonev & Gussev, 2015; Valchev et al., 2015). In Bulgaria the Kaliakra Reserve and protected area Sand Bank Koketrays are marine protected areas. These include 0.2% of the territorial waters of Bulgaria, 0.1% of the shelf zone to 100 m depth and 0.2% of the protected areas of the country (Todorova et al., 2008a).

# References


Abbott R., R. Dance. 1991. Compendium of Seashells. London, Charles Letts & Co. 411 p. [1]

Abatzopoulos T., Beardmore J., Clegg J., Sorgeloos P. 2002. *Artemia*: basic and applied biology. Kluwer Academic Publishers, Dordrecht, 171- 224. [1a]

Althaus B. 1957. Neue Sandbodenrotatorien aus dem Schwarzen Meer. – Wissenschaftliche Zeitschrift der Martin-Luther-Universität Halle-Wittenberg, Mathematisch-naturwissenschaftliche Reihe, 6 (3): 445-458. [2]

Andrássy I. 1958. Erd-und Susswassernematoden aus Bulgarien. – Acta zoologica hungarica, 4: 1-88. [3]

Andreev S. 1997. Brine shrimp (*Artemia salina*) in the Atanasovsko Lake. – In: Michev T. (ed.). Ecology and conservation of Atanasovsko Lake nature reserve. Collection of final scientific reports of experts, Bulgarian-Swiss Biodiversity Conservation Programme, Sofia, 41-47. [3a]

Andreev S. 2003. Invertebrate animals. – In: Michev T. (ed.). 2003. Management Plan of the Atanasovsko Lake. Ministry of Environment and Water & Bulgarian-Swiss Biodiversity Conservation Programme, Sofia, 24-27. Annex 1, 72-75. [3b]

Angelov A. 1976. Revision der Fam. Pisididae Gray, 1857 (Bivalvia – Mollusca) in Bulgarien. – Annuaire de l'Université de Sofia "Kliment Ohridski", Faculté de Biologie, 69 (1 – Zoology): 109–119. (in Bulgarian). [4]

Angelov A. 1983. *Ferrissia wautieri* (Mirolli) (Gastropoda, Ancylidae) – a new representative of the Bulgarian freshwater fauna. – Acta zoologica bulgarica, 21: 95-97. (in Bulgarian). [5]

Angelov A. 2000. Mollusca (Gastropoda et Bivalvia) aquae dulcis. – In: Catalogus faunae bulgaricae. 4. Sofia & Leiden, Pensoft & Backhuys Publishers BV. 57 p. [6]

Anistratenko V., O. Anistratenko. 2001. Fauna of Ukraine. 29 (1-1). Mollusks. Class Polyplacophora or Chitons, class Gastropoda – Cyclobranchia, Scutibranchia and Pectinibranchia (part). Kiev, Veles. 240 p. (in Russian). [7]

Anistratenko V., A. Stadnichenko. 1994. Fauna of Ukraine. 29 (4, 2). Mollusks. Littoriniformes, Rissoiformes. Kiev, Naukova dumka. 174 p. (in Russian). [8]

Apostolov A. 1967. Zwei neue Harpacticoidenarten (Crustacea, Copepoda) aus dem Schwarzmeerbecken. – Zoologischer Anzeiger, 179 (3/4): 303-310. [9]

Apostolov A. 1968. Neue und bemerkenswerte Harpacticoiden der Ruderfuskrebse (Crustacea, Copepoda) aus dem Küstengrundwasser Bulgariens. – Zoologischer Anzeiger, 180 (5/6): 395-402. [10]

Apostolov A. 1969a. *Phyllopodopsyllus ponticus* n. sp., eine neue Art Schwarzmeerharpacticoides. – Fragmenta balcanica Mus. Mac. Sci. nat, 6 (23/158): 209-213. [11]

Apostolov A. 1969b. Harpacticoiden (Crustacea, Copepoda) von der bulgarischen Küste. – Zoologischer Anzeiger, 183 (3/4): 206-267. [12]

Apostolov A. 1969c. Marine Harpacticoiden aus dem Küstensand von Bulgarien. – Acta Mus. Maced. scient. natur. (Skopje), 11 (6): 111-127. [13]

Apostolov A. 1970. Katalog der Harpacticoidenfauna Bulgariens. – Zoologischer Anzeiger, 184 (5/6): 412-427. [14]

Apostolov A. 1971a. Recherches sur la systématique et la distribution de Copepodes harpacticoides de la côte bulgare. – Zoologischer Anzeiger, 186 (5/6): 337-347. [15]

Apostolov A. 1971b. Ein Beitrag zur Kenntnis der harpacticoidenfauna Bulgariens. – Zoologischer Anzeiger, 187 (5/6): 345-356. [16]

Apostolov A. 1972. Catalogue des Copepodes Harpacticoides marins de la Mer Noire. – Zoologischer Anzeiger, 188 (3/4): 202-254. [17]

Apostolov A. 1973a. Sur divers Harpacticoides (Copepodes) de la Mer Noire. – Zoologischer Anzeiger, 190 (1/2): 88-110. [18]

Apostolov A. 1973b. Notes sur les Harpacticoides (Crustacea, Copepoda) de la Mer Noire. – Zoologischer Anzeiger, 190 (3/4): 174-189. [19]

Apostolov A. 1973c. Apporte vers l'étude d'Harpacticoides pontiques habitant les algues marines. – Zoologischer Anzeiger, 191 (3/3): 263-281. [20]

Apostolov A. 1973d. Harpacticoides des eaux saumâtres et des étangs côtières. – Zoologischer Anzeiger, 191 (3/4): 281-294. [21]

Apostolov A. 1973e. Le genre *Schizopera* G. O. Sars (Copepoda, Harpacticoida) de la Mer Noire. – Acta Mus. Mac. Sc. nat., 13 (5): 81-107. [22]

Apostolov A. 1974. Copépodes harpacticoides de la Mer Noire. – Travaux du Museum National d'histoire Naturelle "G. Antipa", 15: 130-139. [23]

Apostolov A. 1975. Les Harpacticoides de la Mer Noire. Description de quelques formes nouvelles. – Vie et Milieu, 25 (1, A): 165-178. [24]

Apostolov A. 1976. New species Copepoda (Harpacticoida) from Bulgaria. – Zoologicheskii zhurnal, 55 (3): 453-458. (in Russian). [25]

Apostolov A. 1977. Harpacticoides nouveaux de la Mer Noire et de la faune bulgare. – Acta zoologica bulgarica, 7: 8-21. [26]

Apostolov A. 1978. Le genre *Robertgurneya* Lang, 1948 (Cop. Harpac.) de la Mer Noire. Description du mâle iconnu de l'especè *Robertgurneya oligochaeta* Noodt. – Acta zoologica bulgarica, 11: 55-59. [27]

Apostolov A. 1980a. Deux formes nouvelles de genre *Nitocra* Boeck (Copepoda, Harpacticoida) de la Mer Noire (côte bulgare). – Acta zoologica bulgarica, 15: 36-42. [28]

Apostolov A. 1980b. Description de deux genres nouveaux de la famille Cletodidae Sars (Copepoda, Harpacticoida)de la Mer Noire. – Fragmenta balcanica mus. Mac. Sci. nat., 10 (19): 167-174. [29]

Apostolov A. 1982. Genre et sous-genre nouveaux de la famille Diossacidae Sars et Cylindropsyllidae Sars (Cop. Harpact.). – Acta zoologica bulgarica, 19: 37-43. [30]

Apostolov A., T. Marinov. 1988. Fauna bulgarica. 18. Copepoda, Harpacticoida. Sofia, Bulgarian Academy of Sciences Publishing House. 384 p. (in Bulgarian). [31]

Arndt W. 1947. Verzeichnis der bisher von der Schwarzen Meer-Küste Bulgariens und Rumäniens angegebenen Schwämme. – Arbeiten aus der Biologischen Meeresstation in Varna, 13: 1-29. [32]

Asem A., Rastegar-Pouyani N., De Los Rios-Escalante P. 2010. The genus *Artemia* Leach, 1819 (Crustacea: Branchiopoda). I. True and false taxonomic descriptions. – Latin American Journal of Aquatic Research, 38 (3): 501-506. [32a]

Băcescu M. 1936. *Hemimysis lamorne* sbsp. *reducta* n. sbsp. et *Hemimysis anomala* dans les eaux roumaines de la Mer Noire. – Ann. Scient. Univ. Jassy, 23: 70-93. [33]

Băcescu M. 1937. Cateva animale noui pentru fauna marina romaneasca si unele date biologice asupra lor. – Bul. Soc. Nat. Bucuresti, 11: 8-19. [34]

Băcescu M. 1940. Les Mysidacés des eaux roumaines. – Ann. Scient. Univ. Jassy, 26 (2): 1-352. [35]

Băcescu M. 1948. Donneés sur la faune carcinologique de la Mer Noire le long de la côte bulgare. – Arbeiten aus der Biologischen Meeresstation in Varna, 14: 1-24. [36]

Băcescu M. 1951a. Fauna Republicii Populare Romîne. 4 (1). Cumacea, Crustacea. Bucharest, Editura Academiei Republicii Populare Romîne, 94 p. (in Romanian). [37]

Băcescu M. 1951b. *Nannastacus euxinicus* n. sp., Cumaceu nou gasit în apele marii negre. Comunicarile Academiei R.P.R, 1 (7): 585-592. (in Romanian). [38]

Băcescu M. 1953. Contributions à la connaissance des Pycnogonides de la Mer Noire. – Academiei Republicii Populare Romîne, Bul. St. Sec. Biol., 5: 263-270. [39]

Băcescu M. 1967. Fauna Republicii Socialiste Romania. 4 (9). Decapoda, Crustacea. Bucuresti, Editura Academiei Republicii Populare Romîne, 351 p. (in Romanian). [40]

Băcescu M., A. Cărăuşu. 1947. *Apseudopsis ostroumowi* n. sp. dans la Mer Noire. – Bull. Sect. Scient. (Bucuresti), 29 (6): 366-383. [41]

Băcescu M., M. Dimitrescu, V. Manea, F. Por and R. Mayer. 1957. Les sables à *Corbulomya* (*Aloides*) *maeotica* Mill., base trophique de prèmier ordre pour les poissons de la Mer Noire. – Travaux du Museum National d'histoire Naturelle "G. Antipa", 1: 305-373. [42]

Băcescu M., G. Müller and M.-T. Gomoiu. 1971. Cercetări de ecologie bentală în Black Sea. Analiza cantitativă, calitativă şi comparată a faunei bentale pontice. Ecologie marină (Bucureşti), 4: 1-357. [43]

Băncilă R., Skolka M., Ivanova P., Surugiu V., Stefanova K., Todorova V. & Zenetos A. 2022. Alien species of the Romanian and Bulgarian Black Sea coast: state of knowledge, uncertainties, and needs for future research. Aquatic Invasions 17 (3): 353-373, https://doi.org/10.3391/ai.2022.17.3.02 [43a]

Bănărescu P. 1990. Zoogeography of Fresh Waters. 1. General Distribution and Dispersal of Freshwater Animals. Wiesbaden, Aula-Verlag. 511 p. [44]

Beshovski V. 1964a. Diptera aus dem litoralen Gebiet der bulgarischen Schwarzmeerküste. – Bulletin de l'Institut de pisciculture et de pêcherie (Varna), 4: 91-98. (in Bulgarian). [45]

Beshovski V. 1964b. Beitrag zum Studium der halobionten und halophilen Diptera Brachycera vom Bulgarischen Küstenbereich des Schwarzen Meeres. – Zoologischer Anzeiger, 172 (4): 261-264. [46]

Beshovski V. 1965. The Insects of the Bulgarian Black Sea Coast. Varna, Darzhavno izdatelstvo. 96 p. (in Bulgarian). [47]

Beshovski V. 1966a. Several families of Diptera unknown to Bulgarian fauna: Dryomyzidae, Chamaemyiidae, Trichoscelidae, Astiidae, Canaceidae and Milichidae. – Bulletin de l'Institut de zoologie et musée, 21: 11-13. (in Bulgarian). [48]

Beschovski V. 1966b. Unbekannte Ephydridae (Dipt.) in der Fauna Bulgariens. – Zoologischer Anzeiger, 176: 270-276. [49]

Beshovski V. 1972a. Ecologic investigations on the representatives of Diptera-Brachycera from the salt biotops along the Bulgarian Seaside. I. Dipteral fauna from the cliffs. – Bulletin de l'Institut de zoologie et musée, 35: 5-21. (in Bulgarian). [50]

Beshovski V. 1972b. The sea-weeds cast ashore around the city of Varna (Bulgaria) and their fauna. – Bulletin de l'Institut de zoologie et musée, 35: 31-40. (in Bulgarian). [51]

Beshovski V. 1973a. Ecologic investigations on the representatives of Diptera-Brachycera from the salt biotops along the Bulgarian Black Sea coast. II. The dipteral fauna from the sea-weeds cas. – Bulletin de l'Institut de zoologie et musée, 38: 5-20. (in Bulgarian). [52]

Beshovski V. 1973b. Ökologische Untersuchungen über Diptera, Brachycera Vertreter aus den salzigen Biotopen des bulgarischen Schwarzmeerstrands. III. Die Diptera-Fauna des von Wellen überfluteten sandigen Strandes. – Bulletin de l'Institut de zoologie et musée, 38: 195-230. (in Bulgarian). [53]

Beshovski V. 1975a. The Black Sea coast inundated by waves and its Dipterous fauna (Diptera-Brachycera). – Hydrobiology, 2: 3-18. (in Bulgarian). [54]

Beschovski V. 1975b. Ökologische Untersuchungen über Vertreter von Diptera Brachycera von den Salzbiotopen der bulgarischen Schwarzmeerküste. – Ecology, 1: 64-74. (in Bulgarian). [55]

Beschovski V. 1976. Diptera-Brachycera in the dunes along the Bulgarian Black Sea coast. – In: Terrestrial fauna of Bulgaria. Materials. 35-59. (in Bulgarian). [56]

Beshovski V. 1978. Invertebrata. – In: Valkanov A., Ch. Marinov, Ch. Danov and P. Vladev (eds.): The Black Sea. Varna, Georgi Bakalov, 218-230. (in Bulgarian). [57]

Beshovski V. 1993. Taxonomic and systematic notes on the genera *Tethina* Haliday, 1838, and *Rhicnoessa* Loew, 1862 (Insecta: Diptera: Tethinidae). – Reichenbachia, 30 (16): 103-107. [58]

Beshovski V. 1994a. Fauna bulgarica. 23. Insecta, Odonata. Sofia, Bulgarian Academy of Sciences Publishing House. 373 p. (in Bulgarian). [59]

Beshovski V. 1994b. Contribution to the study of the West Palaearctic Tethinidae (Diptera). – Acta zoologica bulgarica, 47: 16-29. [60]

Beshovski V. 1996. Qualitative composition and distribution of the Ephydridae species (Diptera) on the moist zones of the Shablenska like system. – Acta entomologica bulgarica, 1: 16-20. (in Bulgarian). [61]

Beshovski V. 2009. Fauna bulgarica. 28. Insecta: Diptera: Ephydridae, Tethinidae, Canacidae. Sofia, Editio Academica "Professor Marin Drinov". 423 p. (in Bulgarian). [62]

Beschovski V., T. Zatwarnicki. 2002. Faunstic Review of the Subfamily Gymnomyzinae (Insecta: Diptera: Ephydridae) in Bulgaria with Some Data from Other European Countries. – Acta zoologica bulgarica 53 (3): 3-18. [63]

Biserkov V., V. Golemanski (Eds.). 2011. Red Data Book of the Republic of Bulgaria. Digital edition. Sofia, Bulgarian Academy of Sciences & Ministry of Environment and Water. [64]

Bogdanova D., A. Konsulov. 1993. On the distribution of the new Ctenophora species *Mnemia maccradyi* in the Black Sea along the Bulgarian coastline in the summer of 1990. – In: Comptes rendus de l'Academie bulgare des Sciences, 46 (3): 71-74. [65]

Borcea I. 1926. Note sur les moules et sur les faciès ou biocénoses à moules de la région littorale roumaine de la Mer Noire. – Annales scientifiques de l'Université de Jassy, 14: 129-139. [66]

Borcea I. 1931. Nouvelles contributions à l'étude dela faune bentique dans la Mer Noire près du littoral roumain. – Annales scientifiques de l'Université de Jassy, 16: 655-750. [67]

Borcea I. 1937. Les résultats de l'expédition de recherches dans la Mer Noire du 28 août – 1 septembre 1935. – Annales scientifiques de l'Université de Jassy, 23: 1-26. [68]

Brandis D., V. Storch and M. Türkay. 2000. Taxonomy and zoogeography of the freshwater crabs of Europe, North Africa, and the Middle East. – Senkerbergiana biologica, 80 (1-2): 5-56. [69]

Bruyne R. 2003. The complete encyclopedia of shells. Lisse, Rebo. 336 p. [70]

Bulgurkov K. 1938a. Study of Rhizocephala and Bopyridae from the Bulgarian Black Sea Coast. – Arbeiten aus der Biologischen Meeresstation in Varna, 7: 69-81. (in Bulgarian). [71]

Bulgurkov K. 1938b. Some freshwater and marine Decapoda drom the Varna and Sozopol. – Arbeiten aus der Biologischen Meeresstation in Varna, 7: 83-116. (in Bulgarian). [72]

Bulgurkov K. 1939. On certain more important fresh- and saltwater Crustacea – Decapoda from Bulgaria. – Travaux de la société bulgare des sciences naturelles, 18: 103-160. (in Bulgarian). [73]

Bulgurkov K. 1961. Systematik, Biologie und zoogeographische Verbreitung der Süsswasserkrebse der Familien Astacidae und Potamonidae in Bulgarien. – Bulletin de l'Institut de zoologie et musée, 10: 165-192. (in Bulgarian). [74]

Bulgurkov K. 1963. Faunistic findings of Crustacea and Mollusca along the western coast of the Black Sea. – Bulletin de l'Institut central de recherche scientifique de pisciculture et de pêcherie (Varna), 3: 265-267. (in Bulgarian). [75]

Bulgurkov K. 1968. *Callinectes sapidus* Rathbun in Black Sea. – Proceedings of the Research Institute of Fisheries and Oceanography (Varna), 9: 97-99. (in Bulgarian). [76]

Bulgurkov K. 1970. On the reproduction and fertility of certain Decapoda (Crustacea) along the Bulgarian coast of the Black Sea. – Proceedings of the Institute of Oceanography and Fisheries (Varna), 10: 65-74. (in Bulgarian). [77]

Bulgurkov K. 1973. Finding of *Alpheus dentipes* Guerin (Crustacea, Decapoda) along the Bulgarian coast of the Black Sea. – Proceedings of the Institute of Oceanography and Fisheries (Varna), 12: 103-105. (in Bulgarian). [78]

Buresch I. 1940. Das tropische Reisen-Wasserinsekt *Belostoma niloticum* Stal. (*Lethoceros cordophanus* Meyr), gefunden in Bulgarien. – Mitteilungen der Bulgarischen Entomologischen Gesellschaft in Sofia, 11: 138-160. (in Bulgarian). [79]

Caraion F. 1960. *Loxoconcha bulgarica* n. sp. a new Ostracod collected in the bulgarien waters of the Black Sea (Sozopol). – Revu de Biol., 5 (3): 21. [80]

Cărăuşu S., A. Cărăuşu. 1942. Amphipodes provenant des dragages effectués dans les eaux roumaines de la Mer Noire (28 août – 1 sept. 1935). – Ann. Sc. Univ. Jassy, 28: 1-18. [81]

Cărăuşu S. 1943. Amphipodes de Roumanie. I. Gamaridés de type caspien. Bucuresti, Instit. de cercet. piscic. al Rômâniei, 293 p. [82]

Cărăuşu S., E. Dobreanu and C. Manolache. 1955. Amphipoda, forme salmastre si di apa dulce. – In: Fauna Republicii Populare Romîne. Crustacea. 4 (4). Bucharest, Editura Academiei Republicii Populare Romîne. 409 p. (in Romanian). [83]

Caspers H. 1951a. Quantitative Untersuchungen über die Bodentierwelt des Schwarzen Meeres im bulgarischen Küstenbereich. – Archiv für Hydrobiologie, 45 (1-2): 1-192. [84]

Caspers H. 1951b. Biozönotische Untersuchungen über die Strandarthropoden im bulgarischen Küstenbereich des Schwarzen Meeres. – Hydrobiologia, den Haag, 3 (2): 131-193. [85]

Caspers H. 1952. Untersuchungen über die Tierwelt von Meeressalinen an der bulgarischen Küste des Schwarzen Meeres. – Zoologischer Anzeiger, 148 (5/8): 243-259. [86]

Černosvitov L. 1935. Über einige Oligochaeten aus dem See- und Brackwasser Bulgariens. – Mitteilungen aus den Königl. Naturwissenschaftlichen Instituten in Sofia, 8: 186-189. [87]

Černosvitov L. 1937. Die Oligochaetenfauna Bulgariens. – Mitteilungen aus den Königl. Naturwissenschaftlichen Instituten in Sofia, 10: 69-92. [88]

Chartosia N., Anastasiadis D., Bazairi H., Crocetta F., Deidun A., Despalatović M., Di Martino V., Dimitriou N., Dragičević B., Dulčić J., Durucan F., Hasbek D., Ketsilis-Rinis V., Kleitou P., Lipej L., Macali A., Marchini A., Ousselam M., Piraino S., Stancanelli B., Theodosiou M., Tiralongo F., Todorova V., Trkov D., & Yapici S. 2018. New Mediterranean Biodiversity Records (July 2018). Mediterranean Marine Science 19 (2): 398-415. https://doi.org/10.12681/mms.18099 [88a]

Chichkoff, G. 1907. Contribution à l'étude de la faune de la Mer Noire. Halacaridae des côtes bulgares. – Archives de zoologie expérimentale et générale, 4-e sér., 7 (6): 247-268. [89]

Chichkoff, G. 1908. Materials to the study of the Fauna of Bulgaria. I. Free-living Copepoda. – Annuaire de l'Université de Sofia. Faculté physico-mathématique, 3-4: 300-350. (in Bulgarian). [90]

Chichkoff, G. 1912. Contribution à l'étude de la faune de la Mer Noire. Animaux récoltés sur les côtes bulgares. –Archives de zoologie expérimentale et générale, 5e sér., 10 (2): 29-39. [91]

Chichkoff, G. 1924. Sur quelques espèces de Triclades d'eau douce de la Bulgarie. – Annuaire de l'Université de Sofia. Faculté physico-mathématique, 20 (2): 113-120. (in Bulgarian). [92]

Cihodaru M. 1937. Notes sur quelques Isopodes du littoral roumain de la Mer Noire. – Annales scientifiques de l'Université de Jassy, 23: 256-260. [93]

Codreanu M., R. Codreanu. 1956. Sur I'*Anisarthrus pelseneeri*, Epicaride parasite abdominal de la crevette *Athanas nitescens*; sa presence dans la Mer Noire et la dispersion du genre *Anisarthrus*. – Bull. biol. France Belgique, 90 (2): 111-121. [94]

Cvetkov L. 1953. New inhabitants of the Black Sea. – Priroda, 4: 66-68. (in Bulgarian). [95]

Cvetkov L. 1955a. Die Chironomidenfauna der Bulgarischen Schwarzmeerseen. – Bulletin de l'Institut zoologique de l'Academie des sciences de Bulgarie, 4-5: 215-249. (in Bulgarian). [96]

Cvetkov L. 1955b. Untersuchungen über das Futter der Fische im Beloslavsko-See. – Bulletin de l'Institut zoologique de l'Academie des sciences de Bulgarie, 4-5: 329-352. (in Bulgarian). [97]

Cvetkov (Zvetkov) L. 1957. Zoobenthos des Beloslav-Sees. – Bulletin de l'Institut zoologique de l'Academie des sciences de Bulgarie, 6: 381-439. (in Bulgarian). [98]

Cvetkov (Tsvetkov) L. 1958. Investigations into the microbenthos of the Bulgarian Black Sea lakes. – Bulletin de l'Institut zoologique de l'Academie des sciences de Bulgarie, 7: 219-250. (in Bulgarian). [99]

Cvetkov L. 1960. *Branchinella spinosa* (M. Edwards) in Bulgaria. – Bulletin de l'Institut zoologique de l'Academie des sciences de Bulgarie, 9: 397-398. (in Bulgarian). [100]

Cvetkov L., T. Gruncharova. 1976. Fouling in the liman of the Ropotamo River. I. Conditions for the development of the fouling biocenosis. – Hydrobiology, 4: 3-18. (in Bulgarian). [101]

Cvetkov L., T. Gruncharova. 1979. Fouling in the liman of the Ropotamo River. III. Composition and quantitative development of the zoocenosis. – Hydrobiology, 8: 26-43. (in Bulgarian). [102]

Cvetkov L., T. Marinov. 1986. Faunistic enrichment of the Black Sea and changes in its benthic ecosystems. – Hydrobiology (Sofia), 27: 3-20 (in Russian). [103]

Czapic A. 1952a. Unterzuchungen über die Infusorien und Rotatorien des Küstengrundwassers und des Sandbodens der Stalinbucht. – Arbeiten aus der Biologischen Meeresstation in Varna, 17: 61-66. [104]

Czapic A. 1952b. Die Infusorien und Rotatorien des Stalin-Sees. – Arbeiten aus der Biologischen Meeresstation in Varna, 17: 67-72. [105]

Czapic A. 1952c. Unterzuchungen über die Lithotelmen der bulgarischen schwarz-Meerküste. – Arbeiten aus der Biologischen Meeresstation in Varna, 17: 73-82. [106]

Dimitrov M. 1962. The benthos of the „Studen Kladenets" Dam. – Rep. Exper. Station of Freshwater Pisciculture (Plovdiv), 1: 81-94. (in Bulgarian). [107]

Dimitrov P., E. Michova and V. Peichev. 1998. Palaeooceanological reconstruction in the West Back sea part during the Quaternary. – Institute of Oceanology, 2: 127-139. (in Bulgarian). [108]

Dimoff I. 1960. Le zooplankton de la Mer Noire devant la cote Bulgare pendant les annes 1954, 1955 et 1956. –Travaux de l'Institut de recherches scientifiques sur la pêche et les industries s'y rattachant (Varna), 2: 85-147. [109]

Dimoff I. 1964. Cladocera et Copepoda d'Eau Douce, Trouvees dans la Mer Noire. – Bulletin de l'Institut central de recherche scientifique de pisciculture et de pêcherie (Varna), 4: 31-37. (in Bulgarian). [110]

Dimoff I. 1966. Zooplankton of the western shores of the Black Sea in 1960-1964. – Bulletin de l'Institut central de recherche scientifique de pisciculture et de pêcherie (Varna), 7: 45-68. [111]

Dimov I., T. Marinov, A. Konsulov. 1970. Some hydrological and hydrobiological particularities of the northwestern section of the Varna Lake and the cooling system of theVarna thermoelectric power station during 1968. – Proceedings of the Institute of Oceanography and Fisheries (Varna), 10: 27-47. (in Bulgarian). [112]

Drensky P. 1947. Synopsis and distribution of freshwater Mollusca in Bulgaria. – Annuaire de l'Université de Sofia. Faculté physico-mathématique, 43 (3): 33-54. (in Bulgarian). [113]

Dumont H., V. Mamaev and Y. Zaitsev. 1999 (Eds.): Black Sea Red Data Book. New York, UNOPS. 413 p. [114]

Earle S., L. Glover. 2009. Ocean an illustrated atlas. Washington, National Geographic. 352 p. [115]

Elder D., J. Pernetta. 1991. The Random House atlas of the oceans. IUCN, New York, Random House. 200 p. [116]

Erk C., Uzunova S., Chartosia N., Kevrekidis K., Almón B. & Schubart Ch. 2024. Further evidence for diversification within the *Diogenes pugilator* complex (Anomura, Diogenidae) in the Mediterranean end Black Seas. – Crustaceana, 97 (5-9): 627-657. [116a]

Evlogiev Y. 2009. Pliocene surface and high Quaternary terraces in the Batova River valley, Northeast Bulgaria. – Review of the Bulgarian Geological Society, 70 (1-3): 91-101. (in Bulgarian). [117]

Gecheva G., Belkinova D., Varadinova E., Mihov S. 2013. Developing a methodology for monitoring in the protected area Atanasovsko Lake. Project LIFE11 NAT/BG000362 – Emergency measures for recovery and long-term conservation of the coastal lagoon Atanasovsko Lake. Final Report. Bulgarian Biodirversity Foundation and Bulgarian Society for the Protection of Birds. Burgas, 63 p. [117a]

Georgiev D. 2012. The freshwater snails (Mollusca: Gastropoda) in Bulgaria. Dr. Sci. thesis, Plovdiv, Plovdiv University "P. Hilendarski", Faculty of Biology, 420 p. (in Bulgarian). [118]

Georgiev B., P. Nikolov. 2010. Invertebrates in the Pomorie Wetland. – In: Collection with reports Pomorie Lake region (Protected Site "Pomorie Lake", Ramsar Site, NATURA 2000 sites "Pomorie" BG 0000620 and "Pomorie Lake" BG 0000152), 72-87. (in Bulgarian). [119]

Georgieva D. 1998. The structure of the zooplankton in the Black Sea in front the Bulgarian coast, formeted in the summer 1990. – Proceedings of the Institute of Oceanology (Varna), 2: 140-145. (in Bulgarian). [120]

Gerlach S. 1951. Freilebende Nematoden aus Varna an der bulgarischen Küste des Schwarzen Meeres. – Archiv für Hydrobiologie, 45: 193-213. [121]

Golemansky V. 2007. Biodiversity and ecology of the Bulgarian Black Sea Invertebrates. – In: V. Fet and A. Popov (Eds.). Biogeography and ecology of Bulgaria. Dordrecht, Springer, 537-553. [122]

Golikov A. 1982. On principles of regionalization and unification of terminology in marine biogeography. – In: Kusakin O. (Ed.): Marine Biogeography, Its Subject, Methods, and Principles of Regionalization. Moscow, Nauka, 94-98 (in Russian). [123]

Golikov A., Y. Starobogatov. 1966. Ponto–Caspian gastropod molluscs in the Azov–Black Sea basin. – Zoologicheskii zhurnal, 45 (3): 352-362. (in Russian). [124]

Golikov A., Y. Starobogatov. 1968. Zoogeographical characteristics of the gastropod mollusks of Black and Azov Seas. – In: Biological Studies of the Black Sea and Its Fishery Resources. Moscow, Nauka, 70-83. (in Russian). [125]

Golikov A., Y. Starobogatov. 1972. Class Gastropoda. – In: Mordukhay-Boltovskogo F. (Ed.): The Keys to the Fauna of Black and Azov Seas. 3. Kiev, Naukova dumka, 65-166. (in Russian). [126]

Gomoiu M., I. Porumb. 1969. *Mya arenaria* L. bivalve presently penetrated into the Black Sea. – Rev. Romaine biol., Ser. zool., 14 (3): 199–202. [127]

Gomoiu M., M. Skolka. 1996. Changements recents dans la biodiversite de la mer Noire dus aux immigrants. GEO-ECO-MARINE, RCGGM, Danube Delta – Black Sea System under Global Changes Impacts, Bucharest – Constanta, 49-65. [128]

Gomoiu M.-T., B. Alexandrov, N. Shadrin and Y. Zaitsev. 2002. The Black Sea – a recipient, donor and transit area for alien species. – In: Geppäkaki E., S. Gellasch and S. Olonin (Eds.): Invasive aquatic species of Europe. Distribution, impacts and management. Dordrecht, Kluwer Academic publishers, 341-350. [129]

Grabowski M., V. Pešić. 2007. New data on the distribution and checklist of fresh- and brackishwater Gammaridae, Pontogammaridae and Behningiellidae (Amphipoda) in Bulgaria. – Lauterbornia, 59: 53-62. [130]

Grinchuk T., P. Michailova. 1979. The karyotypic variation of *Glyptotendipes barbipes* (Staeger) (Diptera, Chironomidae) from various localities. – Tsitologya, 21: 859-964. (in Russian). [131]

Grossu A. 1962. Fauna Republicii Populare Romîne. 3 (3). Mollusca, Bivalvia. Bucharest, Editura Academiei Republicii Populare Romîne. 426 p. (in Romanian). [132]

Grossu A. 1983. Gastropoda Romanie. 4. Ordo Stylommatophora. Bucharest, Litera. 563 p. (in Romanian). [133]

Grossu A. 1986. Gastropoda Romanie. 1. Prosobranchia et Opistobranchia. Bucharest, Litera. 524 p. (in Romanian). [134]

Grossu A. 1987. Gastropoda Romanie. 2. Pulmonata. Bucharest, Litera. 443 p. (in Romanian). [135]

Grossu, A. 1993. The Catalogue of the Molluscs from Romania. – Travaux du Museum National d'histoire Naturelle "G. Antipa", 33: 291-366. [136]

Gruev B. 1995. About the mediterranean faunistic complex in Bulgaria. – Annuaire de l'Université de Sofia "Kliment Ohridski", Faculté de Biologie, 86/87 (1 – Zoology): 75-82. [137]

Gruev B. 2000. About the submediterranean zone of the Palaearctic realm and the submediterranean faunistic element in Bulgaria. – Trav. Sci. Univ. Plovdiv, Animalia, 36 (6): 73-94. (in Bulgarian). [138]

Gruev B., B. Kusmanov. 1994. General biogeography. Sofia, Univ. K. Ohridski, 498 p. (In Bulgarian). [139]

Gruev B., B. Kusmanov. 1999. General biogeography. Plovdiv, Univ. P. Hilendarski, 344 p. (In Bulgarian). [140]

Gruncharova T. 1977. Under-water fouling in the liman of Ropotamo River. II. Bryosoan fauna. – Hydrobiology, 6: 89-100. (in Bulgarian). [141]

Gruncharova T. 1980. *Electra pontica* sp. n. (Bryozoa, Membraniporida) on the Bulgarian Black Sea coasr. – Acta zoologica bulgarica, 14: 88-91. [142]

Guéorguiev V. 1987. Fauna bulgarica. 17. Coleoptera, Hydrocanthares. Sofia, Bulgarian Academy of Sciences Publishing House. 161 p. (in Bulgarian). [143]

Guéorguiev V. 1991. Contribution to the study of family Hydrophilidae (Coleoptera) in Bulgaria. – Acta zoologica bulgarica, 42: 66-69. (in Bulgarian). [144]

Guryanova Ye. 1964. Zoogeographic regionalization of the benthic fauna of the World Ocean. – In: Physico-geographical Atlas of the World, Moscow, Map 68 B, 291-292. (in Russian). [145]

Gussev Ch. 2015. Mediterranean halophytic communities of tall rushes, sedges and grasses. – In: Red Data Book of the Republic of Bulgaria. 3. Natural habitats. Sofia, Bulgarian Academy of Sciences & Ministry of Environment and Water, 47-48. [145a]

Harin N. 1972. Rotatoria. – In: Mordukhay-Boltovskogo F. (ed.): The Key for the Fauna of the Black and Azov Seas. 3. Kiev, Naukova dumka, 183-220. (in Russian). [146]

Hayward P., T. Nelson-Smith and Ch. Shields. 1996. Sea Shore of Britain and Europe. London, Collins. 352 p. [147]

Hesse P. 1916. Mollusken von Varna und Umgebung. – Nachrichtsblatt der Deutschen Malakozoologischen Gesellschaft, Frankfurt am Main, 48 (4): 145-157. [148]

Hook P. 2008. A concise guide to the seashore. Melbourne, Parragon. 256 p. [149]

Hrabě S. 1967. Two new species of the famili Tubificidae from Black Sea with remarks about various species of subfamily Tubificinae. – Spicy Prir. Fac. Univ. Brno, 485: 311-365. [150]

Hristova N. 2012. Hydrology of Bulgaria. Sofia, Tip-Top, 830 p. (In Bulgarian). [151]

Hubenov Z. 2005a. *Dreissena* (Bivalvia: Dreissenidae) – systematics, autochthonous and anthropogenic areas. – Acta zoologica bulgarica, 57 (3): 259-268. [152]

Hubenov Z. 2005b. Malacofaunistic diversity of Bulgaria. – In: Petrova, A. (Ed.): Current state of Bulgarian biodiversity – prblems and perspectives. Sofia, Bulgarian Bioplatform, Drakon, 199-246. (In Bulgarian). [153]

Hubenov Z. 2007a. Fauna and zoogeography of marine, freshwater, and terrestrial mollusks (Mollusca) in Bulgaria. – In: Fet, V., A. Popov (Eds.): Biogeorraphy and ecology of Bulgaria. Springer, Dodrecht, 141-198. [154]

Hubenov Z. 2007b. Faunistic review, distribution and zoogeographical characteristic of the bulgarian Black Sea mollusks (Mollusca: Polyplacophora, Gastropoda et Bivalvia). – Annuaire de l'Université de Sofia "Kliment Ohridski", Faculté de Biologie, 96/98 (1 – Zoology): 17-38. [155]

Hubenov Z. 2014. Zoogeography of the Free-living Metazoan Invertebrate Animals (Metazoa: Invertebrata) from the Bulgarian Sector of the Black Sea and the Coastal Brackish Basins. – Acta zoologica bulgarica, 66 (3): 329-346. [155a]

Hubenov Z. 2015. Species composition of the free living multucellular invertebrate animals (Metazoa: Invertebrata) from the Bulgarian sector of the Black Sea and the coastal brackish basins. – Historia naturalis bulgarica 21: 49-168. [155b]

Hubenov Z. 2024. Species Composition and Distribution of the Mussels (Mollusca: Bivalvia) in the Bulgarian Sector of the Black Sea. – Acta zoologica bulgarica, 76 (2): 159-176. [155c]

Hubenov Z, L. Kenderov, I. Pandourski. 2015. Invertebrate Animals (Metazoa: Invertebrata) of the Atanasovsko Lake, Bulgaria. – Historia naturalis bulgarica, 22: 43-69. [155d]

Ivanov K., A. Sotirov, A. Rozhdestvensky and D. Vodenicharov. 1964. The lakes of Bulgaria. – Proceedings of the Institute of Hydrology and Meteorology, 16: 1-243. (in Bulgarian). [156]

Jacubisiak S. 1938. Les Harpacticoides de la Mer Noire (Côtes Roumaines). – Ann. Sc. Univ. Jassy, 24: 382-402. [157]

Josifov M. 1961. Die halobionten und halophilen Heteropteren an der Schwarzmeerküste. – Bulletin de l'Institut zoologique de l'Academie des sciences de Bulgarie, 10: 5-37. (in Bulgarian). [158]

Josifov M. 1981. Nasekomite ot razred Heteroptera na Balkanskiya poluostrov. Dr. Sci. thesis, Sofia, Bulgarian Academy of Sciences, 31-288. (In Bulgarian) [159]

Kalcheva S. 2017. Shell size and population density of *Cerastoderma glaucum* Poiret 1789 (Mollusca: Bivalvia) in „Pomorie Lake" (Black Sea coast, Bulgaria). ZooNotes 105: 1-4. [159a]

Kalcheva S. 2018. Ecological study of bentic mollusks in Pomorie Lake Protected Area. Ph.D. thesis, Stara Zagora, Trakia University, Faculti of Agriculture, Departmant of Biology and aquaculture. 177 p. (in Bulgarian). [159b]

Kalcheva S. 2019. Ecological study of bentic mollusks in Pomorie Lake Protected Area. Dissertation abstract, Ph.D. thesis, Stara Zagora, Trakia University, Faculti of Agriculture, Departmant of Biology and aquaculture. 36 p. (in Bulgarian). [159c]

Kalcheva S., Georgiev D. & Georgiev D. 2015. Malacofauna (Gastropoda and Bivalvia) of the "Pomoriysko ezero" protected area: preliminary results. Journal of BioScience and Biotechnology, Faculty of Biology, Plovdiv University (Special edition / Online Section "Biodiversity & Ecology", Second National Youth Conference "Biological sciences for a better future", Plovdiv, October 30-31, 2015), 281-183. [159d]

Kamburska L. 2004. The role of nonindigenous ctenophores *Mnemiopsis leidyi* and *Beroe ovata* for the zooplancton alterations along the Bulgarian Black Sea coast. Ph.D. thesis, Sofia, Bulgarian Academy of Sciences. 171 p. (in Bulgarian). [160]

Kamburska L., S. Moncheva. 2003. The least wanted in the Black Sea. International Conference NEAR 2 "The Impact of Global Environmental Problems on Continental & Coastal Marine Waters", 16-18 July, Geneva, Switzerland, Abstract volume, 31-32. [161]

Kamburska L., S. Moncheva, A. Konsulov, A. Krastev and K. Prodanov. 2003. The invasion of *Beroe ovata* in the Black Sea a warning signal for ecosystem concern. – Proceedings of the Institute of Oceanology (Varna), 4: 111-123. [162]

Kamburska L., Valcheva V. 2003. On the peculiarities of the zooplankton spatial distribution in Burgas Bay – May, 1996. – Proceedings of the Institute of Oceanology (Varna), 4, 124-131. (in Bulgarian). [163]

Kaneva-Abadjieva V. 1957. Mollusca and Malacostraca im Varnasee. – Arbeiten aus der Biologischen Meeresstation in Varna, Bulgarien, 19: 127-154. (in Bulgarian). [164]

Kaneva-Abadjieva V. 1958. A new harmful snail on our coast of Black Sea. – Nature (Sofia), 7(3): 89-91. (in Bulgarian). [165]

Kaneva-Abadjieva V. 1960a. Materials to the study of the mollusc fauna in the Back Sea at the Bulgarian shores. –Travaux de l'Institut de recherches scientifiques sur la pêche et les industries s'y rattachant (Varna), 2: 149–172. (in Bulgarian). [166]

Kaneva-Abadjieva V. 1960b. Researches on zoobenthos of the Bay of Varna in respect of Mollusca and Malacostraca. – Travaux de l'Institut de recherches scientifiques sur la pêche et les industries s'y rattachant (Varna), 2: 173-194. (in Bulgarian). [167]

Kaneva-Abadjieva V. 1960c. Higher crustaceans in mussel overgrowths in the gulf of Varna. – Bulletin de l'Institut zoologique de l'Academie des sciences de Bulgarie, 9: 399-403. (in Bulgarian). [168]

Kaneva-Abadjieva V. 1961. Contribution a l'etude des Cumacees de la Mer Noire devant le littoral Bulgare. – Bulletin de l'Institut central de recherche scientifique de pisciculture et de pêcherie (Varna), 1: 221-228. (in Bulgarian). [169]

Kaneva-Abadjieva V. 1962. Répartition des mollusques de la Mer Noire dans les eaux littorales bulgares. – Bulletin de l'Institut central de recherche scientifique de pisciculture et de pêcherie (Varna), 2: 67-79. (in Bulgarian). [170]

Kaneva-Abadjieva V. 1964. On the Amphipod fauna on the Black Sea along the Bulgarian coast and in the area near the Bosphorua. – Bulletin de l'Institut de pisciculture et de pêcherie (Varna), 4: 73-89. (in Bulgarian). [171]

Kaneva-Abadjieva V. 1968a. Distribution des mollusques subfossiles devant le littoral bulgare de la Mer Noire. Report. Comité internationale Mer Méditerranée, 19 (2) : 213-215. [172]

Kaneva-Abadjieva V. 1968b. Une nouvelle espece pour la faune de la Mer Noire: *Cheirocratus sundevalli* (Rathke), Amphipoda, Gammaroidea. – Proceedings of the Research Institute of Fisheries and Oceanography (Varna), 9: 93-96. (in Bulgarian). [173]

Kaneva-Abadjieva V. 1972. A contribution to the study of the Amphipodus Fauna in the Black-Sea-coast lakes and river mouths. – Bulletin de l'Institut de zoologie et musée, 35: 165-178. (in Bulgarian). [174]

Kaneva-Abadjieva V. 1974. A new mussel species for our Black Sea Coast. – Nature (Sofia), 23 (2): 64-65. (in Bulgarian). [175]

Kaneva-Abadjieva V. 1976. On the alimentary zoobenthos dynamics in Bourgas Lake and Mandra Dam during the 1967–1974 period. – Proceedings Institute of Fisheries (Varna), 14: 43-56. (in Bulgarian). [176]

Kaneva-Abadjieva V., T. Marinov. 1960a. Nourriture de quelques poissons benthophaques. – Travaux de l'Institut de recherches scientifiques sur la pêche et les industries s'y rattachant (Varna), 2: 41-72. (in Bulgarian). [177]

Kaneva-Abadjieva V., T. Marinov. 1960b. Zoobenthos distribution in the Bulgarian Black Sea area. – Arbeiten des Zentralen Forschungsinstitutes für Fischzucht und Fischerei Varna, 3: 117-166. (in Bulgarian). [178]

Kaneva-Abadjieva V., T. Marinov. 1965. Fouling organisms along the Bulgarian Black Sea Coast. – Proceedings of the Research Institute of Fisheries and Oceanography (Varna), 6: 137-145. (in Bulgarian). [179]

Kaneva-Abadjieva V., T. Marinov. 1966. Distribution of the zoobenthos in the sand biocoenosis of the Bulgarian Black Sea coast. – Proceedings of the Research Institute of Fisheries and Oceanography (Varna), 7: 69-95. (in Bulgarian). [180]

Kaneva-Abadjieva V., T. Marinov. 1967. Dynamics of the zoobenthos in the lakes along the Bulgarian Black Sea coast in the 1964-1966 period. – Proceedings of the Research Institute of Fisheries and Oceanography (Varna), 8: 177-194. (in Bulgarian). [181]

Kaneva-Abadjieva V., T. Marinov. 1977. The zoobenthos in the cystozyr fouling biocenosis. – Hydrobiology, 6: 76-88. (in Bulgarian). [182]

Kaneva-Abadjieva V., T. Marinov. 1984. A new mollusk species for the Black Sea, *Cunearca cornea* (Reeve). – Nature (Sofia), 33(5): 63-64. (in Bulgarian). [183]

Karachle P., M. C. Foka, F. Crocetta, J. Dulčić, N. Dzhembekova, M. Galanidi, P. Ivanova, N. Shenkar, M. Skolka, E. Stefanova, K. Stefanova, V. Surugiu, I. Uysal, M. Verlaque & A. Zenetos. 2017. Setting-up a billboard of marine invasive species in the ESENIAS area: current situation and future expectancies. – Acta Adriatica, 58 (3): 429-458. [183a]

Karnoschitzky N. 1949. Review of halobiont and halophl Coleopterae of Bulgarian Black Sea shores . – Arbeiten der Biologischen Meeresstation in Varna, 15: 1-66. (in Russian). [184]

Klie W. 1937. Ostracoden und Harpacticoiden aus brackigen Gewässer aus der bulgarischen Küste des Schwarzen Meeres. – Mitteilungen aus den Königl. Naturwissenschaftlichen Instituten in Sofia, 10: 1-42. [185]

Klie W. 1942. Adriatische Ostracoden. – Zoologischer Anzeiger, 138: [186]

Klissurov L. 2008. The Black Sea underwater. Varna, Passat Press. 232 p. [187]

Kolosváry G. 1951. Les Balanides de la Méditerranée. Acta Biologica AcademiaeScientiarum Hungaricae 2: 411-413. [187a].

Konsuloff S. 1912. Materials for Fauna Study of Bulgaria. Rotatoria. – Annuaire de l'Université de Sofia. Faculté physico-mathématique, 7: 1-72. (in Bulgarian). [188]

Konsuloff S. 1930. Die Hummer (*Hommarus vulgaris* M. Edw.) im Schwarzen Meere. – Zoologischer Anzeiger, 87: 11-12. [189]

Konsulov A. 1973. Seasonal and annual dynamics of the zooplankton of the Bourgas Lake and of the Mandra dam during the 1967-1970 period. – Proceedings of the Institute of Oceanography and Fisheries (Varna), 12: 67-85. (in Bulgarian). [190]

Konsulov A. 1974. Seasonal and annual dynamics of zooplankton in the Black Sea coast in the 1967-1970 period. – Proceedings of the Institute of Oceanography and Fisheries (Varna), 13: 64-78. (in Bulgarian). [191]

Konsulov A. 1979a. On zooplankton dynamics in Bourgas Lake. – Institute of Fisheries (Varna), 17: 24-31. (in Bulgarian). [192]

Konsulov A. 1979b. Dynamics of the zooplankton of the Mandra dam. – Institute of Fisheries (Varna), 17: 33-45. (in Bulgarian). [193]

Konsulov A. 1986. Seasonal and annual dynamics of zooplankton in the Black Sea along the Bulgarian coast for the 1974-1984 period. – Ocenology, 16: 23-33. [194]

Konsulov A. 1989. One more new-comer. A new species in the Black Sea - Dangerous or harmless for ecology. – Morski svjat, 2: 8. (in Bulgarian). [195]

Konsulov A. 1990. *Leucothea multicornis* Eschckoltz – a new species for the Black Sea. – Oceanology, 19: 98-99. [196]

Konsulov A. 1991. Biology, ekology and significance of zooplankton along the Bulgarian Black Sea coast. Dr. Sci. thesis, Sofia, Bulgarian Academy of Sciences. 300 p. (in Bulgarian). [197]

Konsulov A. 1998. Black Sea Biological Diversity (Bulgaria). Black Sea Environmental programme. Series. 5. Istanbul, UN Publications. 131 p. [198]

Konsulov A., L. Kamburska. 1997. Sensitivity to anthropogenic factors of the plankton fauna adjacent to the Bulgarian coast of the Black Sea. – In: E. Oszoy and E. Mykaelian (eds.): NATO ASI Series, Environment, Kluwer Academic Publishers, 2/27: 95-104. [199]

Konsulov A., L. Kamburska. 1998. Ekological determination of the new Ctenophora-*Beroe ovata* invasion in the Black Sea. – Institute of Oceanology (Varna), Bulgarian Academy of Sciences, 2: 195-198. [200]

Konsulov A., T. Konsulova. 1993. Biodiversity of the Black Sea benthos and plankton. – In:  Sakalian M. (ed.): The National Biological Diversity Conservation Strategy. 1. Biodiversity Support Program, Sofia, Pensoft, 473-514. (in Bulgarian). [201]

Konsulov A., T. Konsulova. 1998. Biological diversity of the Black Sea zoobenthos and zooplankton. – In:  Meine C. (ed.): Bulgaria's Biological Diversity: Conservation Status and Needs Assessment. 1/2. Biodiversity Support Program, Washington, Pensoft, 319-346. [202]

Konsulova T. 1978. Turbellaria from the Bulgarian Black Sea coast. – Institute of Fisheries (Varna), 16: 87-102. (in Bulgarian). [203]

Konsulova T., K. Stefanova. 2007. Black Sea Species. – In: Evaluation of the alien for the Bulgarian fauna species and measures to reduce their impact on natural ecosystems and native species. National Report, Ministry of Environment and Water of Bulgaria. [204]

Kovachev S., Y. Uzunov. 1981. The macrozoobenthos of the Durankulak Lake. Structure and dynamics of numbers and biomass. – Hydrobiology, 15: 74-80. (in Bulgarian). [205]

Kovachev S., S. Stoichev and V. Hainadjieva. 1999. The zoobenthos of several lakes along the Northern Bulgarian Black Sea coast. – Lauterbornia, 35: 33-38. [206]

Kovachev S., V. Hainadjieva and S. Stoichev. 2002. Quantitative characteristics of the zoobenthos of several coastal lakes in North-East Bulgaria. – Annuaire de l'Université de Sofia "St. Kliment Ohridski". Faculté de Biologie, 93/94 (1 – Zoology): 19-22. [207]

Kovalev A. 2003. Taxonomic composition of zooplankton in the Black Sea. – In: Eremeev V., A. Gaevskaya (Eds.): Modern condition of biological diversity in near-shore zone of Crimea (the Black Sea sector). Institute of Biology of the southern Seas, NAS Ukraine, Sevastopol, Ekosi-Gidrophizika, 44-49. (in Russian). [208]

Kozuharov D. 2007. Catalogue of the Bulgarian Rotifera fauna. – Lauterbornia, 59: 95-114. [209]

Kozuharov D., J. Topalova, S. Andreev, I. Botev, M. Stojneva and S. Stojchev. 1999. Hydrobiological investigations of the protected coastal water basins in the South Dobrudza – North East Bulgaria. – In: Reports of the scientific project supported by Le Balkan foundation, 1-35. [210]

Kumanski K. 1985. Fauna bulgarica. 15. Trichoptera, Annulipalpia. Sofia, Bulgarian Academy of Sciences Publishing House. 244 p. (in Bulgarian). [211]

Kutikova L. 1970. The Rotifera of the USSR Fauna (Rotataoria). Leningrag, Nauka. 742 p. (in Russian). [212]

Lattin G. de. 1967. Grundriss der Zoogeographie. Hochschullehrbücher für Biologie. Band 12. Jena, VEB Gustav Fischer Verlag. 602 p. [213]

Makarov Y. 2004. Fauna Ukraini. 26 (1-2). Desyatinogie rakoobraznie. Kiev, Naukova Dumka. 430 p. (in Russian). [214]

Marinov M. 2000. Pocket field guide of the dragonflies of Bulgaria. Sofia, Eventus. 104 p. (in Bulgarian). [215]

Marinov M. 2003. Chorology, biotopical and habitatic commitments of the Insects from Odonata in Bulgaria. Ph.D. thesis, Sofia, Bulgarian Academy of Sciences. 196 p. (in Bulgarian). [216]

Marinov M. 2005-2009. Dragonflies of Bulgaria. Bulgarian Foundation Biodiversity. http://www.odonata. biodiversity.bg/index.htm [217]

Marinov M., A. Kovachev. 2002. Management plan of the protected area Poda. Sofia, Ministry of Environment and Water, Bulgarian-Swiss Biodiversity Conservation Programme, Bulgarian Society for the Protection of Birds. 75 p. and Annexes. (in Bulgarian). [218]

Marinov T. 1957a. Beitrag zur kenntnis unserer Schwarzmeer-Polychaetenfauna. – Arbeiten aus der Biologischen Meeresstation in Varna, Bulgarien, 19: 105-119. (in Bulgarian). [219]

Marinov T. 1957b. Beitrag zur kenntnis unserer Schwarzmeer-Archiannelidenfauna. – Arbeiten aus der Biologischen Meeresstation in Varna, Bulgarien, 19: 121-126. (in Bulgarian). [220]

Marinov T. 1959a. Beitrag zur Erforschung der Polychätenfauna der Bulgarischen Schwarzmerküste. – Bulletin de l'Institut zoologique de l'Academie des sciences de Bulgarie, 8: 83-104. (in Bulgarian). [221]

Marinov T. 1959b. Sur la faune de polychètes des amas be moules de la Mer Noire. – Comptes rendus de l'Académie bulgare des Sciences, 12 (5): 443-446. [222]

Marinov T. 1959c. Echinoderm animals of the Black Sea. – Priroda, 3: (in Bulgarian). [223]

Marinov T. 1960. On the finding of *Mercierella enigmatica* Fuvel (Polychaeta) along the Bulgarian Black Sea coast. – Bulletin de l'Institut zoologique de l'Academie des sciences de Bulgarie, 9: 405-409. (in Bulgarian). [224]

Marinov T. 1962. Über die Muschelkrebs Fauna des westlichen Schwarzmeerstrandes. – Bulletin de l'Institut central de recherche scientifique de pisciculture et de pêcherie (Varna), 2: 82-108. (in Bulgarian). [225]

Marinov T. 1963a. Über die Polychaeten Faune der Sandzoocenose vor der Bulgarischen Schwarzmeerküste. – Bulletin de l'Institut central de recherche scientifique de pisciculture et de pêcherie (Varna), 3: 61-78. (in Bulgarian). [226]

Marinov T. 1963b. *Pontocytheroma arenaria* n. g. n. sp. Eine neue Ostracoden der Sandbiocönosen in dem Schwarzen Meer. – Comptes rendus de l'Académie bulgare des Sciences, 16 (5): 557 [227]

Marinov T. 1964a. Contribution to Ostracoda fauna of the Black Sea. – Bulletin de l'Institut central de recherche scientifique de pisciculture et de pêcherie (Varna), 4: 39-60. (in Bulgarian). [228]

Marinov T. 1964b. On macrozoobenthic fauna of the Black Sea (Kinorhyncha and Halacaridae). – Bulletin de l'Institut central de recherche scientifique de pisciculture et de pêcherie (Varna), 4: 61-71. (in Bulgarian). [229]

Marinov T. 1964c. Composition and features of the polychaetous fauna of the Bulgarian Black Sea coast. – Bulletin de l'Institut de zoologie et musée, 17: 79-107. (in Bulgarian). [230]

Marinov T. 1964d. *Loxoconcha aestuarii* n. sp. Eine neue Ostracodenart aus dem Schwarzen Meer. – Comptes rendus de l'Académie bulgare des Sciences, 16 (7): 757-67. [231]

Marinov T. 1964e. Untersuchungen über die Ostracodenfauna des Schwarzen Meeres. – Kieler Meeresforsch, 20 (1): 82-91. [232]

Marinov T. 1965. Zwei unbekannte Ostracoda aus dem Schwarzen Meer. – Bulletin de l'Institut de zoologie et musée, 18: 193-198. (in Bulgarian). [233]

Marinov T. 1966a. Uknown Black Sea Polychaetas of the Bulgarian coast. – Bulletin de l'Institut de zoologie et musée, 21: 69-75. (in Bulgarian). [234]

Marinov T. 1966b. Polychaete fauna in the brackish basins along the Bulgarian Black Sea coasts (lakes, marshes, estuaries). – Bulletin de l'Institut de zoologie et musée, 21: 139-152. (in Bulgarian). [235]

Marinov T. 1967a. Species of the genus *Ophelia* Savigny (Opheliidae, Polychaeta) in Black Sea. – Zoologicheskii zhurnal 46 (2), 187–191 (in Russian). [236]

Marinov T. 1967b. Le specie del genera *Leptocythere* (Ostracoda) del littorale Bulgaro del Mar Nero. – Publ. stat. zoolog. Napoli, 35: 273-285. [237]

Marinov T. 1968. *Petitia amphophthalma* Siewing: une nouvelle Polychaete de la Mer Noire. – Proceedings of the Research Institute of Fisheries and Oceanography (Varna), 9: 89-92. (in Bulgarian). [238]

Marinov T. 1971a. Polychètes et Archiannelides des eaux souterraines de plages du littoral bulgare de la Mer Noire. – Rapp. Comm. Int. Mer Médit., 20: 207-209. [239]

Marinov T. 1971b. Harpacticoides of the Bulgarian Black Sea coast. – Proceedings of the Institute of  Oceanography and Fisheries (Varna), 11: 43-87. (in Bulgarian). [240]

Marinov T. 1973a. Quelques espèces de la famille Ameridae (Copepoda, Harpacticoida) dans la Mer Noire. – Crustaceana, 24 (2): 231-241. [241]

Marinov T. 1973b. Quelques harpacticoides psammophiles inconus pour le bassin de la Mer Noire. – Vie et Mil., 22 (2-A): 309-326. [242]

Marinov T. 1974. Supplement to the studi of the Harpacticoid fauna from Bulgarian Black Sea coast. – Proceedings of the Institute of Oceanography and Fisheries (Varna), 13: 71-92. (in Bulgarian). [243]

Marinov T. 1975. Peculiarities of the Meiobenthos from the Sandy pseudolittoral and the Ground water of a Sandy Beaches. – Institute of Fisheries (Varna), 14: 103-135. (in Bulgarian). [244]

Marnov T. 1977. Fauna bulgarica. 6. Polychaeta. Sofia, Bulgarian Academy of Sciences Publishing House. 257 p. (in Bulgarian). [245]

Marinov T. 1978. Qualitative composition and quantitative distribution of the meiobenthos of the Bulgarian Black Sea waters. I. – Proceedings of  the Institute of Fisheries (Varna), 16: 35-49. (in Bulgarian). [246]

Marinov T. 1979. Composition, ecology and economic importance of zoobenthos in the Bulgarian Black Sea shelf. Dr. Sci. thesis, Sofia, Bulgarian Academy of Sciences. 414 p. (in Bulgarian). [247]

Marinov T. 1980. Qualitative composition and quantitative distribution of the meiobenthos of the Bulgarian Black Sea waters. II. – Proceedings of the Institute of Fisheries (Varna), 18: 85-93. (in Bulgarian). [248]

Marinov T. 1990. The Zoobenthos from the Bulgarian Sector of the Black Sea. Sofia, Publishing House of the Bulgarian Academy of Sciences. 195 p. (in Bulgarian). [249]

Marinov T., A. Apostolov. 1982. Contribution à l'étude des copepodes harpacticoides de la mer Adriatique (côte Yougoslave). 2. Sur le meiobenthos du Cap Piran. – Acta zoologica bulgarica, 18: 23-30. [250]

Marinov T., A. Apostolov. 1983. Zoogeography and ekology of the Black Sea Harpacticoids. – Proceedings of the Institute of Fisheries (Varna), 20: 135-144. (in Bulgarian). [251]

Marinov T., V. Golemansky. 1989. Second suplement to the catalogue of the Bulgarian Black Sea fauna. – Acta zoologica bulgarica, 37: 3-34. (in Bulgarian). [252]

Marinov T., S. Stoykov. 1982. Findings of sand gaper (*Mya arenaria*) in front of the Bulgarian Black Sea coast. – Bulletin of the National Oceanographic Committee, 3: 9-16. (in Bulgarian). [253]

Marinov T., S. Stoykov. 1995. Zoobenthos distribution in Bulgarian continental shelf of the Black Sea. – Proceedings of the Institute of Fisheries (Varna), 23: 119-137. (in Bulgarian). [254]

Marinov T., S. Stoykov and M. M'Barek. 1983. The zoobenthos of the sublittoral and muddy bottom of bay of Varna. – Proceedings of the Institute of Fisheries (Varna), 20: 109-133. (in Bulgarian). [255]

Marinova V., T. Trayanov and V. Michneva. 2004. Acoustic and video study on diurnal vertical migration of zooplankton in the central part of the Bulgarian Black Sea shelf. – Comptes rendus d l'Académie bulgare des Sciences, 57 (4): 55-58. [256]

Michev T. (ed.). 1997. Ecology and conservation of Atanasovsko Lake nature reserve. Collection of final scientific reports of experts, Sofia, Bulgarian-Swiss Biodiversity Conservation Programme. 76 p. [257]

Michev T. (ed.). 2003. Management Plan of Atanasovsko Lake. Sofia, Ministry of Environment and Water & Bulgarian-Swiss Biodiversity Conservation Programme. 65 p. and Annexes. (in Bulgarian). [258]

Michev T., M. Stoyneva (eds.). 2007. Inventory of Bulgarian wetlands and their biodiversity. Part 1: Non-lotic wetlands. Sofia, Publ. Hous Elsi-M. 364 p. + CD supplement. [259]

Mihailova-Neikova M. 1961. Hydrobiological research of the Mandra Lake with regard to its importance as a fishing ground. – Annuaire de l'Université de Sofia, Faculté de Biologie, Géologie et Géographie, 53 (1 – Zoology): 57-123. (in Bulgarian). [260]

Mihailova-Neikova M. 1966. *Schizopera gligici* Petkovski, 1957 (Copepoda, Harpacticoida) of Bulgaria. – Fragmenta balcanica, 5 (19): 129-133. [261]

Mihailova-Neikova M. 1968. About the Harpacticoida of Bulgaria. – Annuaire de l'Université de Sofia, Faculté de Biologie, 60 (1 – Zoology): 201-223. [262]

Mihailova-Neikova M., F. Voynova-Stavreva. 1971. Contribution to the study of Harpacticoida (Crustacea, Copepoda) from the Bulgarian Black Sea coast. – Annuaire de l'Université de Sofia, Faculté de Biologie, 63 (1 – Zoology): 37-43. (in Bulgarian). [263]

Michailova P. 1973. Unterzsuchungen über den Chromosomen Polymorphismus bei *Chironomus salinarius* Kieff., *Chironomus valkanovi* Michailova and *Chironomus anchialicus* Michailova (Diptera, Chironomidae) von der bulgarischen Schwarzmerküste. – Zoologischer Anzeiger, 191 (5/6): 348-365. [264]

Michailova P. 1974. Zwei neue Arten der Gattung *Chironomus* (Diptera, Chironomidae) von der bulgarischen Schwarzmeer Küste. – Zoologische Beiträge, 20: 339-357. [265]

Michailova P. 1975. Karyological characteristics of *Cricotopus varians* Staeg., and *Cricotopus vitripennis* Meig. (Diptera, Chironomidae) from Bulgarian Black Sea coast. – Acta zoologica bulgarica, 2: 25-36. (in Bulgarian). [266]

Michailova P. 1980. A review of the genus *Clunio* Haliday, 1855 (Diptera, Chironomidae). – Zoologischer Anzeiger, 200 (5/6): 417-432. [267]

Michailova P. 1982. External morphological and caryological characteristics of *Orthocladius bipunctellus* Zett. (Diptera, Chironomidae). – Zoologischer Anzeiger, 208 (1/2): 82-91. [268]

Mihneva V., K. Stefanova. 2013. The non-native copepod *Oithona davisae* (Ferrari F. D. and Orsi, 1984) in the Western Black Sea: seasonal and annual abundance variability. – Bioinvasions Records, 2 (2): 119-124. [269]

Mischev K., V. Popov. 1978. Morphographische Charakteristik des Schwarzmeerbeckens. – In: The Black Sea. Varna, G. Bakalov, 36-46. (in Bulgarian). [270]

Mitov P. 2015. A new record and redescription of *Acanthochitona crinita* (Pennant, 1777) from the Bulgarian Black Sea coast (Mollusca, Polyplacophora). – Spixiana, 38 (2): 169-185. [270a]

Mitov P. 2016. New data on *Myosotella myosotis* (Draparnaud, 1801) (Pulmonata: Basommatiphora: Ellobiidae) fom Bulgaria. – Acta zoologica bulgarica, 68 (3): 321-330. [270b]

Mitov P. G. 2019. *Eurypanopeus depressus* (Smith, 1869) (Brachyura: Panopeidae) – a new alien species for the Bulgarian Black Sea coast. – Acta Zoologica Bulgarica 71 (2): 253-266. [270c]

Mitov P. G. 2023. The Alien Crab *Eurypanopeus depressus* (Smith, 1869) (Decapoda: Panopeidae) is Already in Front of Turkish Black Sea Territorial Waters. – Acta Zoologica Bulgarica 75 (1): 91-96. [270d]

Mitov P. 2025. A new record for thealien acorn barnacle *Amphibalanus amphitrite* (Darwin, 1854) (Balanidae, Balanomorpha, Crustacea) from the Bulgarian Black Sea Coast. – Acta zoologica bulgarica, 77: [270e].

Mitov P. & Klain S. 2025. New records of the alien bivalve *Arcuatula senhousia* (Benson, 1842) from the Bulgarian Black Sea coast" (Mytiloida: Mytilidae). ........ [270f]

Mitov P., Uzunova S., Kenderov L., Dimov S. & Yanachkov P. 2020. Pacific oyster invasion along Bulgarian Black Sea coast. Poster – Kliment's Days 2020, Faculty of Biology, Sofia University. [270g]

Moncheva S., L. Kamburska. 2002. The control of ballast waters – why a crucial issue to safeguard the Black Sea biodiversity and ecosystem health? MEET/MARIND'2002, Proceedings of First International congress on mechanical and electrical engineering and technology and Fourth International conference on marine industry. 4: 249-257. [271]

Mordukhay-Boltovskogo F. 1960. The Caspian Fauna in the Azov–Black Sea Basin. Moscow & Leningrad, Publishing House of the Academy of Sciences of the USSR. 288 p. (in Russian). [272]

Mordukhay-Boltovskogo F. 1972. General characteristic of the fauna of the Black and Azov Seas – In: Mordukhay-Boltovskogo F. (Ed.): The Key for the Fauna of the Black and Azov Seas. 3. Kiev, Naukova dumka, 316-324. (in Russian). [273]

Motaş C., M. Băcescu. 1938. Sur quelques Cumacés limicoles et maricoles de Roumanie. Bucuresti, Vol. Jubilee "Grigore Antipa", 369-377. [274]

Motaş C., J. Soarec. 1940. Sur quelques Halacarides de la Mer Noire. – Annales scientifiques de l'Université de Jassy, II Section, 26 (1): 140-175. [275]

Munoz J., A. Gomez, A. Green, J. Figuerola, F. Amat, C. Rico. 2010. Evolutionary Origin and Phylogeography of the Diploid Obligate Parthenogen *Artemia parthenogenetica* (Branchiopoda: Anostraca). – PLoS ONE 5 (8): e11932. doi:10.1371/journal.pone.0011932 [275a]

Naidenov W. 1966. Katalog der Copepodenfauna Bulgariens. – Bulletin de l'Institut de zoologie et musée, 21: 109-138. [276]

Naidenow W. 1967. Hydrobiologische Untersuchungen der Wasserbecken an der südlichen Schwarzmeerküste und im Strandža-Gebirge. I. Cladocera, Calanoida, Cyclopoida. – Bulletin de l'Institut de zoologie et musée, 24: 57-95. (in Bulgarian). [277]

Naidenow W. 1981. Struktur und Dynamik des Zooplanktons aus dem See Durankulak. – Hydrobiology, 15: 62-73. (in Bulgarian). [278]

Naidenow W. 1994. Wandel in der Zusammensetzung der Cladocera- und Copepoda-Fauna des Süsßwassers in Bulgarien im letzten Jahrhundert. – Lauterbornia, 19: 95-106. [279]

Naidenow W. 1998. Struktur und Horizontalverteilung des Zooplanktons in zwei Küstenseen am Schwarzen Meer in Nordostbulgarien (Schabla-See und Eserez-See). – In: Golemansky V., W. Naidenov (eds.): Biodiversity of Schabla Lake System, Sofia, "Prof. Marin Drinov" Academic Publishing House, 51-68. [280]

Naumov D. 1968. Coelenterata. – In: Mordukhay-Boltovskogo F. (ed.): The Key for the Fauna of the Black and Azov Seas. 1. Kiev, Naukova dumka, 56-81. (in Russian). [281]

Necrassov O. 1937. Contribution à l'étude des Gastéropodes du littoral roumain de la Mer Noire. – Annales scientifiques de l'Université de Jassy, 23: 117-129. [282]

Nedelkov N. 1912. Sexthe contribution to the entomlogical fauna of Bulgaria. Revue Academy Sciences de Bulgarie, 2: 177-218. (in Bulgarian). [283]

Nesis K. 1982. Zoogeographical position of the Mediterranean Sea. – In: Kusakin O. (ed.): Marine Biogeography, Its Subject, Methods, and Principles of Regionalization. Moscow, Science, 270-299. (in Russian). [284]

Nevesskaya L. 1965. Poznechetvertichnye dvustvorchatye mollyuski Chernogo morya ih sistematika i ekologiya. Moskva, Nauka. 391 p. (in Russian). [285]

Netschaeff A. 1935. The Gebedje Lake as a basin for fishery. – Travaux de la Station Ichthiologique Sozopol, 3: 6-21. (in Bulgarian). [286]

Netschaeff A., W. Neu. 1940. *Aurelia aurita* L. und *Pilema* (*Rhizostoma*) *pulmo* Haeck. im Schwarzen Meer und im Bosphorus. – Zoologischer Anzeiger, 129: 61-63. [287]

Nguen Suan Li. 1984. Qualitative composition and quantitative distribution of zoobenthos in the Bourgas Bay. Ph.D. thesis, Sofia, Bulgarian Academy of Sciences. 177 p. (in Bulgarian). [288]

Nikolov G., A. Atanasov, D. Georgiev and E. Raichev. 2010. Analysis of the plankton in the area around the cape Maslen Nos, Bulgaria: Possibilities for cultivation of mediterranean mussels (*Mytilus galloprovincialis*). – Ecologia Balkanika, 2: 15-18. [289]

Nordsieck F. 1968. Die europäischen Meeres-Gehäuseschnecken (Prosobranchia). Stuttgart, G. Fischer. 273 p. [290]

Nordsieck F. 1969. Die europäischen Meeresmuscheln (Bivalvia). Stuttgart, G. Fischer. 253 p. [291]

Osikovski A., S. Hofman, D. Gweorgiev, S. Kalcheva, A. Falniowski. 2016. *Ecrobia maritima* (Milaschewitsch, 1916) and *E. ventrosa* (Montagu, 1803) in the East Mediterranean and Black Sea. – Annales Zoologici, 66 (3): 447-486. [291a]

Öztürk B., V. Michneva and T. Shigalova. 1911. First records of *Bolinopsis vitrea* (L. Agassiz, 1860) (Ctenophora: Lobata) in the Black Sea. – Aquatic Invasions, 6 (3): 355-360. [292]

Pandourski I. 2001. Recherches hydrobiologiques des zones humides de la côte bulgare de la Mer Noire. I. Le lac de Vaja. – Rivista di Idrobiologia, 40 (2/3): 321-334. [293]

Pandourski I., S. Stoichev. 1999. Sur la faune de l'eau interstitielle littorale de la bande sableuse entre le système lacustre de Chabla-Ezeretz et la Mer Noire. – Historia naturalis bulgarica, 10 : 125-131. [294]

Paspalev G. 1933a. Hydrobiologische Untersuchungen über den Golf von Varna. – Arbeiten aus der Biologischen Meeresstation in Varna, 2: 1-18. [295]

Paspalev G. 1933b. Das Vorkommen von *Hommarus vulgaris* M. Edw. im Schwarzen Meer. – Arbeiten aus der Biologischen Meeresstation in Varna, 2: 29-32. [296]

Paspalev G. 1934. Über das vorkomen von *Thaumantias maeotica* Ostr. in Golf von Varna. – Intern. Revue Hydrobiol. Hydrogr., 31: 373-280. [297]

Paspalev G. 1935. Hydrobiologische Untersuchungen über den Golf von Varna. II. Schiffsbodenbewachsung. – Arbeiten aus der Biologischen Meeresstation in Varna, 4: 1-10. [298]

Paspaleff G. 1936. *Thaumantias maeotica* Ostr. = *Pontia ostroumowi* gen. et sp. nov. (Ammen- und Medusengeneration, Züchtung). – Zoologischer Anzeiger, 115: 311-316. [299]

Paspaleff G. 1937. *Odessia maeotica* n. sp. (Namenberichtigung). – Zoologischer Anzeiger, 118: 112. [300]

Paspalev G. 1938. Über die Entwicklung von *Rhizostoma pulmo* Agass. – Arbeiten aus der Biologischen Meeresstation in Varna, 7: 1-25. [301]

Paspalev G. 1942. Beitrag zur Erforschung des Varna-Golfs und des Varna-Sees. – Arbeiten aus der Biologischen Meeresstation in Varna, 10: 11-14. [302]

Pechlivanov L. 2010. Biological diversity of the Pomorie Lake: Zooplankton. – In: Collection with reports Pomorie Lake region (Protected Site "Pomorie Lake", Ramsar Site, NATURA 2000 sites "Pomorie" BG 0000620 and "Pomorie Lake" BG 0000152), 88-94. (in Bulgarian). [303]

Pesta O. 1926. Carcinologische Mitteilungen. – Achiv für Hydrobiologie, 16: 3. [304]

Petrbok J. 1947. The freshwater molluscs of the lakes of Varna and of Gebedže. – Arbeiten aus der Biologischen Meeresstation in Varna, 13: 71-75 (in Bulgarian). [305]

Petrov B., I. Pandourski. 2015. Sea caves. – In: Red Data Book of the Republic of Bulgaria. 3. Natural habitats. Sofia, Bulgarian Academy of Sciences & Ministry of Environment and Water, 59-60. [305a]

Petrova A. 1972a. Sur la présence d'*Halacarellus suterraneus* Schulz, 1933 et *Halacarellus phreaticus* n. sp. (Halacaridae, Acari) en Bulgarie. – Acarologia, 13 (2): 367-373. [306]

Petrova A. 1972b. Sur quelques Halacariens trouvés dans le littoral de la Mer Noire. – Acarologia, 14 (4): 581-590. [307]

Petrova A. 1976. Une nouvelle espèce d'*Agauopsis* (Prostigmata, Halacaridae) du littoral de la mer Noire. – Hydrobiology (Sofia), 4: 67-70. [308]

Petrova A. 1978. *Pontarachna walkanovi* sp. n. (Acari, Pontarachnidae) du littoral de la mer Noire. – Acta zoologica bulgarica, 10: 60-63. [309]

Petrova E, Stoykov S. 2002. The zoobenthos from the coastal lakes on the northern Bulgarian Black Sea coast during 2000-2001 years. In: Limnological Reports, 34th Conference of the International Association for Danube Research, Tulcea, Romania, August 2002, 34: 337-343. [310]

Raykov V., M. Lepage, R. Pérez-Domínguez. 2010. First record of oriental shrimp, *Palaemon macrodactylus* Rathbun, 1902 in Varna Lake, Bulgaria. – Aquatic Invasions, 5, Supplement 1: S91-S95, doi: 10.3391/ai.2010.5.S1.019. [310a]

Revkov N. 2003. Taxonomical composition of the bottom fauna at the Black Sea Crimean coast. Zoobenthos regional peculiarities. Zoobenthos vertical distribution. – In: Eremeev V., A. Gaevskaya (eds.): Modern condition of biological diversity in near-shore zone of Crimea (the Black Sea sector). Institute of Biology of the southern Seas, NAS Ukraine, Sevastopol, Ekosi-Gidrophizika, 209-222. (in Russian). [311]

Revkov N., V. Abaza, C. Dumitrache, V. Todorova, T. Konsulova, E. Mickashavidze, M. Vershanidze, M. Sezgin, B. Ozturk, M. Chikina and N. Kucheruk. 2008. The state of zoobenthos. – In: T. Oguz (ed.): BSC, 2008. State of the Environment of the Black Sea (2001-2006/7). Istanbul, Publications of the Commission on the Protection of the Black Sea Against Pollution (BSC) 2008-3, 273-320. [312]

Riedl R. 1983. Fauna und Flora des Mittelmeeres. Hamburg, P. Parey. 836 p. [313]

Rudescu L. 1960. Rotatoria. – In: Fauna Republicii Populare Romîne. 2 (2). Bucharest, Editura Academiei Republicii Populare Romîne. 1192 p. (in Romanian). [314]

Rudescu L. 1968. Gastrotricha – In: Mordukhay-Boltovskogo F. (ed.): The Key for the Fauna of the Black and Azov Seas. 1. Kiev, Naukova dumka, 220-236. (in Russian). [315]

Rudescu L. 1972. Tardigrada. – In: Mordukhay-Boltovskogo F. (ed.): The Key for the Fauna of the Black and Azov Seas. 3. Kiev, Naukova dumka, 51-59. (in Russian). [316]

Russev B. 1957. Qualitative und quantitative Untersuchungen des Zooplanktons im Varna-See. – Arbeiten des Zentralen Forschungsinstitutes für Fischzucht und Fischerei (Varna), 1: 127-156. (in Bulgarian). [317]

Russev B. 1965. Midata *Dreissena polymorpha* – golyam vreditel po vodoprovodite i hidrotehnicheskite saorazheniya. – Priroda, 3: 85-87. (in Bulgarian). [318]

Russev B., I. Dimoff. 1957. Qualitative und quantitative Untersuchungen des Zooplanktons im Varna-Bucht. – Travaux de l'Institut de recherches scientifiques sur la pêche et les industries s'y rattachant Varna, 1: 78-109. (in Bulgarian). [319]

Russev B., T. Marinov. 1964. Über die Polychäten- und Hirudinenfauna im bulgarischen Sektor der Donau. – Bulletin de l'Institut et musée de zoologie, 15: 191-197. (in Bulgarian). [320]

Schuchert P. 2008. The European athecate hydroids and their medusae (Hydrozoa, Cnidaria): Filifera Part 4. – Revue suisse de Zoologie, 115 (4): 677-757. [321]

Sergeeva N., V. Zaika. 2013. The Black Sea meiobenthos in permanently hypoxis habitat. – Acta zoologica bulgarica, 65 (2): 139-150. [322]

Shadrin N. 2000. The distant invaders into the Black Sea and Sea of Azov: ecological explosion, their cause, prognosis. – IBSS, Ecology of the Sea (Ecologiya Morya), 51:72-77. [323]

Shigalova T., E. Musaeva, E. Arashkevich, L. Kamburska, K. Stefanova, V. Mihneva, L. Polishchuk, F. Timofte, F. Ustun, T. Oguz, M. Khalvashi and A. Turkan. 2008. The state of zooplankton. – In: T. Oguz (Ed.): BSC, 2008. State of the Environment of the Black Sea (2001-2006/7). Istanbul, Publications of the Commission on the Protection of the Black Sea Against Pollution (BSC) 2008-3, 201-246. [324]

Shopov V. 1993. Stratigraphy of the Quaternary of Bulgarian Black Sea shelf sediments. – Review of the Bulgarian Geological Society, 54 (1): 83-97. (in Bulgarian). [325]

Shopov V. 1996. Quaternary chroostratigraphy of Bulgaria. – Review of the Bulgarian Geological Society, 57 (3): 37-42. (in Bulgarian). [326]

Shornikov E. 1969. Ostracoda Latreille of the Black Sea. – In: Mordukhay-Boltovskogo F. (ed.): The Key for the Fauna of the Black and Azov Seas. 2. Kiev, Naukova dumka, 163-260. (in Russian). [327]

Sinegub I. 1994. New findings of the nudibranch mollusc *Doridella obscura* Verrill in the Black Sea. – Gidrobiol. zhurnal, 30 (3): 107-109. (in Russian). [328]

Skarlato O., Y. Starobogatov. 1972. Class of Bivalve Mollusks – Bivalvia. – In: Mordukhay-Boltovskogo F. (Ed.): The Key for the Fauna of the Black and Azov Seas. 3. Kiev, Naukova dumka, 178-249. (in Russian). [329]

Skolka M. 1970. Contributii la studiul bryozoarelor din Romania. – Comunicări de hidrobiologie, 2/3: 19-26. [330]

Stantschew B. 1940. Das Vorkommen *Podocoryne carnea* Sars im Golf von Varna und sein Verhalten ausserhalb des Wassers. – Arbeiten aus der Biologischen Meeresstation in Varna, 9: 11-18. [331]

Starobogatov Y. 1970. Fauna of Mollusks and Zoogeographic Regionalization of the Continental Water Bodies of the World. Leningrad, Science. 372 p. (in Russian). [332]

Stechow E. 1934. Vorkommen eines nordlischen Hydroiden, *Corymorpha sarsi* Steenstrup, im Schwarzen Meer. – Zoologischer Anzeiger, 107: 197-198. [333]

Stefanova K. 2001. An analysis of the zooplankton state in the lake of Varna. – Proceedings of the Institute of Oceanology (Varna), 3: 155-162. (in Bulgarian). [334]

Stefanova K. 2003. Daily dynamics of zooplankton in the Varna Bay (May 2001). – Proceedings of the Institute of Oceanology (Varna), 4: 133-139. [335]

Stock J. 1968. *Vectoriella marinovi*, un Copepod nouveau parasite d'une annèlide polychète pontique. – Crustaceana, supp. 1, 187-192. [336]

Stoichev S. 1998. The zoobenthos from the lakes Schabla-Ezerets (northern Black Sea coast of Bulgaria). – In: Golemansky V., W. Naidenov (eds.): Biodiversity of Schabla Lake System, Sofia, "Prof. Marin Drinov" Academic Publishing House, 91-99. [337]

Stoykov S. 1974 (1977). Free living Nematodes from the Bulgarian Black Sea Coast. II. – Annuaire de l'Université de Sofia, Faculté de biologie (Livre 1, Zoologie), 68: 57-62. (in Bulgarian). [338]

Stoykov S. 1975. Of the biology of *Orchestia bottae* (M.-Edw.) (Amphipoda, Gammaridae). – Hyrobiology, 2: 80-88. (in Bulgarian). [339]

Stoykov S. 1976. *Bathylaimus filipjevi* n. sp. (Nematoda, Tripyloidae). A new species of free-living Nematoda from Bulgarian Black Sea coast waters. – Acta zoologica bulgarica, 4: 91-93. [340]

Stoykov S. 1977. Free-living Nematodes new for our Black Sea fauna. – Proceedings of the Institute of Fisheries (Varna), 15: 107-113. (in Bulgarian). [341]

Stoykov S. 1978. Qualitative Zusammensetzung und quantitative Verbreitung der freilebenden Meeres-Nematoden der Bulgarischen Schwarzmeerküste. – Proceedings of the Institute of Fisheries (Varna), 16: 103-115. (in Bulgarian). [342]

Stoykov S. 1979. On the zoobenthos dynamics in Bourgas Lake and Mandra dam in the period 1975-1977. – Proceedings of the Institute of Fisheries (Varna), 17: 47-53. (in Bulgarian). [343]

Stoykov S. 1980. Free-living Nematoda of the Bulgarian Black Sea coast. Ph.D. thesis, Sofia, Bulgarian Academy of Sciences. 201 p. (in Bulgarian). [344]

Stoykov S. 1983. Preliminary data on age and growth of the Soft Clam, *Mya arenaria* L., of the Bulgarian coast. – Proceedings of the Institute of Fisheries (Varna), 20: 161-167. (in Bulgarian). [345]

Stoykov S. 1986. Research on the macrobenthos of the Lake Pomorie. – Oceanology, 16: 35-42. (in Bulgarian). [346]

Stoykov S. 1988. Free-living Nematoda from the Bulgarian Black Sea shelf area. – Oceanology, 17: 38-43. [347]

Stoykov S. 1996. Investigation on macrozoobenthos in the shelf along the Bulgarian Black Sea coast during the period 1995-1996. – Proceedings of the Institute of Fisheries (Varna), 24: 117-123. (in Bulgarian). [348]

Stoykov S., S. Uzunova. 1999. Dynamics of macrozoobenthos from the Bourgas Bay (Bulgarian Black Sea coast) during the period 1993-1995. – Proceedings of the Institute of Fisheries (Varna), 25: 153-154. [349]

Studencka B., M. Jasionowski. 2011. Bivalves from the Middle Miocene reefs of Poland and Ukraine: A new approach to Badenian/Sarmatian boundary in the Paratethys. – Acta geologica polonica, 61 (1): 79-114. [350]

Tichiy M. 1911. Zametka o Caprellidae Chernogo morya. – Izvestiya Imperatorskoy akademii nauk, S.-Peterburg, 1125-1134. (in Russian). [351[

Thiel M. 1935. Zur kenntnis der Hydromedusenfauna des Schwarzen Meeres. – Zoologischer Anzeiger, 111: 161-174. [352]

Thinemann A. 1936. Haffmücken und andere Salzwasserchironomiden. – Kieler Meeresforschungen, 1: 167-178. [353]

Todorova E. 2014. Kachestven i kolichestven sastav na nyakoi osnovni grupi bezgrabnachni zhivotni s vazhno nachenie za ezeroto Vaya [Qualitative and quantitative composition of some main groups of invertebrates of importance for the Lake Vaya]. Ph.D. thesis, Sofia University, Faculty of Biology (Department Zoology and Antropology). 276 p. (in Bulgarian). [353a]

Todorova V. 2005. Environmental status of zoobenthos communities of sublittoral sediments of the continental shelf in the northwestern part of the Black Sea and in front of the Bulgarian coast. Ph.D. thesis, Sofia, Bulgarian Academy of Sciences. 206 p. (in Bulgarian). [354]

Todorova V. 2011. Black mussels and / or barnacle communities on mediolittoral rocks. Littoral sands and muddy sands. Underwater „meadows" of sea grasses. In: Red Data Book of the Republic of Bulgaria. Digital edition. Sofia, Bulgarian Academy of Sciences & Ministry of Environment and Water. [354a]

Todorova V. 2015. Black mussels and / or barnacle communities on mediolittoral rocks. Littoral sands and muddy sands. Underwater „meadows" of sea grasses. – In: Red Data Book of the Republic of Bulgaria. 3. Natural habitats. Sofia, Bulgarian Academy of Sciences & Ministry of Environment and Water, 44-47, 51-52. [354a]

Todorova V. 2017. Zoobenthic indicators for assessment of sandy bottom habitats. In: Todorova V. and Milkova T. (Eds). Final project report „Investigations on the State of the Marine Environment and Improving Monitoring Programs developed under MSFD (ISMEIMP)", IO-BAS and BSBD-Varna, 2017, ISBN:978-619-7244-02-1, 23-47. [354b]

Todorova V. (Ed.). 2021. Black Sea monitoring and assessment guideline. ANEMONE Deliverable 1.3, CD Press, 190 p. ISBN 978-606-528-527-9. [354c]

Todorova V., D. Micu, M. Panayotova and T. Konsulova. 2008a. Marine Protected Areas in Bulgaria – Present and Prospects. Varna, HelixPres. 20 p. (in Bulgarian). [355]

Todorova V., Micu D. & Klisurov L. 2009. Unique oyster reefs discovered in the Bulgarian Black Sea. Comptes rendus de l'Académie bulgare des Sciences 62 (7): 871-874. [355a]

Todorova V. & Milkova T. 2017 (Eds.). Investigations on the State of the Marine Environment and Improving Monitoring Programs developed under MSFD. Institute of Oceanology, Bulgarian Academy of Sciences. 595 p. ISBN 978-619-7244-03-8. (in Bulgarian). [355b]

Todorova V. & Moncheva S. 2013 (Eds.). Parvonachalna otsenka na sastoyanieto na morskata okolna sreda [Initial assessment of the state of the marine environment]. Institute of Oceanology, Bulgarian Academy of Sciences, 493 p. [355c]

Todorova V. & Panayotova M. 2011. Infralittoral rocks and other hard substrates. *Cystoseira* spp. on exposed to waves infralittoral bedrock and boulders. Sublittoral sands. Sublittoral mussel beds on sediments. In: Red Data Book of the Republic of Bulgaria. Digital edition. Sofia, Bulgarian Academy of Sciences & Ministry of Environment and Water. [355d]

Todorova V., M. Panayotova. 2015. Infralittoral rocks and other hard substrates. *Cystoseira* spp. on exposed to waves infralittoral bedrock and boulders. Sublittoral sands. Sublittoral mussel beds on sediments. – In: Red Data

Book of the Republic of Bulgaria. 3. Natural habitats. Sofia, Bulgarian Academy of Sciences & Ministry of Environment and Water, 54-58, 60-63. [355e]

Todorova V., Panayotova M., Bekova R. & Prodanov B. 2022. Recovery of *Flexopecten glaber* (Linnaeus, 1758) (Bivalvia: Pectinidae) in the Bulgarian Black Sea Waters: Recent Distribution, Population Characteristics and Future Perspectives for Protection and Commercial Utilisation. Acta zoologica bulgarica 74 (3): 437-444. [355f]

Todorova V., Panayotova M. & Doncheva V. 2023. Distribution and impact on benthic habitats of the non-native invasive species *Rapana venosa* in the Bulgarian Black-Sea. In: Trichkova T., Kalcheva H., Tomov R., Vladimirov V. & Tyufekchieva V. (Eds.). Book of Abstracts. Joint ESENIAS and DIAS Scientific Conference 2023 and 12th ESENIAS workshop, 11-14 October 2023, Varna, IBER-BAS, ESENIAS, DIAS, p. 66. [355g]

Todorova V., A. Trayanova and T. Konsulova. 2008b. Biological monitoring of coastal marine waters and lakes – benthic invertebrate fauna. Report. Varna, Institute of Oceanology, Bulgarian Academy of Sciences, 46 p. [Online: http://www.bsbd.org/UserFiles/Report_zoobenthos_final%20] [356]

Trayanova A. 2003. Ecological status of mscrozoobenthic communities in Beloslav and Varna Lakes during the autumn of 1999. – Proceedings of the Institute of Oceanology (Varna), 4: 140-148. [357]

Trayanova A. 2008. Ecological status of zoobenthos from sediments of Beloslav Lake, Varna Lake and Varna Bay. Ph.D. thesis, Sofia, Bulgarian Academy of Sciences, 200 p. (in Bulgarian). [358]

Trayanova A., Todorova V., Konsulova T., Shtereva G., Hristova O.& Dzhurova B. 2011. Ecological State of Varna Bay in Summer 2009 according to Benthic Invertebrate Fauna. Acta zoologica bulgarica 63 (3): 277-288. [358a]

Triantaphyllidis G., T. Abatzopoulos and P. Sorgeloos. 1998. Review of the biogeography of the genus *Artemia* (Crustacea, Anostraca). – Journal of Biogeography, 25: 213-226. [359]

Trichkova T., D. Kozuharov, Z. Hubenov, S. Cheshmedjiev, I. Botev, M. Zivkov, L. Popa and O. Popa. 2009. Current distribution of *Dreissena* species in the inland waters of Bulgaria. International Scientific Publications: Ecology & Safety, 3 (1): 507-516. (ISSN: 1313-2539) http://www.science-journals.eu/ecology/ISP-ES-3-2009.swf [360]

Trichkova T., Csányi B., Skolka M., Kvach Y., Ivanova P., Kalcheva H. & Paunović M. 2025. Invasive alien species (IAS) in the Danube River Basin and Western Black Sea Coast. In: Bloesch J., Cyffka B., Hein T., Sandu C., Sommerwerk N. (Eds.). The Danube River and the Western Black Sea Coast: Complex Transboundary Management. International Association for Danube Research (IAD), Book Series "Ecohydrology from catchment to coast", Elsevier Publisher. Paperback ISBN: 9780443186868, eBook ISBN: 9780443186875. [360a]

Tzonev R., Ch. Gussev. 2015. Communities of annual halophytes in coastal slt marshes at the Back Sea. – In: Red Data Book of the Republic of Bulgaria. 3. Natural habitats. Sofia, Bulgarian Academy of Sciences & Ministry of Environment and Water, 49-50. [360a]

Uzunov Y. 1974. Two new species of free-living nematodae from Bulgarian Black Sea littoral. – Comptes rendus de l'Académie bulgare des Sciences, 27 (6): 843-845. [361]

Uzunov Y. 1977a. Distribution of interstitial nematodes in the capillary horizon of some Bulgarian Black Sea beach. – Hydrobiology, 15: 183-191. [362]

Uzunov Y. 1977b. New data on interstitial nematodes from Bulgarian Black Sea coast with description of *Camacolaimus pontolittoralis* sp. n. – Acta zoologica bulgarica, 8: 32-37. [363]

Uzunov Y. 1980. Water Oligochets (Oligochaeta Limicola) from some Bulgarian Rivers. Fequency and Domination. – Hydrobiology, 12: 79-89. (in Bulgarian). [364]

Uzunov Y. 2010. Aquatic Oligochets (Oligochaeta Limicola). Annelida: Aphanoneura, Oligochaeta, Branchiobdellea. – In: Catalogus faunae bulgaricae. 7. Sofia, Professor Marin Drinov Academic Publishing House. 120 p. [365]

Uzunov Y., S. Kovachev, K. Kumanski and J. Ludskanova-Nikolova. 1998. Aquatic ecosystems of the Aegean and Black Sea basins. – In: Meine C. (ed.): Bulgaria's Biological Diversity: Conservation Status and Needs Assessment. Volumes I and II. Sofia & Moscow, Pensoft, 292-346. [366]

Uzunova S. 1995. An overview of Crustacea fauna in the bay of Varna. – Proceedings of the Institute of Fisheries (Varna), 23: 158-168. (in Bulgarian). [367]

Uzunova S. 1996. Amphipoda in the biocenosis of *Mytilus galloprovincialis* overgrowths in the Bay of Varna. – Proceedings of the Institute of Fisheries (Varna), 24: 124-131. (in Bulgarian). [368]

Uzunova S. 1999. On the biodiversity of the Ponto-Caspian Amphipoda (Crustacea) from the Bulgarian Black Sea coast. – Proceedings of the Institute of Fisheries (Varna), 25: 175-186. [369]

Uzunova S. 2006. Species composition, distribution and dynamics of the quantitative parameters of Malacostraca (Crustacea) along the Bulgarian coast of the Black Sea. Ph.D. thesis, Sofia, Bulgarian Academy of Sciences. 205 p. (in Bulgarian). [370]

Uzunova S. 2009. Revew of the Decapod fauna from the Bulgarian Black Sea. – Proceedings of the Institute of Fishing Resources (Varna), 27: 27-32. [371]

Uzunova S. 2016. Guide of Decapoda (Crustacea) from the Bulgarian sector of the Black Sea. Varna, Arttreysar. 127 p. (in Bulgarian). [371a]

Valchev V., V. Vassilev and V. Georgiev. 2015. Submerged macrophytic communities in hypersaline water bodies. – In: Red Data Book of the Republic of Bulgaria. 3. Natural habitats. Sofia, Bulgarian Academy of Sciences & Ministry of Environment and Water, 52-54. [371b]

Valkanov A. 1934. Beitrag zur Kenntnis der Hydrofauna Bulgariens. Sofia, Private publication. 32 p. (in Bulgarian). [372]

Valkanov A. 1935. Notizen über die Brackwässer Bulgariens. I. – Annuaire de l'Université de Sofia. Faculté physico-mathématique, 31 (3): 249-303. (in Bulgarian). [373]

Valkanov A. 1936. Notizen über die Brackwässer Bulgariens. II. Versuch einer hydrographischen und biologischen Erforschung derselben. – Annuaire de l'Université de Sofia. Faculté physico-mathématique, 32 (3): 209-341. (in Bulgarian). [374]

Valkanov A. 1937. Die Varnaseen. Beitrag zur Hydrographie und Biologie derselben. – Mitteilungen der Bulgarisches geographischen Gesellschaft, 4: 118-139. (in Bulgarian). [375]

Valkanov A. 1938a. Übersicht der Hydrozoenfamilie Moerisiidae. – Annuaire de l'Université de Sofia. Faculté physico-mathématique, 34: 53-78. (in Bulgarian). [376]

Valkanov A. 1938b. Übersicht der europaischen Vertretern der Gattung *Jaera* Leach 1813 (Isopoda gennisima). – Annuaire de l'Université de Sofia. Faculté physico-mathématique, 34: 1-26. (in Bulgarian). [377]

Valkanov A. 1947. *Protohydra leuckarti* Greeff vom Schwarzen Meer. – Arbeiten aus der Biologischen Meeresstation in Varna, 13: 31-35. (in Bulgarian). [378]

Valkanov A. 1948a. Die Seeschildkröten des Schwarzen Meeres. – Arbeiten aus der Biologischen Meeresstation in Varna, 14: 99-102. (in Bulgarian). [379]

Valkanov A. 1948b. *Thalasomyia frauenfeldi* Schiner vom Schwarzen Meer. – Arbeiten aus der Biologischen Meeresstation in Varna, 14: 103-112. (in Bulgarian). [380]

Valkanov A. 1949a. *Echiniscoides sigismundi* (M. Schulze) vom Schwarzen Meer. – Arbeiten aus der Biologischen Meeresstation in Varna, 15: 181-186. (in Bulgarian). [381]

Valkanov A. 1949b. Gibt es "koloniebildende" Moerisiiden ? – Arbeiten aus der Biologischen Meeresstation in Varna, 15: 187-192. (in Bulgarian). [382]

Valkanov A. 1949c. Notizen über *Ostroumovia maeotica* (Ostr.). – Arbeiten aus der Biologischen Meeresstation in Varna, 15: 193-200. (in Bulgarian). [383]

Valkanov A. 1951a. Autotropismus bei Tieren. – Arbeiten aus der Biologischen Meeresstation in Varna, 16: 1-45. (in Bulgarian). [384]

Valkanov A. 1951b. Untersuchungen über *Centropagis pengoi* (Cladocera, Polyphemidae). – Arbeiten aus der Biologischen Meeresstation in Varna, 16: 65-83. (in Bulgarian). [385]

Valkanov A. 1954a. Betrag zur Kenntnis unserer Schwarzmeerfauna. – Arbeiten aus der Biologischen Meeresstation in Varna, 18: 49-53. (in Bulgarian). [386]

Valkanov A. 1954b. Zwei neue Polychaeten für das Schwarze Meer. – Arbeiten aus der Biologischen Meeresstation in Varna, 18: 55-58. (in Bulgarian). [387]

Valkanov A. 1954c. Die Tardigraden des Schwarzenmeeres. – Arbeiten aus der Biologischen Meeresstation in Varna, 18: 59-61. (in Bulgarian). [388]

Valkanov A. 1957a. Katalog unserer Schwarzmeerfauna. – Arbeiten aus der Biologischen Meeresstation in Varna, 19: 1-62. (in Bulgarian). [389]

Valkanov A. 1957b. Erster Versuch zur erforschung der Gastrotrichen des Schwarzen Meeres. – Annuaire de l'Université de Sofia, Faculté de Biologie, Géologie et Géographie, 50 (1 – Biologie): 383-399. (in Bulgarian). [390]

Valkanov A. 1965. *Microhydrula pontica* n. g. sp., Ein neuer solitärer Vertreter des Hydrozoen. – Zoologischer Anzeiger, 174: 134. [391]

Valkanov A., T. Marinov. 1964. Nachtrag zum Katalog der Bulgarischen Schwarzmeerfauna. – Bulletin de l'Institut de zoologie et musée, 17: 51-59. (in Bulgarian). [392]

Valkanov A., T. Marinov. 1978. Zoobenthos. – In: Valkanov A., Ch. Marinov, Ch. Danov and P. Vladev (Eds.): The Black Sea. Varna, G. Bakalov, 181-190. (in Bulgarian). [393]

Valkanov A, I. Dimov and V. Naidenov. 1978. Zooplankton. – In: Valkanov A., Ch. Marinov, Ch. Danov and P. Vladev (Eds.): The Black Sea. Varna, G. Bakalov, 159-162. (in Bulgarian). [394]

Valkanov A., V. Petrova, A. Roshdestvenski, T. Marinov and V. Naidenov. 1978. Black Sea lakes. – In: Valkanov A., Ch. Marinov, Ch. Danov and P. Vladev (Eds.): The Black Sea. Varna, G. Bakalov, 262-283. (in Bulgarian). [395]

Varadinova E., V. Vassilev, S. Stoichev and Y. Uzunov. 2010. The macrozoobenthos of the Pomorie Lake. – In: Collection with reports Pomorie Lake region (Protected Site "Pomorie Lake", Ramsar Site, NATURA 2000 sites "Pomorie" BG 0000620 and "Pomorie Lake" BG 0000152), 95-100. (in Bulgarian). [396]

Varbanov M. 2002. Lakes and swamps. – In: Geography of Bulgaria. Sofia, ForKom: 237-242. (In Bulgarian) [397]

Vassilev V. 1994. Ecological characteristics of the Pomoriisko Lake. Ph.D. Thesis, Sofia, Bulgarian Academy of Sciences. 218 p. (in Bulgarian). [398]

Vassilev V., A. Konsulov. 1998. Zooplankton in the lake Pomoriysko – composition, dynamics, trophic interactions and secondary assimilation. – Proceedings of the Institute of Oceanology (Varna), 2, 146-153. (in Bulgarian). [399]

Viets K. 1935. Wassermilben aus Bulgarien. – Zoologischer Anzeiger, 109 (1-2): 33-39. [400]

Viets K. 1936. Über eine neue Halacaridae (Acari) aus Bulgarien. – Zoologischer Anzeiger, 115 (11): 199-202. [401]

Viets K. 1940. Hidrachnellae, Porohalacaridae und Halacaridae s. str. (Acari) aus Bulgarien. – Zoologischer Anzeiger, 130 (1-2): 36-41. [402]

Vidinova Y., V. Tyufekchieva, E. Varadinova, S. Stoichev, L. Kenderov, I. Dedov, Y. Uzunov. 2016. Taxonomic list of benthic Macroinvertebrate communities of inland standing water bodies in Bulgaria. – Acta zoologica bulgarica, 68 (2): 147-158. [402a]

Wilke T. 1995. *Hemilepton nitidum* (Turton, 1822) and *Mysella bidentata* (Montagu, 1803), two bivalve species new for the Bulgarian Black Sea coast. – Proceedings of the Institute of Fisheries – Varna, 23: 138-147. [403]

Wilke T. 1996. Annotated check-list of the marine gastropods of the Bulgarian Black Sea coast. – Proceedings of the Institute of Fisheries (Varna), 24: 144-166. [404]

Wilke T. 1997. Gastropods of the Black Sea. Identification of species new for the Bulgarian Coast. – Mitteilungen aus dem Zoologischen Museum in Berlin, 73 (1): 3-15. [405]

Zaitsev Yu., V. Mamaev. 1997. Marine biological diversity in the Black Sea. A study of change and decline. GEF, Black Sea environmental programme. Series. 3. Istanbul, UN Publications. 208 p. [406]

Zaitzev Yu., B. Öztürk. 2001. Exotic species in the Aegean, Marmara, Black, Azov and Caspian Seas, Turkish Marine Foundation, Istanbul. 267 p. [407]

Zaitzev Yu., B. Aleksandrov, N. Berlinskii, Yu. Bogatova, V. Bolshakov., S. Bushuev, E. Volya, G. Garkavaya, M. Gelmboldt, V. Zolotarev, N. Kopitina, I. Kulakova, A. Kurilov, G. Losovskaya, G. Minicheva, D. Nesterova, L. Polishtuk, I. Sinegub, L. Terenko and N. Shurova. 2004. Bazovye biologicheskie issledovaniya Odesskogo morskogo porta (VIII-XII.2001): itogovaiy otchet. – Seriya monografiy Odesskogo demonstratsionnogo tsentra programmy GloBallast, Odessa, № 7. 171 p. (in Russian). [408]

Zapryanov L. 1992. Morfologichna harakteristika na midata *Scapharca inaeqvivalvis* (Brugiere, 1789) (Bivalvia, Anadaridae) ot Cherno more [Morphological characteristics of the clam *Scapharca inaeqvivalvis* (Brugiere, 1789) (Bivalvia, Anadaridae) from the Black Sea]. Izvestiya na Narodniya muzey – Varna 28 (43): 361-365. [408a]

Zaschev G., A. Angelov. 1959. Untersuchungen über den Burgas-See (Waja) in Beziehung zur Verbesserung seiner fischwirtschaftlichen Ausbeutung. – Annuaire de l'Université de Sofia, Faculté de Biologie, Géologie et Géographie, 51 (1 – Biologie): 161-210. (in Bulgarian). [409]

Zernov S. 1913. On the question of life in the Balack sea. – Zap. Imp. Akad. Nauk, S.-Peterb., 32 (1): 299 p. (in Russian). [410]

Zhadin V. 1952. Mollusks of the Fresh and Brakish Waters of the USSR. Moscow & Leningrad, Academy of Sciences of the USSR. 376 p. (in Russian). [411]

Zlateff I. 1936. Über einige Krankheiten unserer Fische. – Arbeiten aus der Biologischen Meeresstation in Varna, 5: 67-80. (in Bulgarian). [412]

**Table 1.** Taxa localities (Symbols: * – highly altered habitat to the time of collection of the material;
[ ] – old geographical names and old data, before 1957).

| Locality | Number | Old number | Number of species |
|---|---|---|---|
| **Sea localities** | | | |
| Durankulak [Blatnitsa] (Durankulak north – Durankulak Lake – Krapets) | **1** | 1 | 131 |
| Shabla (Shabla Lake – Shabla Tuzla – Cape Shabla) | **2** | 2 | 153 |
| Cape Kaliakra (Rusalka – Bolata – Cape Kaliakra) | **3** | 3 | 230 |
| Kavarna (Cape Chairburun – Cape Chirakman – Cape Kalkanburun) | **4** | 4 | 138 |
| Balchik (Balchik Tuzla – Balchik) | **5** | 5 | 172 |
| Batova (Albena – Kranevo) | **6** | 6 | 141 |
| Varna (Golden Sands – Evksinograd – Varna – Cape Galata – Pasha Dere River) | **7** | 7 | 601 |
| Kamchiya (Cape Ilandzhik – Camping Ray – Kamchiya River – Shkorpilovtsi) | **8** | 8 | 139 |
| Byala (Cape Cherni – Byala – Dvoynitsa River – Obzor) | **9** | 9 | 128 |
| Cape Emine (Irakli – Cape Emine – Cocketrice sandy bank) | **10** | 10 | 176 |
| Nesebar (Elenite – Sunny Beach – Nesebar – Ravda) | **11** | 11 | 256 |
| Pomorie (Aheloy – Pomorie – Camping Evropa – Cape Lahna) | **12** | 12 | 220 |
| Burgas (Saraphovo – Burgas – Kraymorie – Park Rosenets – Cape Chukalya) | **13** | 13 | 264 |
| Sveta Anastasiya Island [Bolshevik Island] | **14** | 14 | 136 |
| Chernomorets (Cape Atiya – Chernomorets – Cape Chervenka [Cape Hrisotira]) | **15** | 15 | 134 |
| Sozopol (Camping Gradina – Sozopol – Kavatsite – Dyuni) | **16** | 15 | 274 |
| Cape Maslen Nos (Alepu Marsh – Ropotamo River – C. Maslen Nos – Stomoplo Marsh) | **17** | 16 | 184 |
| Primorsko (Stomoplo Marsh – Primorsko – Dyavolska Reka River) | **18** | 17 | 133 |

| | | | |
|---|---|---|---|
| Kiten [Urdoviza] (International Youth Centre – Kiten – Karaagach River – Lozenets) | **19** | 18 | 133 |
| Tsarevo [Michurin, Vasiliko] (Cape Arapya – Tsarevo – Varvara) | **20** | 19 | 177 |
| Ahtopol (Varvara – Achtopol – Veleka River) | **21** | 20 | 143 |
| Sinemorets (Sinemorets – Silistar River – Rezovo) | **22** | 20 | 140 |
| *Zostera* overgrowths (0-6 m) | **23** | | 45 |
| Rocky sublittoral, *Cystoseira* and other algae, *Mytilus* (from 0.5-1 m to 15-25 m) | **24** | | 257 |
| Sandy sublittoral (1-25 m); clean sand – to 17 m, with *Branchiostoma* – to 20 m | **25** | | 396 |
| Coastal silt (from 15-20 m to 30-40 m); dominated by *Melinna* – to 25-30 m | **26** | | 115 |
| *Mytilus* silt (from 15-20 m to 60-80 m) | **27** | | 148 |
| *Phaseolina* silt (from 65 m to 140-180 m) | **28** | | 104 |
| Black Sea, pelagic in front of the Bulgarian coast | **29** | | 55 |
| **Localities along the sea coast** | | | |
| Lithotelms, Shabla – Cape Kaliakra, Varna | **32** | 21 | 15 |
| Lithotelms, Ravda, Sozopol – Cape Maslen Nos | **33** | 22 | 40 |
| Basins of Varna Aquarium | **34** | 25 | 11 |
| Subterranean (ground) waters of sandy beach, interstitial, mesopsamal | **35** | 26 | 146 |
| Sunny Beach, coastal zone, sand bottom and floating algae | **36** | 29 | 4 |
| Arkutino, coastal zone | **37** | 27 | 4 |
| Camping (Residence) Perla, coastal zone, sand bottom and floating algae | **38** | 30 | 4 |
| Kiten, coastal zone, sand bottom and floating algae | **39** | 28 | 5 |
| Lozenets, coastal zone | **40** | 31 | 4 |
| Small saltwater marshes along the coast | **41** | 72 | 8 |
| Small freshwater marshes along the coast | **42** | 73 | 26 |
| Mouths of small streams | **43** | 74 | 3 |

| | | | |
|---|---|---|---|
| Temporary salty puddles and floods around the coastal basins | **44** | 79 | 4 |
| Rocks along the entire coast, rocky supralittoral | **45** | 76 | 14 |
| Algae washed ashore along the coast, supralittoral | **46** | 77 | 23 |
| Salty soils around coastal basins | **47** | 78 | 28 |
| Terrestrial coastal zone with halophilic plants (to 50-100 m from the sea) | **48** | 75 | 24 |
| Sea coastal zone; littoral (medio- or pseudolittoral) | **49** | | 24 |
| Rocky littoral (medio- or pseudolittoral, enteromorpha zone) | **50** | | 16 |
| Sandy littoral (medio- or pseudolittoral) | **51** | | 64 |
| Sandy supralittoral | **52** | | 8 |
| Springs and wells with brackish water along the coast, Durakulak – Cape Kaliakra | **53** | | 1 |
| Springs along the coast, Sozopol – Cape Maslen Nos | **54** | | 2 |
| **Coastal basins (lakes, swamps, firths and river floods)** | | | |
| Durankulak [Blatnitsa] Lake: [0-5‰, 3.4 km$^2$, depth 4 m], 1-4‰, average salinity – 2‰ | **58** | 41 | 113 |
| Ezerets Lake: [1-2‰, average salinity – 1.6‰], 0.58-0.79‰, 0.72 km$^2$, depth 9.0 m | **59** | 42 | 101 |
| Shabla Lake: [0.1-2‰, 0.6-1.6‰], 0.52-0.60‰, 0.79 km$^2$, depth 9.5 m | **60** | 42 | 137 |
| *Schabla Tuzla: [10-30‰], 22-200‰, 0.19 km$^2$, depth 0.6 m | **61** | 43 | 17 |
| Nanevska Tuzla [Tauk Liman]: 1-90‰ (often about 20‰) 0.10 km$^2$, depth 0.3 m | **62** | | 3 |
| *Bolata River Mouth: 0.1‰ | **63** | 44 | 13 |
| *Balchik Tuzla: [80-150‰], 35-160‰, 0.14 km$^2$, depth 0.5-0.8 m | **64** | 45 | 7 |
| Batova River Mouth and Swamp: 0.03-6‰, depth 0.5-1 m, | **65** | 46 | 9 |
| *Golden Sands Marshes: [0-60‰] | **66** | 47 | 7 |
| *Sindel [Sultanlar] Swamp: [0‰] | **67** | 48 | 6 |
| *Beloslav [Devnya, Gebedzhe] Lake: 0.1-15.6‰, 3.90 km$^2$, depth 3.5 m, sea canal – 1923 | **68** | 49 | 160 |
| *Varna Lake: [5-14‰], 6.5-8-16.8‰, 17.40 km$^2$, depth 19 m, sea canals – 1909, 1976 | **69** | 50 | 264 |

| | | | |
| --- | --- | --- | --- |
| Pasha Dere [Chatal Dere, Novata Voda] River Mouth: [0-7‰] | 70 | 51 | 10 |
| Kamchiya River Mouth – Swamps: [0.4-0.7‰], average 0.1‰ | 71 | 52 | 53 |
| Fandakliyska Reka [Shkorpilova] River Mouth: [0.1‰] | 72 | 53 | 11 |
| *Dvoynitsa [Cherta, Suha Kamchiya] River Mouth | 73 | | 8 |
| *Hadzhiyska River Mouth [Nesebar Marsh]: [1-10‰] | 74 | 54 | 30 |
| Aheloy River Mouth | 75 | | 9 |
| Pomorie Lake: 30-70 to 140‰, 8.50 km$^2$, depth 1.4 m | 76 | 55 | 106 |
| Atanasovsko Lake: 1-250‰, average 50-60‰, 16.90 km$^2$, depth 0.3-0.8 m | 77 | 56 | 113 |
| *Burgas [Vaya] Lake: [9-20‰], 1.8-45‰, average 10.6‰, 27.60 km$^2$, depth 1.3 m | 78 | 57 | 89 |
| *Mandra Dam [Mandra Lake to 1963]: [0.1-12‰, max. 30‰, 14.00 km$^2$, depth 1.1-5 m] | 79 | 58 | 94 |
| Uzungeren-Poda Complex: 0.1-32‰, 3.12 km$^2$ | 80 | | 80 |
| Tsiganski Skelet Marsh [Chengene Skele Marsh]: [7-20‰] | 81 | 59 | 11 |
| Alepu Marsh: 4-11-27‰ (usually 3.5-7.2‰), 0.14 km$^2$, depth 0.6-1 m | 82 | 60 | 14 |
| Arkutino Marsh: 0.1-1‰, 0.03 km$^2$, depth 0.5 m | 83 | | 24 |
| Ropotamo River Mouth: 5-15‰ | 84 | 61 | 97 |
| Stomoplo Marsh: [2-25‰], 1.5-4‰, 6-14‰, 0.06 km$^2$, depth 0.5 m | 85 | 62 | 20 |
| *Dyavolsko Blato Swamp: [1-20‰], 6-14‰, 0.80 km$^2$, depth 1 m | 86 | 63 | 78 |
| *Dyavolska Reka River Mouth: depth 4 m | 87 | | 10 |
| Karaagachka Reka [Kitenska, Oryashka] River Mouth and Swamp: [5-15‰] | 88 | 64 | 68 |
| Tsarevska Reka [Michurinska (small)] River: [3‰] | 89 | 65 | 10 |
| Izgrevsko Dere [Michurinska Reka (great)] River: [5-10‰] | 90 | 66 | 14 |
| Puddles and mouths of streams between Tsarevo and Ahtopol | 91 | 67 | 12 |
| Veleka River Mouth: [0-0.5‰] | 92 | 68 | 59 |
| Butamyata [Potamyata] River Mouth: [12‰] | 93 | 69 | 22 |
| Silistar River Mouth: [5-15‰] | 94 | 70 | 25 |

| | | | |
|---|---|---|---|
| Rezovska Reka [Rezvaya] River Mouth: [0-1.45‰] | **95** | 71 | 30 |
| Black Sea coastal lakes and swamps | **96** | | 71 |

**Table 2.** Taxonomic diversity of the invertebrate animals from the Bulgarian Black Sea.

| Types | Classes | Orders | Families | Species |
|---|---|---|---|---|
| Porifera | 2 | 6 | 13 | 23 |
| Cnidaria | 3 | 10 | 25 | 38 |
| Ctenophora | 2 | 3 | 3 | 4 |
| Plathelminthes | 1 | 5 | 16 | 29 |
| Nemertini | 2 | 3 | 9 | 26 |
| Gastrotricha | 1 | 2 | 6 | 13 |
| Nematoda | 2 | 8 | 34 | 112 |
| Cephalorhyncha | 1 | 2 | 3 | 4 |
| Rotifera | 1 | 4 | 22 | 125 |
| Annelida | 5 | 16 | 49 | 175 |
| Tardigrada | 2 | 3 | 5 | 5 |
| Arthropoda | 4 | 28 | 206 | 813 |
| Mollusca | 3 | 20 | 66 | 155 |
| Bryozoa | 2 | 3 | 9 | 19 |
| Phoronida | 1 | 1 | 1 | 1 |
| Entoprocta | 1 | 1 | 1 | 1 |
| Chaetognatha | 1 | 1 | 1 | 1 |
| Echinodermata | 2 | 3 | 3 | 6 |
| Chordata | 3 | 4 | 6 | 8 |
| **Total** | **39** | **123** | **477** | **1558** |

**Table 3.** Distribution of the invertebrate animals by categories according to depth (Note. Only marine and brackish species, for which data are available, are included.).

| Types | Epibathic | Hypobathic | Mesobathic | Eurybathic |
| --- | --- | --- | --- | --- |
| Porifera | 2 | | 4 | 9 |
| Cnidaria | 7 | 1 | 2 | 10 |
| Ctenophora | 1 | | | 1 |
| Plathelminthes | 13 | | | |
| Nemertini | 5 | 4 | 6 | 5 |
| Gastrotricha | 4 | | | |
| Nematoda | 21 | 14 | 11 | 29 |
| Cephalorhyncha | | 1 | 1 | 1 |
| Rotifera | 12 | | | |
| Annelida | 81 | 3 | 16 | 19 |
| Tardigrada | 5 | | | |
| Arthropoda | 239 | 21 | 36 | 74 |
| Mollusca | 69 | 2 | 33 | 15 |
| Bryozoa | 5 | 1 | 3 | |
| Phoronida | | | | 1 |
| Entoprocta | 1 | | | |
| Chaetognatha | 1 | | | |
| Echinodermata | | 3 | | 3 |
| Chordata | 1 | 2 | 3 | 1 |
| **Total** | **467** | **52** | **115** | **168** |

257

**Table 4.** Zoogeographical characteristic of the marine and brackish Invertebrate fauna from the Bulgarian Black Sea coast.

| Zoogeographical scheme of the used categories and main taxa | Total | | Benthos | | Plankton | | Marine | | Brakish | |
|---|---|---|---|---|---|---|---|---|---|---|
| | Number | % | number | % | number | % | number | % | Number | % |
| **COSMOPOLITAN TYPE** | **121** | **11.48** | **105** | **10.74** | **25** | **21.19** | **119** | **11.63** | **18** | **11.69** |
| **Arctic-Antarctic-Atlantic-Indian-Pacific** | **42** | **3.98** | **36** | **3.68** | **7** | **5.93** | **42** | **4.10** | **7** | **4.54** |
| Cosmopolitan | 15 | 1.42 | 10 | 1.02 | 5 | 4.24 | 15 | 1.47 | 5 | 3.25 |
| Subcosmopolitan | 10 | 0.95 | 10 | 1.02 | 1 | 0.85 | 10 | 0.98 | 2 | 1.30 |
| Arctic-Antarctic-Atlantic-Mediterranean-Indo-Pacific | 4 | 0.38 | 4 | 0.41 | | | 4 | 0.39 | | |
| Arctic-Atlantic-Mediterranean-Indo-Pacific | 6 | 0.57 | 6 | 0.61 | | | 6 | 0.59 | | |
| Arctic-Atlantic-Mediterranean-Indo-North Pacific | 1 | 0.09 | 1 | 0.10 | | | 1 | 0.09 | | |
| Arctic-North Atlantic-Mediterranean-Indo-Pacific | 1 | 0.09 | 1 | 0.10 | | | 1 | 0.09 | | |
| Arctic-North Atlantic-Mediterranean-Indo-North Pacific | 2 | 0.19 | 2 | 0.20 | | | 2 | 0.19 | | |
| Antarctic-Atlantic-Mediterranean-Indo-Pacific | 2 | 0.19 | 1 | 0.10 | 1 | 0.85 | 2 | 0.19 | | |
| Antarctic-Pontian-Indo-Pacific | 1 | 0.09 | 1 | 0.10 | | | 1 | 0.09 | | |
| **Atlantic-Indian-Pacific** | **79** | **7.49** | **69** | **7.05** | **18** | **15.25** | **77** | **7.53** | **11** | **7.14** |
| **HOL- AND EAST ATLANTIC-INDIAN-PACIFIC** | **58** | **5.50** | **49** | **5.01** | **15** | **12.71** | **57** | **5.58** | **10** | **6.49** |
| Atlantic-Mediterranean-Indo-Pacific | 37 | 3.51 | 31 | 3.17 | 12 | 10.17 | 36 | 3.52 | 7 | 4.54 |

| | | | | | | | | | | |
| --- | --- | --- | --- | --- | --- | --- | --- | --- | --- | --- |
| Atlantic-Mediterranean-Indo-West Pacific | 4 | 0.38 | 2 | 0.20 | 2 | 1.69 | 4 | 0.39 | 1 | 0.65 |
| Atlantic-Mediterranean-Indo-Southwest Pacific | 3 | 0.28 | 3 | 0.31 | | | 3 | 0.29 | | |
| Atlantic-Mediterranean-Indo-New Zealand | 4 | 0.38 | 4 | 0.41 | | | 4 | 0.39 | 2 | 1.30 |
| Atlantic-Mediterranean-Indo-North Pacific | 2 | 0.19 | 2 | 0.20 | | | 2 | 0.19 | | |
| Atlantic-Mediterranean-Indo-Northwest Pacific | 3 | 0.28 | 3 | 0.31 | | | 3 | 0.29 | | |
| Atlantic-Mediterranean-Red Sea-Pacific | 2 | 0.19 | 1 | 0.10 | 1 | 0.85 | 2 | 0.19 | | |
| East Atlantic-Mediterranean-Indo-Pacific | 2 | 0.19 | 2 | 0.20 | | | 2 | 0.19 | | |
| East Atlantic-Mediterranean-Indo-Southwest Pacific | 1 | 0.09 | 1 | 0.10 | | | 1 | 0.09 | | |
| **TROPICAL AND SUBTROPICAL ATLANTIC-INDIAN-PACIFIC** | **6** | **0.57** | **5** | **0.51** | **2** | **1.69** | **6** | **0.59** | | |
| Circum[sub]tropical | 3 | 0.28 | 2 | 0.20 | 1 | 0.85 | 3 | 0.29 | | |
| Lusitanian-Mediterranean-West Indo-West Pacific | 1 | 0.09 | 1 | 0.10 | | | 1 | 0.09 | | |
| Mediterranean-Indo-West Pacific | 1 | 0.09 | 1 | 0.10 | | | 1 | 0.09 | | |
| Pontian-Indo-New Zealand | 1 | 0.09 | 1 | 0.10 | | | 1 | 0.09 | | |
| **NORTH ATLANTIC-INDIAN-PACIFIC** | **15** | **1.42** | **15** | **1.53** | **1** | **0.85** | **14** | **1.37** | **1** | **0.65** |
| North Atlantic-Mediterranean-Indo-Pacific | 3 | 0.28 | 3 | 0.31 | 1 | 0.85 | 2 | 0.19 | 1 | 0.65 |
| North Atlantic-Mediterranean-Indo-West Pacific | 1 | 0.09 | 1 | 0.10 | | | 1 | 0.09 | | |
| North Atlantic-Mediterranean-Indo-Malayan | 2 | 0.19 | 2 | 0.20 | | | 2 | 0.19 | | |
| North Atlantic-Mediterranean-Indo-New Zealand | 5 | 0.47 | 5 | 0.51 | | | 5 | 0.49 | | |

| | | | | | | | | | | |
|---|---|---|---|---|---|---|---|---|---|---|
| North Atlantic-Mediterranean-Red Sea-Northeast Pacific | 1 | 0.09 | 1 | 0.10 | | | 1 | 0.09 | | |
| Northeast Atlantic-Mediterranean-Indo-New Zealand | 3 | 0.28 | 3 | 0.31 | | | 3 | 0.29 | | |
| **ATLANTIC-INDIAN TYPE** | **72** | **6.83** | **67** | **6.85** | **10** | **8.47** | **70** | **6.84** | **7** | **4.54** |
| **Arctic-Antarctic-Atlantic-Indian** | **4** | **0.38** | **4** | **0.41** | | | **4** | **0.39** | | |
| Arctic-Atlantic-Mediterranean-Indian | 1 | 0.09 | 1 | 0.10 | | | 1 | 0.09 | | |
| Arctic-Atlantic-Mediterranean-North Indian | 1 | 0.09 | 1 | 0.10 | | | 1 | 0.09 | | |
| Arctic-North Atlantic-Mediterranean-Red Sea | 1 | 0.09 | 1 | 0.10 | | | 1 | 0.09 | | |
| Antarctic-Atlantic-Mediterranean-Indian | 1 | 0.09 | 1 | 0.10 | | | 1 | 0.09 | | |
| **Atlantic-Indian** | **68** | **6.45** | **63** | **6.44** | **10** | **8.47** | **66** | **6.45** | **7** | **4.54** |
| **TROPICAL AND SUBTROPICAL ATLANTIC-INDIAN** | **7** | **0.66** | **5** | **0.51** | **2** | **1.69** | **7** | **0.68** | **2** | **1.30** |
| Lusitanian-Mediterranean-Indian | 1 | 0.09 | 1 | 0.10 | | | 1 | 0.09 | | |
| Lusitanian-Mediterranean-Mauritanian-West Indian | 1 | 0.09 | | | 1 | 0.85 | 1 | 0.09 | | |
| Lusitanian-Mediterranean-West Indian | 1 | 0.09 | 1 | 0.10 | | | 1 | 0.09 | 1 | 0.65 |
| Lusitanian-Mediterranean-Northeast Indian | 2 | 0.19 | 2 | 0.20 | | | 2 | 0.19 | 1 | 0.65 |
| Mediterranean-North Indian | 1 | 0.09 | 1 | 0.10 | | | 1 | 0.09 | | |
| Mediterranean-Red Sea | 1 | 0.09 | | | 1 | 0.85 | 1 | 0.09 | | |
| **HOL- AND NORTH ATLANTIC-INDIAN** | **36** | **4.42** | **33** | **3.37** | **5** | **4.24** | **34** | **3.32** | **3** | **1.95** |
| Atlantic-Mediterranean-Indian | 6 | 0.57 | 6 | 0.61 | | | 5 | 0.49 | 1 | 0.65 |
| Atlantic-Pontian-Indian | 1 | 0.09 | 1 | 0.10 | | | 1 | 0.09 | | |
| Atlantic-Mediterranean-North Indian | 1 | 0.09 | 1 | 0.10 | | | 1 | 0.09 | | |

| | | | | | | | | | | |
|---|---|---|---|---|---|---|---|---|---|---|
| Atlantic-Mediterranean-Northeast Indian | 2 | 0.19 | 2 | 0.20 | | | 2 | 0.19 | | |
| Atlantic-Mediterranean-West Indian | 2 | 0.19 | 2 | 0.20 | 1 | 0.85 | 2 | 0.19 | | |
| Atlantic-Mediterranean-Red Sea | 2 | 0.19 | 2 | 0.20 | | | 2 | 0.19 | | |
| North Atlantic-Mediterranean-Indian | 7 | 0.66 | 7 | 0.72 | 3 | 2.54 | 6 | 0.59 | | |
| North Atlantic-Mediterranean-North Indian | 6 | 0.57 | 5 | 0.51 | 1 | 0.85 | 6 | 0.59 | | |
| NorthAtlantic-Mediterranean-Northeast Indian | 5 | 0.47 | 5 | 0.51 | | | 5 | 0.49 | 1 | 0.65 |
| North Atlantic-Pontian-Northeast Indian | 1 | 0.09 | 1 | 0.10 | | | 1 | 0.09 | 1 | 0.65 |
| North Atlantic-Mediterranean-West Indian | 1 | 0.09 | 1 | 0.10 | | | 1 | 0.09 | | |
| North Atlantic-Mediterranean-Red Sea | 2 | 0.19 | | | 2 | 1.69 | 2 | 0.19 | | |
| **EAST AND NORTHEAST ATLANTIC-INDIAN** | **25** | **2.37** | **25** | **2.56** | **3** | **2.54** | **25** | **2.44** | **2** | **1.30** |
| East Atlantic-Mediterranean-Indian | 4 | 0.38 | 4 | 0.41 | 1 | 0.85 | 4 | 0.39 | 1 | 0.65 |
| East Atlantic-Mediterranean-West Indian | 1 | 0.09 | 1 | 0.10 | | | 1 | 0.09 | | |
| East Atlantic-Mediterranean-Southwest Indian | 1 | 0.09 | 1 | 0.10 | | | 1 | 0.09 | | |
| East Atlantic-Mediterranean-Red Sea | 2 | 0.19 | 2 | 0.20 | | | 2 | 0.19 | | |
| Celtic-Lusitanian-Mediterranean-Indian | 7 | 0.66 | 7 | 0.72 | | | 7 | 0.68 | 1 | 0.65 |
| Celtic-Lusitanian-Mediterranean-West Indian | 5 | 0.47 | 5 | 0.51 | 2 | 1.69 | 5 | 0.49 | | |
| Celtic-Lusitanian-Mediterranean-Northwest Indian | 2 | 0.19 | 2 | 0.20 | | | 2 | 0.19 | | |
| Celtic-Lusitanian-Mediterranean-Northeast Indian | 1 | 0.09 | 1 | 0.10 | | | 1 | 0.09 | | |
| Celtic-Pontian-Northeast Indian | 2 | 0.19 | 2 | 0.20 | | | 2 | 0.19 | | |
| **ATLANTIC-PACIFIC TYPE** | **120** | **11.38** | **94** | **9.61** | **33** | **27.97** | **118** | **11.53** | **28** | **18.18** |

| | | | | | | | | | | |
|---|---|---|---|---|---|---|---|---|---|---|
| **Arctic-Antarctic-Atlantic-Pacific** | **23** | **2.18** | **20** | **2.04** | **4** | **3.39** | **23** | **2.25** | **3** | **1.95** |
| Arctic-Antarctic-Atlantic-Mediterranean-Boreal Pacific | 1 | 0.09 | 1 | 0.10 | 1 | 0.85 | 1 | 0.09 | | |
| Arctic-North Atlantic-Mediterranean-Pacific | 2 | 0.19 | 2 | 0.20 | | | 2 | 0.19 | | |
| Arctic-North Atlantic-Mediterranean-North Pacific | 5 | 0.47 | 4 | 0.41 | 1 | 0.85 | 5 | 0.49 | | |
| Arctic-North Atlantic-Mediterranean-Northeast Pacific | 4 | 0.38 | 4 | 0.41 | | | 4 | 0.39 | | |
| Arctic-North Atlantic-Pontian-Northeast Pacific | 1 | 0.09 | | | 1 | 0.85 | 1 | 0.09 | 1 | 0.65 |
| Arctic-Boreal Atlantic-Mediterranean-Boreal Pacific | 3 | 0.28 | 3 | 0.31 | | | 3 | 0.29 | | |
| Arctic-Boreal Atlantic-Pontian-Boreal Pacific | 3 | 0.28 | 2 | 0.20 | 1 | 0.85 | 3 | 0.29 | 2 | 1.30 |
| Arctic-Boreal Atlantic-Pontian-Northeast Pacific | 1 | 0.09 | 1 | 0.10 | | | 1 | 0.09 | | |
| Arctic-Atlantic-Mediterranean-Southwest Pacific | 1 | 0.09 | 1 | 0.10 | | | 1 | 0.09 | | |
| Arctic-Celtic-Mediterranean-New Zealand | 1 | 0.09 | 1 | 0.10 | | | 1 | 0.09 | | |
| Antarctic-Atlantic-Mediterranean-Pacific | 1 | 0.09 | 1 | 0.10 | | | 1 | 0.09 | | |
| **Atlantic-Pacific** | **97** | **9.20** | **74** | **7.57** | **29** | **24.58** | **95** | **9.27** | **25** | **16.23** |
| **HOL- AND NORTH ATLANTIC-PACIFIC** | **26** | **2.47** | **22** | **2,25** | **6** | **5.08** | **26** | **2.54** | **3** | **1.95** |
| Atlantic-Mediterranean-Pacific | 14 | 1.33 | 13 | 1.33 | 3 | 2.54 | 14 | 1.37 | | |
| Atlantic-Mediterranean-North Pacific | 1 | 0.09 | | | 1 | 0.85 | 1 | 0.09 | | |
| North Atlantic-Mediterranean-Pacific | 6 | 0.57 | 4 | 0.41 | 2 | 1.69 | 6 | 0.59 | 2 | 1.30 |
| Boreal Atlantic-Pontian-Pacific | 1 | 0.09 | 1 | 0.10 | | | 1 | 0.09 | 1 | 0.65 |

| | | | | | | | | | | |
|---|---|---|---|---|---|---|---|---|---|---|
| Atlantic-Mediterranean-Japonic | 1 | 0.09 | 1 | 0.10 | | | 1 | 0.09 | | |
| Atlantic-Mediterranean-Northeast Pacific | 3 | 0.28 | 3 | 0.31 | | | 3 | 0.29 | | |
| **TROPICAL AND SUBTROPICAL ATLANTIC-PACIFIC** | **7** | **0.66** | **6** | **0.61** | **1** | **0.85** | **7** | **0.68** | | |
| Lusitanian-Mediterranean-West Pacific | 1 | 0.09 | 1 | 0.10 | | | 1 | 0.09 | | |
| Mediterranean-Japonic | 2 | 0.19 | 2 | 0.20 | | | 2 | 0.19 | | |
| Mediterranean-Mauritanian-Guinean-Tasmanian | 1 | 0.09 | 1 | 0.10 | | | 1 | 0.09 | | |
| Lusitanian-Mediterranean-New Zealand | 2 | 0.19 | 1 | 0.10 | 1 | 0.85 | 2 | 0.19 | | |
| Mediterranean-New Zealand | 1 | 0.09 | 1 | 0.10 | | | 1 | 0.09 | | |
| **NORTH ATLANTIC-PACIFIC** | **19** | **1.80** | **19** | **1.94** | | | **18** | **1.76** | **1** | **0.65** |
| North Atlantic-Mediterranean-North Pacific | 3 | 0.28 | 3 | 0.31 | | | 3 | 0.29 | | |
| Boreal Atlantic-Mediterranean-Boreal Pacific | 2 | 0.19 | 2 | 0.20 | | | 2 | 0.19 | | |
| Boreal Atlantic-Pontian-Boreal Pacific | 1 | 0.09 | 1 | 0.10 | | | 1 | 0.09 | | |
| Circumboreal-Mediterranean | 2 | 0.19 | 2 | 0.20 | | | 2 | 0.19 | | |
| North Atlantic-Mediterranean-Northeast Pacific | 2 | 0.19 | 2 | 0.20 | | | 2 | 0.19 | | |
| Boreal Atlantic-Mediterranean-Northeast Pacific | 2 | 0.19 | 2 | 0.20 | | | 2 | 0.19 | | |
| Northeast Atlantic-Mediterranean-North Pacific | 2 | 0.19 | 2 | 0.20 | | | 2 | 0.19 | | |
| Northeast Atlantic-Mediterranean-Aleutian | 1 | 0.09 | 1 | 0.10 | | | 1 | 0.09 | | |
| Northeast Atlantic-Mediterranean-Japonic | 2 | 0.19 | 2 | 0.20 | | | 2 | 0.19 | | |
| Celtic-Pontian-Japonic | 1 | 0.09 | 1 | 0.10 | | | | | 1 | 0.65 |
| Pontian-Northeast Pacific | 1 | 0.09 | 1 | 0.10 | | | 1 | 0.09 | | |

| | 28 | 2.66 | 17 | 1.74 | 12 | 10.17 | 27 | 2.64 | 15 | 9.74 |
|---|---|---|---|---|---|---|---|---|---|---|
| **NORTH AND SOUTH ATLANTIC-PACIFIC** | **28** | **2.66** | **17** | **1.74** | **12** | **10.17** | **27** | **2.64** | **15** | **9.74** |
| Circumboreal-Mediterranean-Australian | 1 | 0.09 | 1 | 0.10 | 1 | 0.85 | 1 | 0.09 | 1 | 0.65 |
| North Atlantic-Mediterranean-South Pacific | 2 | 0.19 | 1 | 0.10 | | | 2 | 0.19 | 1 | 0.65 |
| North Atlantic-Mediterranean-Southeast Pacific | 1 | 0.09 | 1 | 0.10 | | | 1 | 0.09 | | |
| North Atlantic-Mediterranean-Southwest Pacific | 3 | 0.28 | 2 | 0.20 | 2 | 1.69 | 3 | 0.29 | 2 | 1.30 |
| Boreal Atlantic-Mediterranean-Southwest Pacific | 4 | 0.38 | 2 | 0.20 | 2 | 1.69 | 4 | 0.39 | 1 | 0.65 |
| North Atlantic-Mediterranean-New Zealand | 9 | 0.85 | 4 | 0.41 | 5 | 4.24 | 8 | 0.78 | 6 | |
| Carolinian-Celtic-Pontian-New Zealand | 1 | 0.09 | 1 | 0.10 | | | 1 | 0.09 | 1 | 0.65 |
| Northeast Atlantic-Mediterranean-Southwest Pacific | 2 | 0.19 | 2 | 0.20 | | | 2 | 0.19 | 1 | 0.65 |
| Northeast Atlantic-Mediterranean-New Zealand | 1 | 0.09 | 1 | 0.10 | | | 1 | 0.09 | | |
| Celtic-Lusitanian-Mediterranean-New Zealand | 3 | 0.28 | 1 | 0.10 | 2 | 1.69 | 3 | 0.29 | 2 | 1.30 |
| Celtic-Lusitanian-Pontian-New Zealand | 1 | 0.09 | 1 | 0.10 | | | 1 | 0.09 | | |
| **HOL- AND SOUTH ATLANTIC-PACIFIC** | **9** | **0.85** | **4** | **0.41** | **6** | **5.08** | **9** | **0.88** | **5** | **3.25** |
| Atlantic-Mediterranean-Southwest Pacific | 4 | 0.38 | 2 | 0.20 | 3 | 2.54 | 4 | 0.39 | 2 | 1.30 |
| Atlantic-Pontian-Southwest Pacific | 1 | 0.09 | | | 1 | 0.85 | 1 | 0.09 | 1 | 0.65 |
| Atlantic-Mediterranean-New Zealand | 4 | 0.38 | 2 | 0.20 | 2 | 1.69 | 4 | 0.39 | 2 | 1.30 |
| **EAST AND WEST ATLANTIC-PACIFIC** | **8** | **0.76** | **6** | **0.61** | **4** | **3.39** | **8** | **0.78** | **1** | **0.65** |
| East Atlantic-Mediterranean-Pacific | 1 | 0.09 | 1 | 0.10 | | | 1 | 0.09 | | |
| Northeast Atlantic-Mediterranean-East Pacific | 1 | 0.09 | 1 | 0.10 | | | 1 | 0.09 | | |

| | | | | | | | | | | |
|---|---|---|---|---|---|---|---|---|---|---|
| Atlantic-Mediterranean-West Pacific | 4 | 0.38 | 2 | 0.20 | 3 | 2.54 | 4 | 0.39 | 1 | 0.65 |
| North Atlantic-Mediterranean-West Pacific | 2 | 0.19 | 2 | 0.20 | 1 | 0.85 | 2 | 0.19 | | |
| **ATLANTIC TYPE** | **740** | **70.20** | **712** | **72.80** | **50** | **42.37** | **716** | **69.99** | **101** | **65.58** |
| **Arctic-Antarctic-Atlantic** | **43** | **4.08** | **41** | **4.19** | **7** | **5.93** | **42** | **4.11** | **5** | **3.25** |
| Arctic-Antarctic-North Atlantic-Mediterranean | 1 | 0.09 | 1 | 0.10 | | | 1 | 0.09 | | |
| Arctic-Atlantic-Mediterranean | 4 | 0.38 | 4 | 0.41 | | | 4 | 4.11 | | |
| Arctic-North Atlantic-Mediterranean | 12 | 1.14 | 11 | 1.12 | 2 | 1.69 | 11 | 1.07 | 1 | 0.65 |
| Arctic-North Atlantic-Pontian | 1 | 0.09 | 1 | 0.10 | | | 1 | 0.09 | | |
| Arctic-Boreal Atlantic-Mediterranean | 7 | 0.66 | 7 | 0.72 | 1 | 0.85 | 7 | 0.68 | 3 | 1.95 |
| Arctic-Boreal Atlantic-Pontian | 3 | 0.28 | 3 | 0.31 | | | 3 | 0.29 | 1 | 0.65 |
| Arctic-Circumeuropean-Mauritanian | 1 | 0.09 | 1 | 0.10 | | | 1 | 0.09 | | |
| Arctic-Circumeuropean | 4 | 0.38 | 3 | 0.31 | 2 | 1.69 | 4 | 0.39 | | |
| Arctic-Celtic-Pontian | 1 | 0.09 | 1 | 0.10 | | | 1 | 0.09 | | |
| Circumeuropean-Mauritanian | 1 | 0.09 | 1 | 0.10 | | | 1 | 0.09 | | |
| Circumeuropean | 7 | 0.66 | 7 | 0.72 | 2 | 1.69 | 7 | 0.68 | | |
| Antarctic-Celtic-Lusitanian-Mediterranean | 1 | 0.09 | 1 | 0.10 | | | 1 | 0.09 | | |
| **Atlantic** | **697** | **66.13** | **671** | **68.61** | **43** | **36.44** | **674** | **65.88** | **96** | **62.34** |
| **HOL- AND NORTH ATLANTIC** | **99** | **9.39** | **95** | **9.71** | **8** | **6.78** | **98** | **9.58** | **13** | **8.44** |
| Holatlantic-Mediterranean | 13 | 1.23 | 12 | 1.23 | 2 | 1.69 | 12 | 1.17 | 5 | 3.25 |
| Atlantic-Mediterranean | 5 | 0.47 | 5 | 0.51 | | | 5 | 0.49 | | |
| Atlantic-Pontian | 2 | 0.19 | 2 | 0.20 | | | 2 | 0.19 | | |
| North Atlantic-Mediterranean | 27 | 2.56 | 25 | 2.56 | 3 | 2.54 | 27 | 2.64 | 3 | 1.95 |
| North Atlantic-Pontian | 5 | 0.47 | 5 | 0.51 | | | 5 | 0.49 | 1 | 0.65 |
| Boreal-Antiboreal Atlantic-Pontian | 1 | 0.09 | 1 | 0.10 | | | 1 | 0.09 | | |
| Boreal Atlantic-Mediterranean | 23 | 2.18 | 22 | 2.25 | 2 | 1.69 | 23 | 2.25 | 3 | 1.95 |
| Boreal Atlantic-Pontian | 13 | 1.23 | 13 | 1.33 | 1 | 0.85 | 13 | 1.27 | 1 | 0.65 |
| Virginian-Celtic-Lusitanian-Mediterranean | 1 | 0.09 | 1 | 0.10 | | | 1 | 0.09 | | |

265

| | | | | | | | | | | |
|---|---|---|---|---|---|---|---|---|---|---|
| Carolinian-Celtic-Lusitanian-Mediterranean | 4 | 0.38 | 4 | 0.41 | | | 4 | 0.39 | | |
| Carolinian-Celtic-Pontian | 1 | 0.09 | 1 | 0.10 | | | 1 | 0.09 | | |
| Caribbean-Celtic-Lusitanian-Mediterranean | 4 | 0.38 | 4 | 0.41 | | | 4 | 0.39 | | |
| **TROPICAL AND SUBTROPICAL ATLANTIC** | **109** | **10.34** | **99** | **10.12** | **8** | **6.78** | **108** | **10.56** | **9** | **5.84** |
| Tropical Atlantic-Mediterranean | 1 | 0.09 | 1 | 0.10 | | | 1 | 0.09 | | |
| Virginian-Carolinian-Caribbean | 2 | 0.19 | 1 | 0.10 | 1 | 0.85 | 2 | 0.19 | | |
| Carolinian-Lusitanian-Mediterranean | 2 | 0.19 | 2 | 0.20 | | | 2 | 0.19 | | |
| Carolinian-Lusitanian-Pontian | 1 | 0.09 | 1 | 0.10 | | | 1 | 0.09 | | |
| Caribbean-Lusitanian-Mediterranean | 5 | 0.47 | 4 | 0.41 | | | 5 | 0.49 | | |
| Caribbean-Mediterranean-Mauritanian | 1 | 0.09 | 1 | 0.10 | | | 1 | 0.09 | | |
| Lusitanian-Mediterranean-Mauritanian-Guinean | 5 | 0.47 | 5 | 0.51 | | | 5 | 0.49 | | |
| Lusitanian-Mediterranean-Mauritanian | 11 | 1.04 | 8 | 0.82 | 2 | 1.69 | 11 | 1.07 | | |
| Lusitanian-Mediterranean-South African | 1 | 0.09 | 1 | 0.10 | | | 1 | 0.09 | | |
| Lusitanian-Mediterranean | 79 | 7.49 | 74 | 7.57 | 5 | 4.24 | 78 | 7.62 | 9 | 5.84 |
| Lusitanian-Pontian | 1 | 0.09 | 1 | 0.10 | | | 1 | 0.09 | | |
| **EAST AND NORTHEAST ATLANTIC** | **251** | **23.81** | **244** | **24.94** | **16** | **13.56** | **247** | **24.14** | **32** | **20.78** |
| East Atlantic-Mediterranean | 10 | 0.95 | 10 | 1.02 | | | 10 | 0.98 | | |
| Celtic-Lusitanian-Mediterranean-Mauritanian | 26 | 2.47 | 26 | 2.66 | 1 | 0.85 | 26 | 2.54 | 1 | 0.65 |
| Celtic-Lusitanian-Mediterranean | 148 | 14.04 | 147 | 15.03 | 7 | 5.93 | 147 | 14.37 | 21 | 13.64 |
| Celtic-Lusitanian-Pontian | 15 | 1.42 | 12 | 1.23 | 4 | 4.24 | 13 | 1.27 | 5 | 3.25 |
| Celtic-Mediterranean | 6 | 0.57 | 6 | 0.61 | | | 6 | 0.59 | | |
| Celtic-Pontian-Caspian | 2 | 0.19 | 1 | 0.10 | 1 | 0.85 | 2 | 0.19 | 2 | 1.30 |
| Celtic-Pontian | 44 | 4.17 | 42 | 4.29 | 3 | 2.54 | 43 | 4.20 | 3 | 1.95 |
| **MEDITERRANEAN-PONTIAN-CASPIAN** | **238** | **22.58** | **233** | **23.82** | **11** | **9.32** | **221** | **21.60** | **42** | **27.27** |
| Holomediterranean | 23 | 2.18 | 22 | 2.25 | 1 | 0.85 | 23 | 2.25 | 3 | 1.95 |
| Mediterranean | 28 | 2.66 | 28 | 2.86 | | | 28 | 2.74 | | |

266

| | | | | | | | | | | |
|---|---|---|---|---|---|---|---|---|---|---|
| East Mediterranean | 9 | 0.85 | 9 | 0.92 | 1 | 0.85 | 9 | 0.88 | | |
| North Mediterranean | 3 | 0.28 | 3 | 0.31 | | | 3 | 0.29 | | |
| Adriatic-Aegean-Pontian | 3 | 0.28 | 3 | 0.31 | | | 3 | 0.29 | | |
| Adriatic-Pontian | 10 | 0.95 | 10 | 1.02 | | | 10 | 0.98 | | |
| Aegean-Pontian | 14 | 1.33 | 14 | 1.43 | 1 | 0.85 | 13 | 1.27 | | |
| Pontian-Caspian-Aral | 2 | 0.19 | 2 | 0.20 | 1 | 0.85 | 2 | 0.19 | 1 | 0.65 |
| Pontian-Caspian | 28 | 2.66 | 27 | 2.76 | 2 | 1.69 | 16 | 1.56 | 27 | 17.53 |
| Pontian | 118 | 11.19 | 115 | 11.76 | 5 | 4.24 | 114 | 11.14 | 11 | 7.14 |
| **CASPIAN RELICT** | **41** | **3.89** | **39** | **3.99** | **4** | **3.39** | **21** | **2.05** | **38** | **24.67** |
| **Total** | **1055** | | **978** | **92.88** | **118** | **11.21** | **1023** | **97.15** | **154** | **14.62** |

**Table 5.** Zoogeographical characteristic of the brackish, freshwater and terrestrial Invertebrate fauna from the Bulgarian Black Sea coast.

| Zoogeographical scheme of the used categories and main taxa | Total | | Brakish (rare marine) | | Freshwater | | Terrestrial | |
|---|---|---|---|---|---|---|---|---|
| | number | % | number | % | number | % | number | % |
| **Species distributed in Palaearctic and out of it** | **296** | **58.27** | **127** | **73.84** | **271** | **65.14** | **74** | **36.81** |
| **NORTHERN TYPE** | **282** | **55.51** | **126** | **73.25** | **262** | **62.98** | **63** | **31.34** |
| Cosmopolitan | 75 | 14.76 | 54 | 31.40 | 74 | 17.79 | 2 | 0.99 |
| Subcosmopolitan | 41 | 8.07 | 22 | 12.79 | 39 | 9.37 | | |
| Holarctic-Paleotropical-Neotropical | 5 | 0.98 | 2 | 1.16 | 4 | 0.96 | 1 | 0.50 |
| Holarctic-Paleotropical-Australian | 2 | 0.39 | | | 1 | 0.24 | 1 | 0.50 |
| Holarctic-Paleotropical | 3 | 0.59 | 2 | 1.16 | 3 | 0.72 | | |
| Holarctic-Neotropical-Oriental-Australian | 3 | 0.59 | 1 | 0.58 | 3 | 0.72 | 1 | 0.50 |
| Holarctic-Neotropical-Oriental | 8 | 1.57 | 4 | 2.32 | 7 | 1.68 | 4 | 1.99 |
| Holarctic-Neotropical-Afrotropical-Australian | 3 | 0.59 | | | 2 | 0.48 | 1 | 0.50 |
| Holarctic-Neotropical-Afrotropical | 6 | 1.18 | 3 | 1.74 | 5 | 1.20 | 1 | 0.50 |
| Holarctic-Neotropical-Australian | 5 | 0.98 | 2 | 1.16 | 4 | 0.96 | 1 | 0.50 |
| Holarctic-Afrotropical-Australian | 3 | 0.59 | 1 | 0.58 | 3 | 0.72 | | |
| Holarctic-Oriental-Australian | 3 | 0.59 | 1 | 0.58 | 3 | 0.72 | | |
| Holarctic-Neotropical | 5 | 0.98 | 3 | 1.74 | 5 | 1.20 | 1 | 0.50 |
| Holarctic-Afrotropical | 7 | 1.38 | 3 | 1.74 | 5 | 1.20 | 2 | 0.99 |
| Holarctic-Oriental | 8 | 1.57 | 2 | 1.16 | 7 | 1.68 | 4 | 1.99 |

| | | | | | | | | |
|---|---|---|---|---|---|---|---|---|
| Holarctic-Australian | 2 | 0.39 | 1 | 0.58 | 2 | 0.48 | | |
| Palearctic-Paleotropical-Australian | 3 | 0.59 | | | 3 | 0.72 | | |
| Palearctic-Afrotropical-Australian | 2 | 0.39 | | | 1 | 0.24 | 1 | 0.50 |
| Palearctic-Oriental-Australian | 2 | 0.39 | | | 2 | 0.48 | 2 | 0.99 |
| Palearctic-Paleotropical | 1 | 0.20 | 1 | 0.58 | 1 | 0.24 | 1 | 0.50 |
| Palearctic-Afrotropical | 8 | 1.57 | 2 | 1.16 | 8 | 1.92 | 1 | 0.50 |
| Palearctic-Oriental | 9 | 1.77 | 3 | 1.74 | 9 | 2.16 | 6 | 2.98 |
| West Palearctic-Paleotropical | 1 | 0.20 | | | 1 | 0.24 | 1 | 0.50 |
| Transpalaearctic-Oriental | 2 | 0.39 | | | 2 | 0.48 | 2 | 0.99 |
| West and Central Palaearctic-Oriental | 1 | 0.20 | | | 1 | 0.24 | 1 | 0.50 |
| West Palearctic-Afrotropical | 3 | 0.59 | | | 1 | 0.24 | 2 | 0.99 |
| West Palearctic-Oriental | 2 | 0.39 | 1 | 0.58 | 2 | 0.48 | 2 | 0.99 |
| Holarctic | 68 | 13.39 | 17 | 9.88 | 63 | 15.14 | 23 | 11.44 |
| European-Australian | 1 | 0.20 | | | | | 1 | 0.50 |
| **SOUTH TYPE** | **14** | **2.76** | **1** | **0.58** | **9** | **2.16** | **11** | **5.47** |
| Paleotropical-South Palearctic | 2 | 0.39 | | | 2 | 0.48 | 1 | 0.50 |
| Paleotropical-Mediterranean-Central Asian | 1 | 0.20 | | | 1 | 0.24 | 1 | 0.50 |
| Paleotropical-Mediterranean (ptm) | 2 | 0.39 | | | 1 | 0.24 | 2 | 0.99 |
| Afrotropical-Mediterranean (atm) | 3 | 0.59 | | | 1 | 0.24 | 2 | 0.99 |
| Oriental-Mediterranean-Central Asian-Australian | 1 | 0.20 | 1 | 0.58 | | | | |
| Oriental-Mediterranean-Central Asian | 3 | 0.59 | | | 2 | 0.48 | 3 | 1.49 |
| Oriental-Mediterranean | 2 | 0.39 | | | 2 | 0.48 | 2 | 0.99 |
| **Species with Palaearctic distribution** | **205** | **40.35** | **45** | **26.16** | **145** | **34.85** | **127** | **63.18** |
| **PALAEARCTIC TYPE** | **79** | **15.55** | **14** | **8.14** | **65** | **15.62** | **52** | **25.87** |

| | | | | | | | | |
|---|---|---|---|---|---|---|---|---|
| Holopalaearctic | 10 | 1.97 | 1 | 0.58 | 10 | 2.40 | 4 | 1.99 |
| Transpalaearctic | 21 | 4.13 | 1 | 0.58 | 18 | 4.33 | 17 | 8.46 |
| West and Central Palaearctic | 14 | 2.76 | 2 | 1.16 | 13 | 3.12 | 11 | 5.47 |
| West Palaearctic | 18 | 3.54 | 2 | 1.16 | 14 | 3.36 | 10 | 4.97 |
| Disjunct Palaearctic | 2 | 0.39 | 1 | 0.58 | 1 | 0.24 | 1 | 0.50 |
| Eurosiberian-Central Asian | 1 | 0.20 | | | 1 | 0.24 | 1 | 0.50 |
| European-Central Asian | 3 | 0.59 | 1 | 0.58 | 2 | 0.48 | 1 | 0.50 |
| European-West Central Asian | 1 | 0.20 | | | 1 | 0.24 | 1 | 0.50 |
| European-Iran-Turanian | 1 | 0.20 | 1 | 0.58 | 1 | 0.24 | | |
| European-Turanian | 2 | 0.39 | 1 | 0.58 | 1 | 0.24 | 1 | 0.50 |
| European-North African | 6 | 1.18 | 4 | 2.32 | 3 | 0.72 | 5 | 2.49 |
| **EUROSIBERIAN TYPE** | **55** | **10.83** | **12** | **6.98** | **48** | **11.54** | **18** | **8.95** |
| Holoeurosiberian | 2 | 0.39 | 1 | 0.58 | 3 | 0.72 | 3 | 1.49 |
| West and Central Eurosiberian | 7 | 1.38 | | | 5 | 1.20 | 2 | 0.99 |
| West Eurosiberian-Anatolian | 2 | 0.39 | | | 2 | 0.48 | 1 | 0.50 |
| West Eurosiberian | 6 | 1.18 | | | 5 | 1.20 | 2 | 0.99 |
| European-Anatolian | 1 | 0.20 | 1 | 0.58 | 1 | 0.24 | | |
| Central and Southeast European-Anatolian | 1 | 0.20 | 1 | 0.58 | 1 | 0.24 | | |
| European | 31 | 6.10 | 9 | 5.23 | 26 | 6.25 | 8 | 3.98 |
| Central and South European | 2 | 0.20 | | | 1 | 0.24 | 1 | 0.50 |
| Central and Southeast European | 4 | 0.79 | | | 4 | 0.96 | 1 | 0.50 |
| **MEDITERRANEAN TYPE** | **71** | **13.98** | **19** | **11.05** | **32** | **7.69** | **57** | **28.36** |
| Mediterranean-Central Asian | 5 | 0.98 | 1 | 0.58 | 3 | 0.72 | 5 | 2.49 |
| Mediterranean-West Central Asian | 3 | 0.59 | 2 | 1.16 | 2 | 0.48 | 2 | 0.99 |

| | | | | | | | | |
|---|---|---|---|---|---|---|---|---|
| North Mediterranean-West Central Asian | 1 | 0.20 | | | 1 | 0.24 | | |
| East Mediterranean-Central Asian | 1 | 0.20 | | | | | 1 | 0.50 |
| Northeast Mediterranean-Iran-Turanian | 1 | 0.20 | | | 1 | 0.24 | 1 | 0.50 |
| Central and Southeast European-Iran-Turanian | 1 | 0.20 | | | | | 1 | 0.50 |
| Central and South European-North African | 2 | 0.39 | | | 1 | 0.24 | 2 | 0.99 |
| Holomediterranean | 19 | 3.74 | 3 | 1.74 | 10 | 2.40 | 16 | 7.96 |
| Atlantomediterranean | 3 | 0.59 | | | | | 3 | 1.49 |
| North Mediterranean | 4 | 0.79 | | | 1 | 0.24 | 4 | 1.99 |
| Atlantic-South European | 1 | 0.20 | 1 | 0.58 | | | 1 | 0.50 |
| South European | 3 | 0.59 | 1 | 0.58 | 2 | 0.48 | 3 | 1.49 |
| Southeast European-Pontian-Caspian | 1 | 0.20 | 1 | 0.58 | 1 | 0.24 | | |
| Southeast European-Pontian | 1 | 0.20 | | | 1 | 0.24 | | |
| Southeast European-Anatolian | 1 | 0.20 | | | 1 | 0.24 | | |
| Southeast European | 1 | 0.20 | | | | | 1 | 0.50 |
| East Mediterranean | 4 | 0.79 | 2 | 1.16 | 3 | 0.72 | 4 | 1.99 |
| Northeast Mediterranean | 5 | 0.98 | 1 | 0.58 | | | 4 | 1.99 |
| Pontomediterranean | 3 | 0.59 | 1 | 0.58 | 1 | 0.24 | 2 | 0.99 |
| Pontian endemic | 3 | 0.59 | 3 | 1.74 | | | 2 | 0.99 |
| Balkan endemic | 2 | 0.39 | | | 1 | 0.24 | 1 | 0.50 |
| Bulgarian endemic | 1 | 0.20 | 1 | 0.58 | 1 | 0.24 | | |
| Regional Bulgarian endemic | 4 | 0.79 | 2 | 1.16 | 1 | 0.24 | 4 | 1.99 |
| Local Bulgarian endemic | 1 | 0.20 | | | 1 | 0.24 | | |
| **Total** | **528** | | **172** | **34.33** | **416** | **83.03** | **201** | **40.12** |

271

**Table 6.** Alien invertebrate animals, recorded from the Bulgarian Black Sea coast (Note. The years in brackets show finding of the species in Bulgaria, * – invasive species).

| Taxa | Year of introduction or finding | Donor region |
| --- | --- | --- |
| **Coelenterata** | | |
| *Blackfordia virginica* Mayer, 1910 | 1925 (1935) | North Atlantic and North America |
| *Rathkea octopunctata* (Sars, 1835) | (1957) | Atlantic Ocean and Mediterranean Sea |
| *Diadumene lineata* (Verrill, 1869) | (1960) | Indo Northwest Pacific Ocean |
| *Bougainvillia muscus* (Allman, 1863) | 1960 | Atlantic |
| *Calyptospadix cerulea* (Clarke, 1882) | 1933 (1933)1959 | North Atlantic and North America |
| **Ctenophora** | | |
| **Mnemiopsis leidyi* Agassiz, 1865 | 1982 (1986) | Western Atlantic Ocean |
| **Beroe ovata* Bruguière, 1789 | 1997 (1997) | North Atlantic Ocean |
| *Bolinopsis vitrea* (L. Agassiz, 1860) | 2010 (2010) | Tropical and Subtropical |
| **Annelida: Polychaeta** | | |
| **Ficopomatus enigmaticus* (Fauvel, 1923) | 1929 (1935) | Indian Ocean |
| *Nephtys ciliata* (Müller, 1788) | 1977 | Cosmopolitan |
| *Glycera capitata* Örsted, 1842 | 1970 | Cosmopolitan |
| *Capitellethus dispar* (Ehlers, 1907) | 1977 | Indo-Pacific |
| *Hesionides arenaria* Friedrich, 1937 | 1950-те (1954) | Pacific Ocean or Atlantic Ocean |
| *Streblospio shrubsolii* (Buchanan, 1890) | 1957 (1957) | North Atlantic Ocean |
| *Streptosyllis varians* Webster & Benedict, 1887 | 1966 (1966) | North Atlantic or Pacific Ocean |
| **Polydora cornuta* Bosc, 1802 | 2005 (2008) | North and West Atlantic Ocean |
| **Dipolydora quadrilobata* (Jacobi, 1883) | (1990) | Atlantic Ocean and Pacific Ocean |

| Arthropoda: Crustacea | | |
|---|---|---|
| *Acartia tonsa* Dana, 1849 | 1976 (2000) | Indian Ocean and Pacific Ocean |
| *Oithona davisae* Ferrari F. D. & Orsi, 1984 | 2001 (2009) | Northwest Pacific Ocean |
| *Triconia minuta* (Giesbrecht, 1893 [1892]) | (2000) | Indian Ocean and Pacific Ocean |
| *Monstrilla grandis* Giesbrecht, 1891 | 1912, 1980 | North Atlantic |
| *Amphibalanus eburneus* Gould, 1841 | 1892 (1933) | North America |
| * *Amphibalanus improvisus* Darwin, 1854 | 1844 (1912) | North America |
| *Alpheus dentipes* Guerin, 1832 | 1884 | Atlantic |
| *Palaemon macrodactylus* Rathbun, 1902 | (2009) | Pacific Ocean and Souteast Asia |
| *Eurypanopeus depressus* Smith, 1869 | (2017) | Atlantic |
| *Rhithropanopeus harrisii* (Gould, 1841) | 1937 (1934, 1953) | North America |
| *Callinectes sapidus* Rathbun, 1896 | 1967 (1968) | North Atlantic Ocean |
| *Eriocheir sinensis* (H. Milne Edwards, 1854) | 2006 (2005) | East and Southeast Asia |
| Mollusca | | |
| *Potamopyrgus  antipodarium* (J. E. Gray, 1843) | 1952 (2008) | coast of New Zeland |
| *Neptunea arthritica* (Valenciennes, 1858) | (2000) | Pacific, Atl |
| **Rapana venosa* (Valenciennes, 1846) | 1946 (1956) | Sea of Japan |
| *Neptunea arthritica* (Valenciennes, 1858) | (2000) | Indian Ocean and Pacific Ocean |
| *Corambe obscura* (Verrill, 1870) | 1980 (1986) | North Atlantic Ocean |
| *Ferrissia fragilis* (Tryon, 1862 | (1983) | North America |
| *Physella acuta* (Draparnaud, 1805) | (1927) | North America |
| **Anadara kagoshimensis* (Tokunaga, 1906) | 1982 (1984) | Indian Ocean and Pacific Ocean |
| *Arcuatula senhousia* (Benson, 1842) | (2017, 2018) | Northwest Pacific |
| *Magallana gigas* (Thunberg, 1793) | 2010 (2020) | Northwest Pacific Ocean |
| **Mya arenaria* Linnaeus, 1758 | 1966 (1973) | Circumboreal |
| *Teredo navalis* Linnaeus, 1758 | 750-500 B.C. | Atlantic Ocean and Pacific Ocean |

**Table 7.** Conservation status of invertebrare animals of the Bulgarian Black Sea coast.

| Taxa | Black Sea Red Data Book | Ecological data, Bulgarian Red Data Book, IUCN and European category | Distribution (area and depth) |
|---|---|---|---|
| *Halichondria panicea* (Pallas, 1766) | VU | M, bt, eb, lt | K, 2-65 |
| *Odessia maeotica* (Ostroumoff, 1896) | VU | M-B, 25‰, bt-p | lm, ? Rc |
| *Hesionides arenaria* Friedrich, 1937 | EN, VU | M, bt, sep, sls, ps, gw | aminwp, 0-10 |
| *Ophelia bicornis* Savigny, 1918 | EN | M, bt, sep, l-sl, ps | bam, 0.5-1.5 |
| *Hirudo verbana* Carena, 1820 | | L, bt, ph, 10‰, ▲-VU | (cse) |
| *Halacarellus procerus* (Viets, 1927) | EN | M-B, bt, l-sl, ps, gw | Bam |
| *Branchinella spinosa* (H. Milne Edwards, 1840) | EN | B-M, 30‰, p | lm, (atm) |
| *Centropages ponticus* Karavaev, 1895 | EN | M, p | Mrs |
| *Anomalocera patersoni* Templeton, 1837 | EN | M, p, et | Amip |
| *Labidocera brunescens* (Czerniavsky, 1868) | EN | M, p, eh, th | Lmm |
| *Pontella mediterranea* (Claus, 1863) | EN | M, p | Lmm |
| *Oithona minuta* (Krichagin, 1877; Scott, 1894) | EN | M-B-L, p, eh | amip, 20 |
| *Hemimysis anomala* G. O. Sars, 1907 | EN | B-L, bt, ep, lt | pc, clm, (h), -20 |
| *Chaetogammarus ischnus* (Stebbing, 1899) | VU | M-B, eh, sep, ro, l-sl | pc, cpc, Rc |
| *Dikerogammarus villosus* (Sowinsky, 1894) | VU | B-L, bt, ep, ro, ph | pc, cpc, Rc, (ean) |
| *Iphigenella andrussowi* (G. O. Sars, 1896) | LR | B-L, bt, ep, ps, ps-s | pc, Rc |
| *Shablogammarus shablensis* (Carausu, 1943) | VU | B-L, bt, ep, ps, ps-s | pc, Rc |
| *Apseudopsis ostroumovi* Bacescu & Carausu, 1947 | LR | M, bt, sg, ms, phs | p, 27, -100 |
| *Upogebia pusilla* (Petagna, 1792) | EN | M, bt, ep (p), ps-pe | clmm, -22 |
| *Diogenes pugilator* (Roux,1829) | EN | M, bt, mb, ps, sg | eam, ? eami, 0-40 |
| *Carcinus aestuarii* Nardo, 1847 | VU | M, bt, ps, zc, ro, r | lm, ? amp, 0-70 |

274

| Species | Status | Characters | Distribution |
|---|---|---|---|
| *Liocarcinus navigator* (Herbst, 1794) | VU | M, bt, eb, ps, sg, s | acem, 3-80 |
| *Eriphia verrucosa* (Forskål, 1775) | EN | M, bt, ep, sp-l-sl, lt, ▲-VU | lmm, ?lmmg, 0-30 |
| *Pilumnus hirtellus* (Linnaeus, 1761) | VU | M, bt, mb, ph, mc, lt | clmm, 0-40 |
| *Xantho poressa* (Olivi, 1792) | VU | M, bt, eb, ro | lmm, 1-15, 100 |
| *Potamon ibericum* (Bieberstein, 1808) | DD, EN | L, eu | (nmwca) |
| *Pachygrapsus marmoratus* (Fabricius, 1787) | VU | M, bt, sep, spr-slr, lt | cmm, 0-7 |
| *Calopteryx splendens* (Harris, 1782) | LR, VU | L-TL, ♦-LC | (wcp) |
| *Calopteryx virgo* (Linnaeus, 1758) | VU | L-TL, ♦-LC | (tp) |
| *Epallage fatime* (Charpentier, 1840) | NE, DD, VU | L-TL, ▲-VU, ♦-NT | (om) |
| *Lestes viridis* (Vander Linden, 1825) | | L-TL, ▲-VU, ♦-LC | (wp) |
| *Anax imperator* Leach, 1815 | NE, DD, VU | L-TL, ♦-LC | (wppt) |
| *Sympetrum depressiusculum* (Sélys, 1841) | | L-TL, ▲-VU, ♦-VU | (tp) |
| *Sympetrum vulgatum* (Linnaeus, 1758) | | L-TL, ▲-VU, ♦-LC | (ewca) |
| *Patella ulyssiponensis* Gmelin, 1791 | EN | M, bt, sep, lt-ro, l-sl, r | clm, 0-10 |
| *Theodoxus pallasi* Lindcholm, 1924 | | M-B, eh-14‰, ro, ▲-EX | pc, Rc, Sf, +, 4-10 |
| *Hauffenia lucidula* Angelov, 1967 | | L, bt, cr, 1‰, ▲-CR, ♦-CR | (El) |
| *Valvata cristata* O. F. Müller, 1774 | | L, 0.5‰, bt, ph, s, r, ♦-LC | (wes) |
| *Valvata piscinalis* (O. F. Müller, 1774) | | L, 0.4‰, bt, sw, ph, pe, x, ♦-LC | (wcp, i - h), 3, 80 |
| *Hydrobia acuta* (Draparnaud, 1805) | | B-M, 60‰, bt, ep, sw, ph, ro, s, ♦-LC | lm, ? hm, 0-22 |
| *Bithynia tentaculata* (Linnaeus, 1758) | | L, bt, ph, ro, s, ps, ♦-LC | (wp, ? h - i), 5 |
| *Turricaspia lincta* (Milaschewitch, 1908) | | M-B, 8‰, bt, s, sw, ♦-LC | pc, Rc, Sf, +, -148 |
| *Acroloxus lacustris* (Linnaeus, 1758) | | L, bt, ph, sw, ♦-LC | (wes, ? hoes) |
| *Lymnaea stagnalis* (Linnaeus, 1758) | | L, 7‰, bt, ph, pe, ♦-LC | (h), 0-4 |
| *Stagnicola corvus* (Gmelin, 1791) | | L, bt, sw, ph, ♦-LC | (hop, ? e), 0-50 |
| *Radix auricularia* (Linnaeus, 1758) | | L, 6‰, bt, ph, rh, s, ♦-LC | (h, ? hop), 0.2-25 |
| *Radix balthica* (Linnaeus, 1758) | | L, 3-10‰, bt, ph, eu, ♦-LC | (hop) |

| | | | |
|---|---|---|---|
| *Ferrissia fragilis* (Tryon, 1862 | | L, bt, eu, th, ph, r, is, ♦-DD | (h, sk - i), 0-8 |
| *Planorbis carinatus* O. F. Müller, 1774 | | L, bt, sw, ph, pe, r, ♦-LC | (wes, ? h), -10, -18 |
| *Planorbis planorbis* (Linnaeus, 1758) | | L, bt, 2‰, sw, ph, pe, ♦-LC | (h) |
| *Anisus septemgyratus* (Rossmaessler, 1835) | | L, 8‰, sw, ph, α-β, r, ♦-LC | (wes, ? e) |
| *Anisus vortex* (Linnaeus, 1758) | | L, bt, 8‰, ph, α-β, r, ♦-LC | (wces) |
| *Anisus vorticulus* (Troschel, 1834) | | L, bt, pe, ph, rh, sw, r, ♦-NT, HD | (wces, ? wes) |
| *Gyraulus crista* (Linnaeus, 1758) | | L, 1.5‰, eu, ph, α-β, ♦-LC | (h) |
| *Hippeutis complanatus* (Linnaeus, 1758) | | L, bt, ph, s-ar, sw, α, r, ♦-LC | (wces, ? wcp) |
| *Segmentina nitida* (O. F. Müller, 1774 | | L, bt, sw, ph, α-β, ♦-LC | (wcp) |
| *Planorbarius corneus* (Linnaeus, 1758) | | L, 5‰, sw, po, α-β, ♦-LC | (wces), -9 |
| *Physella acuta* (Draparnaud, 1805) | | L, bt, pe, tx, α-β, is, ♦-LC | (na, sk - i) |
| *Physa fontinalis* (Linnaeus, 1758) | | L, bt, sw, po, ph, β, r, ♦-LC | (tp, ? h) |
| *Branchiostoma lanceolatum* (Pallas, 1774) | VU | M, bt, ep, ps | amip, 17, 21 |

**Table 8.** Conservation significance of Bivalvia from the Bulgarian Black Sea coast

| Taxa | Ecological data, IUCN and European category, BSZL, BU-II, Black Sea Red Data Book | Distribution (area and depth) |
|---|---|---|
| *Striarca lactea* (Linnaeus, 1758) | M, bt, ep, lt, r; BU-II | (-20 m); eami, Sf, + |
| *Modiolus adriaticus* Lamarck, 1819 | M, bt, mb, pe, s-ps, BU-II | (20-36, 75 m); lm, ?clm |
| *Flexopecten glaber* (Linnaeus, 1758) | M, bt, ep, ps, ps-s; BU-II | (3-40, 70 m); lm |
| *Anomia ephippium* Linnaeus, 1758 | M, bt, ep, ro; BU-II | (-30 m); amep |
| *Ostrea edulis* Linnaeus, 1758 | M, bt, sep, ro, r, BU-II, BSZL, ■ – EN, VU | (7-20, 65 m); anamnep |
| *Loripes orbiculatus* Poli, 1795 | M, bt, sep, ps, zc; BSZL | (-25 m); clmm |
| *Hypanis plicata* (Eichwald, 1829) | B, bt, pe, ar, s-ps, eb; BSZL | (35-92 m); pc, Rc, Sf, + |
| *Gastrana fragilis* (Linnaeus, 1758) | M, bt, ep, ar, ar-ps, sg; BSZL | (-36 m); clm |
| *Moerella donacina* (Linnaeus, 1758) | M, bt, sep, ps-pe; r, BU-II, BSZL | (-20, 30 m); clmm |
| *Fabulina fabula* (Gmelin, 1791) | M, bt, ep, ps, ps-s; BU-II; BSZL | (-10, 40 m); clm |
| *Donax trunculus* Linnaeus, 1758 | M, bt, sep, ps; BSZL | (2-15 m); clm |
| *Donacilla cornea* (Poli, 1795) | M, bt, sep, ps, BU-II, BSZL, ■ – EN | (0.2, -2 m); lm |
| *Pitar rudis* (Poli, 1795) | M, bt, mb, ps-pe, sg; BSZL | (11-40-200 m); lmmg |
| *Irus irus* (Linnaeus, 1758) | M, bt, sep, lt, slr; BU-II, BSZL | (0-10 m); eamiswp |
| *Polititapes aureus* (Gmelin, 1791) | M, bt, mb, ps-pe, s; BU-II; BSZL | (11-60 m, 2-65 m); ce |
| *Petricola lithophaga* (Retzius, 1788) | M, bt, sep, lt; BU-II, BSZL | (-10, 26 m); lmi |
| *Solen marginatus* Pulteney, 1799 | M, bt, sep, ps, r, BU-II; BSZL, ■ – EN | (0-10 m); clmm |
| *Pholas dactylus* Linnaeus, 1758 | M, bt, sep, lt; r, BA-II – SPA/BD (Annex II – 1999, 2017), BC-II, BU-II, BSZL | (-15 m); amrs |
| *Barnea candida* (Linnaeus, 1758) | M, bt, ep, lt, eh; r, BU-II | (-15 m); namni |

www.ingramcontent.com/pod-product-compliance
Lightning Source LLC
Chambersburg PA
CBHW070748160726
48004CB00001B/101